Geospatial Modeling for Environmental Management

Geospatial Modeling for Environmental Management
Case Studies from South Asia

Edited by
Shruti Kanga, Suraj Kumar Singh,
Gowhar Meraj, and Majid Farooq

CRC Press
Taylor & Francis Group
Boca Raton London New York

CRC Press is an imprint of the
Taylor & Francis Group, an **informa** business

First edition published 2022
by CRC Press
6000 Broken Sound Parkway NW, Suite 300, Boca Raton, FL 33487-2742

and by CRC Press
2 Park Square, Milton Park, Abingdon, Oxon, OX14 4RN

Library of Congress Cataloging-in-Publication Data
Names: Kanga, Shruti, 1984- editor. | Singh, Suraj Kumar, 1983- editor. |
Meraj, Gowhar, editor. | Farooq, Majid, editor.
Title: Geospatial modeling for environmental management : case studies from
South Asia / edited by Shruti Kanga, Suraj Singh, Gowhar Meraj, and Majid Farooq.
Description: First edition. | Boca Raton, FL : CRC Press, 2022. |
Includes bibliographical references and index.
Identifiers: LCCN 2021041803 (print) | LCCN 2021041804 (ebook) |
ISBN 9780367702892 (hbk) | ISBN 9780367705916 (pbk) | ISBN 9781003147107 (ebk)
Subjects: LCSH: Geospatial data—Himalayan Mountains Region. |
Environmental management—Geographic information systems. |
Himalaya Mountains Region—Remote sensing. | Himalaya Mountains
Region—Environmental conditions. | Geoinformatics.
Classification: LCC GE160.H5 G46 2022 (print) | LCC GE160.H5 (ebook) |
DDC 333.7095496—dc23/eng/20211029
LC record available at https://lccn.loc.gov/2021041803
LC ebook record available at https://lccn.loc.gov/2021041804

ISBN: 978-0-367-70289-2 (hbk)
ISBN: 978-0-367-70591-6 (pbk)
ISBN: 978-1-003-14710-7 (ebk)

DOI: 10.1201/9781003147107

Typeset in Times
by codeMantra

Contents

Preface...ix
Editors...xv
Introduction..xvii
Editorial Advisory Board...xix
List of Contributors..xxi

PART A   Geospatial Modeling in Hydrological Studies

Chapter 1   Flood Vulnerability and Risk Assessment with Parsimonious
Hydrodynamic Modeling and GIS 3

Sudhakar B. Sharma, Anupam K. Singh, and Ajay S. Rajawat

Chapter 2   Estimation of Parameters in Ungauged Catchment Using Map-
Correlation Method: A Case Study on Krishna-Godavari Basin....... 19

Laxmipriya Mohanty, Anindita Swain, and Bharadwaj Nanda

Chapter 3   Soil and Water Assessment Tool for Simulating the Sediment
and Water Yield of Alpine Catchments: A Brief Review...................37

*Ishfaq Gujree, Fan Zhang, Gowhar Meraj, Majid Farooq,
Muhammad Muslim, and Arfan Arshad*

Chapter 4   Temporal Assessment of Sedimentation in Siruvani Reservoir
Using Remote Sensing and GIS 59

J. Brema and A. Tamilarasan

Chapter 5   Review of Conceptual Models of Estimating the Spatio-
Temporal Variations of Water Depth Using Remote Sensing and
GIS for the Management of Dams and Reservoirs 83

Abdulkadir Isah Funtua

PART B   Geospatial Modeling in Landslide Studies

Chapter 6   Geospatial Modeling in Landslide Hazard Assessment: A Case
Study along Bandipora-Srinagar Highway, N-W Himalaya,
J&K, India 113

Ahsan Afzal Wani, Bikram Singh Bali, Sareer Ahmad,
Umar Nazir, and Gowhar Meraj

Chapter 7   Causes, Consequences, and Mitigation of Landslides in the
Himalayas: A Case Study of District Mandi, Himachal Pradesh 127

Krishan Chand and D.D. Sharma

Chapter 8   Landslide Hazard and Exposure Mapping of Risk Elements in
Lower Mandakini Valley, Uttarakhand, India 157

Habib Ali Mirdda, Masood Ahsan Siddiqui,
Somnath Bera, and Bhoop Singh

PART C   Geospatial Modeling for Climate Change Studies

Chapter 9   Crop Response to Changing Climate, Integrating Model
Approaches: A Review 175

Arul Prasad S., Maragatham M., Vijayashanthi V.A., and Naveen

Chapter 10   Snow and Glacier Resources in the Western
Himalayas: A Review 195

Asif Marazi and Shakil Ahmad Romshoo

Chapter 11   Detecting Vegetation and Timberline Dynamics in Pinder
Watershed Central Himalaya Using Geospatial Techniques............ 213

N. C. Pant, Meenaxi, Anand Kumar,
and Upasana Choudhury

Chapter 12   Climate Change Studies, Permanent Forest Observational Plots
and Geospatial Modeling 223

Swayam Vid and Shanti Kumari

Chapter 13   Analyzing the Relationship of LST with MNDWI and NDBI in Urban Heat Islands of Hyderabad City, India 241

Subhanil Guha and Himanshu Govil

PART D   Geospatial Modeling in Change Dynamics Studies

Chapter 14   Assessment of the Visual Disaster of Land Degradation and Desertification Using TGSI, SAVI, and NDVI Techniques 261

B. Pradeep Kumar, K. Raghu Babu, M. Rajasekhar, and M. Ramachandra

Chapter 15   Dynamics of Forest Cover Changes in Hindu Kush-Himalayan Mountains Using Remote Sensing: A Case Study of Indus Kohistan Palas Valley, Pakistan.. 281

Noor ul Haq, Fazlur Rahman, and Iffat Tabassum

Chapter 16   Remote Sensing and Geographic Information System for Evaluating the Changes in Earth System Dynamics: A Review...... 309

Asraful Alam and Nilanjana Ghosal

PART E   Geospatial Modeling in Policy and Decision-Making

Chapter 17   Sustainable Livelihood Security Index: A Case Study in Chirrakunta Rurban Cluster... 325

Supratim Guha, Dillip Kumar Barik, and Venkata Ravibabu Mandla

Chapter 18   Carrying Capacity of Water Supply in Shimla City: A Study of Sustainability and Policy Framework........................... 347

Rajesh Kumar and M. L. Mankotia

Index.. 365

Preface

Chapter 1 demonstrates the flood frequency analysis of the Tapi river carried out using an integrated geospatial, hydrological, and hydrodynamic modeling approach. The study discusses how the causal factors of flooding in Lower Tapi Basin, i.e., heavy rainfall and the associated high discharge from the Ukai dam, get further aggravated by various other elements on the ground to induce severe flooding in the region. The authors carried out an exhaustive study wherein parameters causing inundation, such as the river's inadequate drainage capacity, congestion at the confluence point, excessive silt load, and high tide due to proximity to the Gulf of Khambhat, were studied. Using the HEC-RAS hydrodynamic model, river cross-sections were examined for understanding anomalies along and across the river. Flood risk analysis has also been carried out using vulnerability, land use, and population density maps. The risk-prone areas were classified into very high, high, moderate, and low-risk zones based on GIS-based weighted overlay. Overall, the study highlighted the role of geospatial modeling in understanding the reasons behind the drastically reduced water carrying capacity at Surat, India. In hydrological studies, calibration and validation of simulation outputs are critical stages of analysis in order to use the model outputs for decision and policy-making. Both of them depend upon the reliability of the observational data at the gauging stations. **Chapter 2** demonstrates how the indirect assessment of daily streamflow time series data at any ungauged station requires selecting a reference stream-gauge for calibration purposes. Therefore, the accuracy of simulation of daily streamflow at such ungauged stations directly depends on the degree of correlation of the reference stream gauge. In this chapter, the authors used the map correlation method (MCM) to identify the most correlated stream gauge station as the reference of an ungauged station. The efficiency of this method is demonstrated through a case study on the Krishna-Godavari river basin in southern India to predict the catchment parameters for different ungauged sites. The study shows how streamflow estimation in all stream gauges gets improved using this method of calibrating gauging sites.

Chapter 3 reviewed the latest research using the Soil and Water Assessment (SWAT) model Soil and Water Assessment Tool (SWAT) for hydrological studies. SWAT model works on local-scale watersheds to larger river basins to simulate the ground and surface water to assess the environmental impacts resulting from land use, land management practices, and climate change. The SWAT application categories include river discharge calibration and related hydrologic evaluations, comparisons with other models, sensitivity analyses, and calibration techniques. The United States and China have done the majority of studies reporting SWAT since 1999. However, there are still gaps in SWAT research, such as bias correction of GCMs and regional climate estimations, SWAT+ and SWAT advancement for exceptional flow simulations in various arid/nonarid basins, and integration of machine learning in the SWAT model. The chapter discusses a detailed review of multiple SWAT applications in hydrologic studies. **Chapter 4** discusses how sedimentation in the water reservoir is critical for understanding the reservoir function life cycle.

Using normalized difference water index (NDWI) for the water-spread area and universal soil loss equation (USLE) for soil deposition of the reservoir for 2011–2018, the study showed that soil loss has increased by 5.27% catchment. The integrated geospatial assessment approach provided a holistic understanding of the impact of sedimentation on the usability and life cycle of the reservoir. **Chapter 5** reviews various geospatial models used for assessing the impacts of multiple parameters on water depth data. Then, it evaluates the prospects of the models as aid and tools in the management of dams and reservoirs and water resources management in general. Next, some examples of relevant conceptual models and their application scenarios (case studies) and a comparative presentation of their respective significances have been discussed. Finally, the chapter concludes with a representation of the significant findings that show the influence of the physical characteristics of the water, topography, satellite image resolution, and selected band combination, on the accuracy of the extracted water depth data.

Frequent slope failure incidences have been increasing in the last couple of decades in the Himalayan ranges, mainly due to human activities like deforestation and road construction on unstable slopes. Landslide incidence infrequently involves a single type of movement resulting from various kinds of processes: geology, gravity, weather, groundwater, wind action, and human influence. In this context, **Chapter 6** mapped the landslide-prone areas based on integrated field and geospatial modeling approaches. In this chapter, the authors have shed light on the data inputs and data product terminologies associated with the GIScience. The study used the dip, strike of the joints and faults, fold altitudes, and dip strike of the bedding planes to map the areas that require immediate interventions for the landslide management on the Bandipora-Srinagar National highway. This zone is one of the most vulnerable land sliding areas in Kashmir, the Indian Himalayas. More specifically, data products such as slope angle, aspect, land use/land cover, and contours have been used in the integrated geospatial model for landslide mapping. Data inputs used included ASTER 30 m DEM, georeferenced topo sheets, and Landsat satellite imagery.

Similarly, **Chapter 7** demonstrates integrated geospatial modeling and field-based approach for landslide hazard vulnerability zonation in the Mandi District, Himachal Pradesh, Indian Himalayas. The study used a GIS-based model to assess the landslide zonation using terrain evaluation, geomorphic forms, hydro-meteorological elements, hydro-geological components, and natural and anthropogenic component activities and defined low, medium, and high-risk zones. Further, **Chapter 8** analyzed landslide hazard, exposure, and risk elements in Lower Mandakini Catchment using remote sensing multitemporal data from 1990 to 2015. The study used several causative factors for spatial probability mapping through a logistic regression model in the geospatial environment, while the temporal probability of landslide is calculated using the Poisson distribution model. Spatial probability and temporal probability have been validated by the seed cell area index (SCAI) method to prepare the landslide probability map showing different zones of landslide probability. The study is one of the best vulnerability and risk assessments to plan future loss in probable landslide zones. These studies discuss how the research on the landslide hazard vulnerability for a rugged mountainous area is imperative for a foolproof warning alert system for timely rescue of life and property in such areas.

Climate change assessments from global to local depend on different tools and methods to determine anthropogenic activities' impact on the human-climate system. **Chapter 9** reviewed a vast set of scientific literature focused on the modeling approach for climate change studies that specifically addressed the impact of climate change on agricultural productivity. Land use performance and crop simulation models have been used to understand the role of anthropogenic influences on agricultural productivity. The authors have discussed how impact assessments use process-based crop models to predict crop yields under current and future climate conditions. The role of simulation modeling in the global climate change analysis and the potential use of crop simulation models is indispensable in suggesting that adaptation and mitigation strategies support the regional agricultural production. Further, the review concludes by discussing how geospatial simulation modeling has become the most reliable for assessing the impact of climate change on agriculture could be used in policy and decision-making. **Chapter 10** discusses how the snow and glacier resources in the Himalayan region primarily regulate the water flow in major rivers like Indus, Ganges, and Brahmaputra that drain the South Asian region and are therefore crucial for sustainable development economic development of the people living in the area. Providing an example of Kashmir valley, the study shows the similarity between the cryospheric resources of this region and the rest of the Himalayas, is almost the same and is losing mass at an accelerated rate at nearly the same time rates. The study reviews various studies focusing on cryosphere quantification, analyzing the changing climate, and the consequential impacts on the streamflow components in the Himalayas, specifically focusing on the Kashmir Himalayan region. Such assessments are significant inputs for prudent management of the water resources in the region.

Climate change in high alpine regions is one of the reasons behind the shifting tree lines. **Chapter 11** studies the variations in vegetation line and timberline altitude using the normalized difference vegetation index (NDVI) in Pinder Watershed, Uttarakhand (India). The study uses Landsat images of three periods, 1990, 1999, and 2011, to quantify the change in the vegetation line and timberline altitude. Due to climate change, the study suggests that the vegetation line in this region has shifted 51m toward higher elevation at an average rate of 2.42 m/year during the last 21 years. In comparison, the timberline has gone about 40m to higher elevation at an average rate of 1.90 m/year in the previous 21 years (1990–2011). **Chapter 12** emphasizes the importance of long-term monitoring using Predictive spatial modeling and species distribution models to assess the impacts of climate change on biodiversity distribution and apply fact-based models to expand future dispersals under overall global change. The study has focused on the Himalayan Region and Eastern states of India, i.e., Bihar, Jharkhand, and West Bengal. The study concludes that these approaches have become helpful for producing biogeographical data connected over an expansive scope of fields, including conservation science, ecology, and evolutionary biology.

Chapter 13 focuses on assessing the effects of urban growth on increased temperatures of Hyderabad city using MODIS-based land surface temperature (LST) data. The study models the relationship between modified normalized difference water index (MNDWI) and normalized difference built-up index (NDBI) and tries to identify urban heat islands (UHIs) and urban hot spots (UHSs) influencing the urban climate of the region. The validation of the models used indicated that landscape heterogeneity reduced the relationship between LST with MNDWI and NDBI. Thus, there are some levels of uncertainty in using these indices for climate change studies.

Chapter 14 demonstrates a desertification study in the semi-arid part of the Kanekal area in Anantapur, India. The study uses Landsat TM/ETM+ data for the past 29 years to develop a quantitative geospatial model for assessing desertification and land cover changes in the area. The model used SAVI (Soil Adjusted Vegetation Index) and TGSI (Topsoil Grain Size Index) to identify the degradation and desertification conditions in the study area. The study has proven that the multitemporal analysis of data from the TGSI, SAVI, and NDVI indices can provide a sophisticated measure of ecosystem health and variability. The study is essential to help address the impact of land degradation and desertification on the global socio-economic and biophysical environment. **Chapter 15** studied forest cover changes in the Indus Kohistan Palas valley, Pakistan, primarily due to rapid population growth and road network expansion. Hindu Kush-Himalayan mountains experience a drastic change due to intensive human activities, population growth, proximate causes, accessibility, unstable political situations, government policy failure, and poverty. Geospatial techniques are used to monitor the impact of population growth on forest cover changes. The study demonstrated the role of geospatial technologies in assessing the effects of human developmental works on the fragile Himalayan ecosystem systems. **Chapter 16** reviews up-to-date literature on the role of remote sensing (RS) and geographical information systems (GIS) in assessing various dynamics related to the Earth system. The chapter discusses how Remote sensing and GIS technologies are well-established tools to work hand-in-hand in mapping, analyzing, and disseminating spatial information. The subject of correctly combining both technologies and assessing different earth system processes has been the focus of this chapter.

Chapter 17 is a policy-related geospatial modeling approach wherein the importance of the sustainable livelihood security index (SLSI) in Shyama Prasad Mukherjee Rurban Mission (SPMRP) has been assessed. This SLSI tool has the potential to identify the current sustainability condition and future needs to achieve and hold the sustainability tag. The study focuses on the Chirrakunta urban cluster. Three different indices viz. ecological security index (ESI), economic efficiency index (EEI), and social equity index (SEI) have been dealt with, using both spatial and nonspatial data. The study shows that the only area with a very low SLSI score is Ada Panchayat, which covers about 8.57% of the total area and 4.91% of the total population in this cluster. Hence, it achieves the tag of Highly Sustainable (HS). The other nine Panchayats cover 75.34% of the total area, and 52.38% of the total population comes under the Moderately Sustainable (MS) category. The remaining 16.09% of the total area and 42.71% of the total population come under Low Sustainable (LS) category in this cluster. The study concludes that there is an urgent need to reorient the development policies. **Chapter 18 is** a policy-oriented geospatial modeling study in Shimla, India, using demand-supply analysis of drinking water supplies. The study examines the carrying capacity of water supply in Shimla city, north-western Himalayas. The study reveals that total water demand (2014) is 58.46 million liters per day (MLD) against a system carrying capacity of 54.54 MLD. The gap between demand and supply further increases with system inefficiencies. As a result, some consumers are getting water supply only for about 45 minutes a day. The present deficit of 3.92 MLD (2014) may amplify demand 51.01 MLD in 2041 and 72 MLD in 2071. The study suggested that the sustained strategy must consider identifying the source of supply, collection, transmission, distribution, and other related aspects in Shimla city. The studies focus on the need for the investment for the development on a priority basis with proper resources and opportunities to facilitate their ecological security, economic efficiency, and social equity, which may ascertain sustainable livelihood security.

Editors

Dr. Gowhar Meraj is currently working as Young Scientist Fellow, under the Department of Science and Technology, Government of India's Scheme for Young Scientist and Technologists (SYST) at the Department of Ecology, Environment and Remote sensing, Government of Jammu and Kashmir. Previously, he has worked as Programme Officer on ENVIS scheme of Ministry of Environment, Forests, and Climate Change (MoEF&CC) for five years. He has also worked as a consultant with World Bank Group, New Delhi for its South Asia Water Initiative Program. He has done his M Sc., M Phil. and Ph. D. in Environmental Sciences with special research focus on spatial information sciences. He has also done Post Graduate Diploma in Remote sensing and GIS. He has more than 9 years of experience in research and teaching in various fields related to satellite remote sensing and geographic information science (GIS) such as hazard vulnerability assessments, hydrological modeling, flood assessment, watershed management, natural resource management, ecosystem services, and modeling. He has more than 42 publications in various international and national journals. He is on reviewer boards of various SCI journals such as Information Fusion, Natural Hazards, Arabian Journal of Geosciences, Geocarto International, etc.

Google scholar profile: https://scholar.google.com/citations?user=tLe6BeEAΛA ΛJ&hl=en&oi=ao

Dr. Shruti Kanga is currently working as Associate Professor and Coordinator, Centre for Climate Change and Water Resources, Suresh Gyan Vihar University, Jaipur. She has previously worked as Assistant Professor in the Centre for Land Resource Management, Central University of Jharkhand (CUJ), Ranchi. She had also served as Research Associate in Ministry of Environment, Forest and Climate Change (MoEF), GoI funded project on Forest Fires, and has teaching experience in different universities, i.e., Central University of HNB Garhwal and the University of Jammu. She did her Ph.D. degree in Technology (Geoinformatics) from the Department of Remote Sensing, Birla Institute of Technology, Ranchi, in 2013. She has worked in the area of forest fires risk modeling and management, tourism, resource management, etc. She has attended and organized around 35 national and international conferences. She was also a course coordinator of different modules of EDUSAT based distance learning program organized by IIRS (Indian Institute of Remote Sensing), ISRO (Indian Space Research Organization), Govt. of India. She has more than 12 years of teaching and research experience.

Google scholar profile: https://scholar.google.co.in/citations?user=RSkmve8AAA AJ&hl=en&oi=ao

Dr. Suraj Kumar Singh is currently working as Associate Professor and Coordinator, Centre for Sustainable Development, Suresh Gyan Vihar University, Jaipur. Previously, he had worked as an Assistant Professor in the Centre for Land

Resource Management, Central University of Jharkhand (CUJ), Ranchi. He has also served as a Research Associate in National Remote Sensing Centre (NRSC), Indian Space Research Organisation (ISRO), GoI. Sponsored projects, i.e., National Wasteland Mapping and National Urban Information System for Northern parts of India. He did his Ph.D. degree in Technology (Geoinformatics) from the Department of Remote Sensing, Birla Institute of Technology, Ranchi, in 2012. He has worked in the area of waterlogging and flood hazards, geospatial applications in water resources, disaster management, hydrogeomorphology, urban planning & wasteland mapping. He was also a course coordinator of different modules of EDUSAT-based distance learning program organized by IIRS (Indian Institute of Remote Sensing), ISRO (Indian Space Research Organization), Govt. of India. He has 12 years of teaching and research experience. He had the responsibility of an NSS program officer at the Central University of Jharkhand and Coordinator of Centre for Sustainable development, Suresh Gyan Vihar University, Jaipur.

Google Scholar profile: https://scholar.google.co.in/citations?user=wXlG45AAA AAJ&hl=en

Dr. Majid Farooq is currently working as Scientist-D at Department of Ecology, Environment and Remote Sensing, Government of Jammu and Kashmir, India. He has more than 15 years of experience in research, teaching, and consultancy related to remote sensing and GIS such as climate change vulnerability assessments, flood modeling, ecosystem assessment, watershed management, natural resource management, ecosystem services, and modeling. He has done his M Tech. Ph. D. in Remote sensing and GIS. He has more than 35 publications in various international and national journals. Besides, he has also served as a reviewer in various SCI journals such as Environmental Monitoring and Assessment (Springer), International Journal of Remote Sensing (Taylor and Francis), etc.

Google Scholar profile: https://scholar.google.co.in/citations?user=6X2n7iMAA AAJ&hl=en

Introduction

Protecting the environment in which we live and breathe is perhaps the most pressing concern for all of us today. Environmental changes mainly due to human activity are among the many problems wreaking havoc on the planet. Issues such as global warming, glacier recession, lake eutrophication, unprecedented rainfall, land use-landcover changes, forest degradation, flooding, landslides, and decreased agricultural production have endangered the sustainability of future human generations. Moreover, climate change with unpredictable weather patterns influencing everything from food output to glacier melting has further aggravated the situation. Hence, there's a lot to be concerned about, and quick action is necessary.

It is not that the world has not prepared to take remedial action; nonetheless, an all-inclusive understanding of the system is a must. Effective environmental monitoring and modeling require holistic knowledge of the environment for understanding its current situation and future possibilities. Under present circumstances, environmental modeling is needed to evaluate, monitor, and mitigate environmental issues. Such as during the disaster response, forest fire management, natural resource management, wastewater management, and wastewater spills.

Geospatial modeling using satellite remote sensing and geographic information system (RSGIS) is one such tool that has improved the performance of every sector of the human-environment system. The most successful application of RSGIS is for environmental data analysis and forecasting. It provides a better perspective and comprehension of physical characteristics and the linkages that impact any environmental state. Environmental characteristics and efficient assessments may be determined by viewing and overlaying n number of parameters such as slope steepness, aspect, vegetation, elevation, soil, socio-economic factor, so on and so forth governing an environmental process.

Managing environmental risks and dangers requires completing the data analysis effectively. For example, hazard and risk assessment – a basis of planning choices for mitigation measures in coping with various environmental problems, is aided by RSGIS. The latter can help mitigate the danger of a disaster to a significant extent, whether it is modeling through early warning systems or utilizing decision-making tools to see which disaster would impact or affect which region. The use of RSGIS can improve planning, lead efforts, and response. It allows the response teams to obtain awareness of the problem, connect with the public and understand its impact. GIS will make it easier and faster to identify the places and persons in need.

Geospatial Modeling for Environmental Management: Case Studies from South Asia is a comprehensive book focusing on managing earth resources using innovative techniques of spatial information sciences and satellite remote sensing. The enormous stress on the environment over the years due to anthropogenic activities for commercialization and livelihood needs has increased manifold. The only solution to this problem lies in the awareness of the stakeholders, which can be addressed through scientific means alone. The understanding is the basis of the sustainable development concept, which involves optimal management of natural resources, subject to reliable, accurate, and timely information from the global to local scales.

The geospatial approach involves the use of satellite remote sensing (RS), geographical information system (GIS), and global positioning system (GPS) technology for environmental protection, natural resource management, and sustainable development and planning. Being a powerful and proficient tool for mapping, monitoring, modeling, and managing natural resources can understand the earth's surface and dynamics at different observational scales. Through the spatial understanding of the problem concerning land resources, policymakers can make prudent decisions for restoration and conservation of critically endangered resources, such as water bodies, lakes, rivers, air, forests, wildlife, and biodiversity.

The book comprises research-based chapters from eminent researchers and experts, mainly from South Asia. We have divided the book into five major parts, geospatial modeling in hydrological studies, geospatial modeling in landslide studies, geospatial modeling for climate change studies, geospatial modeling in change dynamics studies, and concluding with geospatial modeling in policy and decision making. The primary focus is to replenish the gap in the available literature on the subject by bringing the concepts, theories, and experiences of the specialists and professionals in this field jointly. The editors have worked hard to get the best literature in this field in a book form for helping the students, researchers, and policymakers develop a complete understanding of the vulnerabilities and solutions to the whole environmental system.

We hope this book shall do service to humanity at large and South-Asian communities in particular. We want to acknowledge the help of all the reviewers who tirelessly read chapters and sent suggestions to authors that greatly enhanced their quality and prospective reach. Special thanks are to Dr. Neelu Gera, Dr. Muzamil Amin, Dr. Ishtiyaq Ahmad Rather, and Dr. Muhammad Muslim for their valuable suggestions during the book's editing. Finally, we would like to thank our families who supported us in thick and thin at each stage of our lives; nothing would have been possible without their help and support.

Editors
Shruti Kanga
Suraj Kumar Singh
Gowhar Meraj
Majid Farooq

Editorial Advisory Board

List of Contributors

Sareer Ahmad
Department of Earth Sciences
University of Kashmir
Srinagar, Jammu and Kashmir, India

Asraful Alam
Department of Geography, Serampore
    Girls' College
University of Calcutta
Kolkata, West Bengal, India

Arfan Arshad
Department of Irrigation and Drainage,
    Faculty of Agricultural Engineering
University of Agriculture
Faisalabad, Pakistan

Arul Prasad S.
Department of Agroclimate Research
    Center
Tamil Nadu Agricultural University,
    Krishi Vigyan Kendra
Tiruvallur, Tamil Nadu, India

K. Raghu Babu
Department of Geology
Yogi Vemana University
Kadapa, Andhra Pradesh, India

Bikram Singh Bali
Department of Earth Sciences
University of Kashmir
Srinagar, Jammu and Kashmir, India

Dillip Kumar Barik
School of Civil Engineering
Vellore Institute of Technology
Vellore, Tamil Nadu, India

Somnath Bera
Centre for Geoinformatics
Tata Institute of Social Sciences (TISS)
Mumbai, Maharashtra, India

J. Brema
Department of Civil Engineering
Karunya Institute of Technology and
    Sciences
Coimbatore, Tamil Nadu, India

Krishan Chand
Aryabhatta Geo-Informatics & Space
    Application Centre
Shimla, Himachal Pradesh, India

Upasana Choudhury
Centre for Climate Change and Water
    Research
Suresh Gyan Vihar University
Jaipur, Rajasthan, India

Majid Farooq
Department of Ecology, Environment
    and Remote Sensing
Government of Jammu and Kashmir
Srinagar, Jammu and Kashmir, India

Abdulkadir Isah Funtua
Department of Surveying and
    Geoinformatics
Abubakar Tafawa Balewa
    University
Bauchi, Nigeria

Nilanjana Ghosal
Jadavpur University
Kolkata, West Bengal, India

Himanshu Govil
Department of Applied Geology
National Institute of Technology
Raipur, Chhattisgarh, India

Subhanil Guha
Department of Applied Geology
National Institute of Technology
Raipur, Chhattisgarh, India

Supratim Guha
Department of Civil Engineering
Indian Institute of Technology Ropar
Rupnagar, Punjab, India
School of Civil Engineering
Vellore Institute of Technology
Vellore, Tamil Nadu, India

Ishfaq Gujree
Key Laboratory of Tibetan Environment
    Changes and Land Surface Processes
Institute of Tibetan Plateau Research,
    Chinese Academy of Sciences
Beijing, China
Key Laboratory of Alpine Ecology and
    Biodiversity
Institute of Tibetan Plateau Research,
    Chinese Academy of Sciences
Beijing, China
CAS Center for Excellence in Tibetan
    Plateau Earth Sciences
Beijing, China
University of Chinese Academy of
    Sciences
Beijing, China

Noor ul Haq
Department of Geography
University of Peshawar
Peshawar, Pakistan

Anand Kumar
Centre for Climate Change and Water
    Research
Suresh Gyan Vihar University
Jaipur, Rajasthan, India

B. Pradeep Kumar
Department of Geology
Yogi Vemana University
Kadapa, Andhra Pradesh, India

Rajesh Kumar
Department of Geography
Government College Haripur Guler
Kangra, Himachal Pradesh, India

Shanti Kumari
Indian Institute of Remote Sensing,
    ISRO
Dehradun, Uttarakhand, India

Venkata Ravibabu Mandla
Centre for Geo-informatics
    Applications in Rural Development
    (CGARD)
National Institute of Rural Development
    & Panchayati Raj, Ministry of
    Rural Development, Government
    of India
Hyderabad, Telangana, India

M. L. Mankotia
Department of Geography
Government Degree College Theog
Shimla, Himachal Pradesh, India

Maragatham M.
Department of Agroclimate Research
    Center
Tamil Nadu Agricultural University
Coimbatore, Tamil Nadu, India

Asif Marazi
Department of Earth Sciences
University of Kashmir
Srinagar, Jammu and Kashmir,
    India

Meenaxi
Department of Geography
S.S.J. University
Almora, Uttarakhand, India

Gowhar Meraj
Department of Ecology, Environment
  and Remote Sensing
Government of Jammu and
  Kashmir
Srinagar, Jammu and Kashmir, India

Habib Ali Mirdda
Department of Geography
Jamia Millia Islamia
New Delhi, India

Laxmipriya Mohanty
Department of Civil Engineering
Veer Surendra Sai University of
  Technology
Burla, Odisha, India

Muhammad Muslim
Department of Environmental
  Science
University of Kashmir
Srinagar, Jammu and Kashmir, India

Bharadwaj Nanda
Department of Civil Engineering
Veer Surendra Sai University of
  Technology
Burla, Odisha, India

Naveen
Department of Agroclimate Research
  Center
Tamil Nadu Agricultural University
Coimbatore, Tamil Nadu, India

Umar Nazir
Department of Earth Sciences
University of Kashmir
Srinagar, Jammu and Kashmir,
  India

N. C. Pant
Department of Geography
S.S.J. University
Almora, Uttarakhand, India

Fazlur Rahman
Department of Geography
University of Peshawar
Peshawar, Pakistan

M. Rajasekhar
Department of Geology
Yogi Vemana University
Kadapa, Andhra Pradesh, India

Ajay S. Rajawat
Geoscience Division
Space Application Centre
  (SAC/ISRO)
Ahmedabad, Gujarat, India

M. Ramachandra
Department of Geology
Yogi Vemana University
Kadapa, Andhra Pradesh, India

Shakil Ahmad Romshoo
Department of Earth Sciences
University of Kashmir
Srinagar, Jammu and Kashmir,
  India

D.D. Sharma
Department of Geography
Himachal Pradesh University
Shimla, Himachal Pradesh, India

Sudhakar B. Sharma
Institute of Science
Nirma University
Ahmedabad, Gujarat, India

Masood Ahsan Siddiqui
Department of Geography
Jamia Millia Islamia
New Delhi, India

Anupam K. Singh
Institute of Engineering and Technology
Indus University
Ahmedabad, Gujarat, India

Bhoop Singh
Department of Science and
   Technology
Natural Resource Data Management
   System
New Delhi, India

Anindita Swain
Department of Civil Engineering
Veer Surendra Sai University of
   Technology
Burla, Odisha, India

Iffat Tabassum
Department of Geography
University of Peshawar
Peshawar, Pakistan

A. Tamilarasan
Department of Civil
   Engineering
Karunya Institute of Technology and
   Sciences
Coimbatore, Tamil Nadu, India

Swayam Vid
Institute of Forest Productivity,
   ICFRE
Ranchi, Jharkhand, India

Vijayashanthi V.A.
Department of Agroclimate Research
   Center
Tamil Nadu Agricultural University,
   Krishi Vigyan Kendra
Tiruvallur, Tamil Nadu, India

Ahsan Afzal Wani
Department of Earth Sciences
University of Kashmir
Srinagar, Jammu and Kashmir, India

Fan Zhang
Key Laboratory of Tibetan Environment
   Changes and Land Surface Processes
Institute of Tibetan Plateau Research,
   Chinese Academy of Sciences
Beijing, China
Key Laboratory of Alpine Ecology and
   Biodiversity
Institute of Tibetan Plateau Research,
   Chinese Academy of Sciences
Beijing, China
CAS Center for Excellence in Tibetan
   Plateau Earth Sciences
Beijing, China
University of Chinese Academy of
   Sciences
Beijing, China

Part A

Geospatial Modeling in Hydrological Studies

1 Flood Vulnerability and Risk Assessment with Parsimonious Hydrodynamic Modeling and GIS

Sudhakar B. Sharma
Nirma University

Anupam K. Singh
Indus University

Ajay S. Rajawat
Space Application Centre (SAC/ISRO)

CONTENTS

1.1   Introduction 3
1.2   Study Area 6
1.3   Methodology 7
    1.3.1   Flood Frequency Analysis 8
    1.3.2   Hydrodynamic Modeling 9
    1.3.3   Flood Vulnerability Analysis 10
    1.3.4   Flood Risk Assessment 10
1.4   Results and Discussion 11
    1.4.1   Land Use Under Risk 14
    1.4.2   Villages Under Risk 14
1.5   Conclusion 15
References 16

1.1   INTRODUCTION

The recent advances in the geographical information system (GIS) and remote sensing (RS) has opened new vistas on flood assessment and forecasting. GIS together with RS is a useful tool to integrate and manipulate not only technical information but also a communication tool for interaction with the public and decision-makers

DOI: 10.1201/9781003147107-2

(Gujree et al., 2020; Hassanin et al., 2020; Kanga et al., 2021). It is at the strategic planning level that the possibility of utilizing information available in cartographic forms assumed significant importance for urban flood mapping (Singh & Sharma, 2009). However, a significant contribution to flood mapping and planning is derived from the information made available by the use of RS technologies and its integration with a GIS (Prasad, 2006). Flooding is considered the world's most costly type of natural disaster in terms of both human causalities and property damage (E.S.A., 2006; Joy et al., 2019). There was a considerable increase in the occurrence of flood and flood-related damages in 2006 globally. The flood events accounted for nearly 55% of all disasters registered and approximately 72.5% of total economic losses worldwide. The annual disaster review indicates that flood occurrence has increased almost ten-fold during the last 45 years, merely 20 events in 1960 to 190 events in 2005 (Scheuren et al., 2007). These extreme flood events have caused significant damage to property, agriculture productivity, industrial production, communication network and infrastructure parallax mainly in the downstream catchment.

In India, the area affected by floods is an average of 4.942 million ha from the year 2000–2009. During this period, the monetary damage to the tune of Indian rupees (INR) 36,004.75 million (Central Water Commission, 2009) is observed. About 40 million ha of land is flood-prone, which is about 12% of the total geographical area of 328 million ha of the country (National Disaster Management Authority, 2007). The floods in India occur typically during the monsoon season (from July to September) caused by the formation of heavy tropical storms over the watershed, increase in imperviousness of urban land surfaces and unregulated construction practices in urban areas.

The river floodplains/river beds are being consistently encroached resulting in ever decreasing channel capacity and sometimes due to tidal backwater effect from the sea for cities situated on estuaries. The Indian subcontinent in general and the Western Peninsula including river basins of Sabarmati, Mahi, Narmada, Tapi, Godavari, Krishna and Kaveri have experienced heavy floods during August 2006. The arid regions of Rajasthan and Gujarat to humid regions in the South were caught by surprise and incurred a heavy loss of personal and property. It is estimated that a single flood during the 7–14 August 2006 event in the LTB in Gujarat state – a river stretch between the Ukai Dam and the Arabian Sea – resulted in 300 people being killed and approximately Rs. 20,000 crore, approximately US$ 4.5 billion, of property damage (National Disaster Management Authority, 2007). The human life was stand-still for almost 2 weeks in Surat and Hazira cities, as well as tens of rural villages along the LTB. The Tapi River in Surat has recorded the highest water level during the last 35-year return period. The flood frequency analysis of the Ghala station shows that there is a likelihood of a flood hazard 1 in 8 years. Therefore, it is important to identify flood risk zones for minimizing loss of life and property.

In light of the above-mentioned scenarios and recent developments in space technology, flood mapping, monitoring and forecasting are expected to contribute significantly to flood disaster management. Early warning of floods is one of the most critical requirements. There is a need to understand the potentials of different hydrological models in conjunction with RS and GIS techniques for improved flood forecast and carry out vulnerability and risk assessment predictions for different return periods of flooding. It would lead to effective flood disaster management and reduce the loss of lives and property.

National Flood Risk Mitigation Project has been envisaged for mitigation or reduction in risk, severity or consequences of floods (National Disaster Management Division, Government of India, 2007). The UN/ISDR (United Nations/International Strategy for Disaster Reduction, 2004) determined that the risk to a particular system has two factors. One factor is the "hazard", and the second factor is the "vulnerability" of a system (e.g., an urban area), which denotes the relationship between the severity of the hazard and the degree of damage caused.

Samantaray et al. (2015) have estimated flood risk through hydrodynamic flood-plain modeling in the flood-prone delta region of the Mahanadi river basin in India. They used the MIKE FLOOD model based on river cross-sections and floodplain elevation model derived from freely available SRTM DEM. Flood hazard maps obtained from the MIKE FLOOD model along with the flood vulnerability data for different rice varieties are used to develop an optimal rice allocation model for the study area. Saini et al. (2012) assessed the risk and vulnerability of flood hazards in part of Ghaggar Basin, Kaithal, Haryana, India. The approach resulted in three classes of flood risk mapping ranging from low to highly vulnerable areas. GIS techniques were utilized to produce flood risk map which was validated with flood that occurred during July 2010. Their findings are that the GIS-based long-term inundation maps can offer a cost-effective solution for planning mitigation measures and preparedness in flood-prone areas.

Sinha et al. (2008) have integrated hydrological analysis with a GIS-based flood risk mapping in the Kosi river basin. They used parameters such as land use topography, population density and geomorphology and proposed a flood risk index based on the Analytical Hierarchy Process (AHP). Flood risk maps were validated with long-term flood events and offered a cost-effective solution for planning mitigation measures in the flood-prone area. A multicriteria analysis (MCA) approach has been utilized in the GIS environment for disaster risk reduction at the village level in Kopili River Basin (KRB) of Assam State, India (Sharma et al., 2017). The proposed approach can be applied elsewhere in other river basins to estimate flood risk useful for mitigating disaster risk posed by the floods.

Khalid et al. (2019) introduced a probabilistic approach for flood risk assessment in a reach of the Richelieu River, south of Saint-Jean-Sur-Richelieu, Quebec, Canada. The approach is based on a combination of three simple modules: (i) flood frequency analysis (frequency and peak discharge), (ii) estimation of inundation depth and (iii) damage and loss estimation. The damage was calculated accounting for the building location and altitude within the flooded zone in combination with hydrologic modeling, hydraulic modeling and stage-damage relationship.

Winter et al. (2018) studied the uncertainties in various flood risk models. Sensitivity analysis of the flood risk model was performed with respect to the five variables. They concluded that the result of a flood risk model is subjective to the selected values of variables, and the better estimation of such values indicates a better decision-making process.

Rincón et al. (2018) developed an updated and accurate flood risk map in the Don River Watershed within the Great Toronto Area (GTA). The risk maps use GIS and

MCA along with the application of AHP methods to define and quantify the optimal selection of weights for the criteria that contribute to flood risk.

Scheuer et al. (2011) assessed flood risk by considering social, economic and ecological dimensions of flood risk. Sharma et al. (2012) have estimated the flood risk index for the Nagaon district of the Assam state by deriving flood hazard zonation from the historical flood inundations observed during 1998–2007 with population density, infrastructure and land use as parameters.

Khan et al. (2011) have estimated the risk of flood in river Indus using historical data of maximum peak discharges and observed which dam/barrage reservoir need to be updated in capacity and whether there are more dams/barrages needed. Guhathakurta et al. (2011) brought out some of the interesting findings that are very useful for hydrological forecasting and disaster management. According to the authors, extreme rainfall and flood risk are increasing significantly in the country except for some parts of central India.

Taubenböck et al. (2011) presented a spatial risk analysis using past sample flood events for the city of Gorakhpur in India and utilized capabilities of multisource RS data for supporting decision-making before, during and after a flood event. Digital Elevation Model and discharge data are the primary parameters to delineate floodplain mapping and risk assessment for Pune City in Maharashtra, India (Pawar et al., 2016).

One of the initial efforts was to identify the flood risk zones which will be useful for future planning of the area (Gogoi et al., 2013). Flood risk zone mapping has been done for the Subansiri subbasin in Assam, India. Sisir and Balan (2014) have prepared flood risk maps of Kadalundi river basin, Kerala based on MCA using GIS and RS. Flood intensity curves have also been generated for the different return periods.

1.2 STUDY AREA

The Tapi River is the second-largest westward draining inter-state river in India after the mighty Narmada River. Tapi river basin covers a geographical area of 65,145 km^2, which is almost 2% of the geographical area of the country. Tapi being an inter-state river basin shares its geographical area with states such as Maharashtra (51,504 km^2) 79%, Madhya Pradesh (9,804 km^2) 15% and Gujarat (3,837 km^2) 5% state. The river rises at Multai (752m MSL), travels 724km westward and has Purna, Girna, Panjhara, and Varekhadi as major tributaries. The Tapi River and its tributaries flow over the plains of Vidharbha and Khandesh and later to Gujarat.

The river basin can be classified into three zones, viz. upper Tapi basin, middle Tapi basin and LTB. The portion between the Ukai Dam and the Arabian Sea has been considered as LTB, mainly in the Gujarat state occupying Surat-Hazira twin city along with tens of small towns and villages along the river course. The LTB with a geographical area of 1,919 km^2 lies between 72° 40′ and 73° 40′ east longitudes and 21° 02′ to 21° 27′ north latitudes situated in the Deccan Plateau in India (Figure 1.1). It is a lower portion of the Tapi River basin (between Ukai Dam to Arabian Sea) coming under the administrative boundary of Gujarat state in Western India. The Total Geographical area under LTB is derived through watershed delineation from DEM. In the study area, there are 385 villages and four urban towns, i.e., Mandvi, Kamrej, Songarh and Surat city.

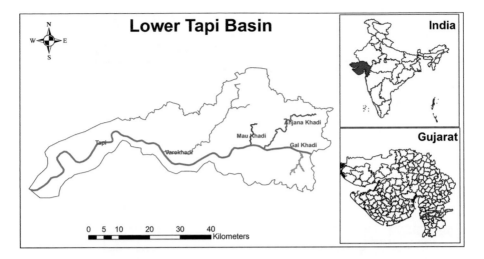

FIGURE 1.1   Location of study area LTB.

The basin has its outlet in the Arabian Sea at Hazira approximately 26 km downstream near Surat city. Surat city is a major commercial hub in the state of Gujarat and is bounded on the three sides by the hill ranges. There are 11 discharge gauge stations and three flood forecasting stations operated by Central Water Commission. The length between two-flood forecasting stations is approximately 180 km and an estimated travel time of 16 hours but LTB does not have any flood forecasting station for almost a stretch of 106 km river length.

The study area experiences a moderate but humid climate having extreme day temperatures ranging from 37.8°C to 44.4°C resulting in hot summers. In winters, night temperatures drop to around 15.5°C. The region receives an average annual rainfall of 1,376 mm associated with heavy daily downpours resulting in flood and water loggings mainly between Ukai Dam and Hazira townships. Most of the rainfall occurs between June and September, and it is observed that 80% of rainfall occurs in July and August.

The main reasons for flooding in LTB are heavy rainfall and discharge due to high water levels from the Ukai Dam. Various issues related to flooding in 106 km stretch of LTB are inundation due to overflowing of the banks, inadequate drainage capacity of the river, congestion at the point of confluence, excessive silt load factor in river discharge and high tide due to proximity to Gulf of Khambhat.

1.3   METHODOLOGY

Flood management is more and more adopting a risk-based approach, whereby flood risk is the product of the probability and consequences of flooding. One of the most common approaches to flood risk assessment is to estimate the damage that would occur for floods of several return periods (Ward et al., 2011; Pandey et al., 2010). Flood risk assessment is an important part of flood risk management; particularly, areas with a high concentration of people and goods are vulnerable to floods. Flood

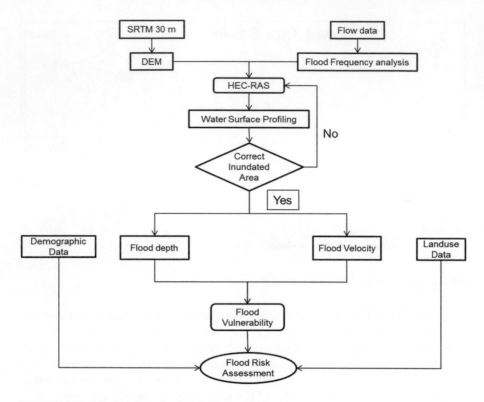

FIGURE 1.2 Methodology for flood risk mapping,

risk management policies have not been developed for the LTB region, and there is an inadequate spatial planning approach. In this chapter, an integrated hydrological modeling approach for flood vulnerability and risk assessment for LTB is described using the inundated scenarios (Figure 1.2).

Flood vulnerability has been assessed based on flood depth and flood velocity in flood inundation scenarios of different return periods. Subsequently, flood risk assessment has been carried out based on the integration of spatial layers of land use/land cover and demographic data of the identified vulnerable classes.

1.3.1 FLOOD FREQUENCY ANALYSIS

The annual peak flow data of the river for 30 years were collected for flood frequency analysis, and the annual peak flow values have been calculated for 100 successive years. Several methods exist for the determination of maximum flood magnitudes. The historical hydrological data recorded during the past events were more reliable for the estimation of the maximum probable flood. Six flood frequency distribution methods for fitting the extreme events are tested: (i) extreme value type I distribution (EVI), (ii) generalized extreme value distribution (GEV), (iii) log-normal distribution parameter two (LN2), (iv) log-normal distribution parameter three (LN3), (v) Pearson type III distribution (P3) and (vi) log Pearson type III distribution (LP3).

Out of these, the LP3 method was selected because it gave the largest discharge compared to others (worst case scenario). This method is the most widely used probability distribution function for extreme values in hydrological studies for the prediction of flood peaks.

Distribution function: The log-Pearson Type III distribution is a refinement to the Pearson Type III distribution. The distribution function is as follows:

$$f(x) = \frac{\left(\ln x - x_0\right)^{-1} \exp\left\{\frac{-\left(\ln x - x_0\right)}{\beta}\right\}}{\beta^\gamma \prod(\gamma)}$$

where x_0, b and g are control location, scale and shape, respectively. The lower bound of the distribution is $\exp(x_0)$. It is difficult to calculate exactly by hand the magnitude of the event with a return period T. The simplest approach is to estimate the T-year event from the sample mean, standard deviation and a factor K_T:

$$x_T = \bar{x} + \sigma K_T$$

The factor K_T varies with return period T and sample skewness g and can be read from tables. Sample skewness g can be calculated directly from the sample or estimated from the parameter g using

$$g = \frac{2}{\sqrt{\gamma}}$$

Parameters of fitted distributions for LP3-III are as follows:
$x_0 = 5.560$ $\beta = 0.756$ $\gamma = 2.534$

1.3.2 HYDRODYNAMIC MODELING

The HEC-RAS is an integrated software system for one-dimensional flow modeling and can be used for steady flow calculation to determine water surface profile. HEC-RAS has capabilities of creating, storing and analyzing geometry data in the GIS environment. In the present analysis, the HEC-RAS software is used for hydrodynamic analyses with a set of parameters (e.g., Manning n) and input data (e.g., flows) that are obtained from the flood frequency analysis. The channel roughness usually specified by using Manning's n also has a significant impact on hydraulic simulations. Manning's n is a commonly assigned range from 0.035 to 0.065 in the main channel and 0.080 to 0.150 in the floodplains. This model uses as input six future scenarios for 2-, 5-, 10-, 25-, 50- and 100-year return periods. The results (e.g., water surface elevations) from HEC-RAS are exported into GIS for the development of corresponding floodplain maps and estimating flood inundation area for LTB. Therefore, the HEC-GeoRAS tool not only increases the precision of hydrodynamic modeling but also generates floodplain maps for various return periods. This analysis assumes that natural channel meets uniform flow condition; it means, the energy gradient is approximately equal to the average channel bed slope and that water surface elevation can be obtained from normal depth calculations.

Flood plain boundary and flood depth maps have been prepared for the different return periods. The grid cell size is 30×30 m based on the input DEM resolution for creating various geometry layers, i.e., river channel, banks, and cross-sections. Flood simulations have been generated over a 30×30 m cell size for the six flood scenarios (i) $T = 2$ years, (ii) $T = 5$ years, (iii) $T = 10$ years, (iv) $T = 20$ years, (v) $T = 50$ years, (vi) $T = 100$ years.

1.3.3 FLOOD VULNERABILITY ANALYSIS

HEC-RAS model has been utilized to simulate flood inundation for different return periods of flooding. Detailed water surface profiling at various locations in the Tapi river was carried out and utilized to improve the prediction of river discharge and depth downstream, and river gauge data were utilized to validate the accuracy. The spatial extent of flood inundation for different return periods could be generated. Flood depth and velocity maps were also generated for LTB. The flood depth map is the water surface elevation grid minus the grid representing the ground elevation or DEM. The depth values for each depth cell are calculated by subtracting the ground elevation value from the water surface elevation value for each return period or flood scenario computed. The topographic data used for the development of any depth grid should be the same source as used to generate the effective floodplain boundaries to ensure consistency of the same projection system and grid size for accurate results. In the present study, 30m grid size elevation data have been taken. Any negative values from the resulting depth grids were removed from the output result by either removing the cells or setting them to depths of zero.

A flood velocity map is a digital representation of flood velocity distribution throughout the floodplain by mapping the velocity output data obtained from hydrodynamic models. Any point on the grid describes the average flood velocity for that floodplain location for a given flood frequency. The hydrodynamic model provides velocity grids by connecting specific velocity values in between cross-sections. Velocity distributions and values generated from 1-D models are typically linearly interpolated from cross-section to cross-section, whereas there is likely more variation of flood velocities in reality. The velocity grid can provide a general awareness of areas within the floodplain where flood velocities are likely to be higher than their surrounding areas (Altaf et al., 2013, 2014; Meraj et al., 2015, 2016).

1.3.4 FLOOD RISK ASSESSMENT

The process of defining the nature and impacts of hazards involves hazard analysis. Hazard analysis can be divided into four steps:

1. Hazard identification
2. Vulnerability analysis
3. Risk analysis
4. Reality check

Hazard identification provides information on the nature and characteristics of the hazardous event to the community. It characterizes the extent of the potential risk to

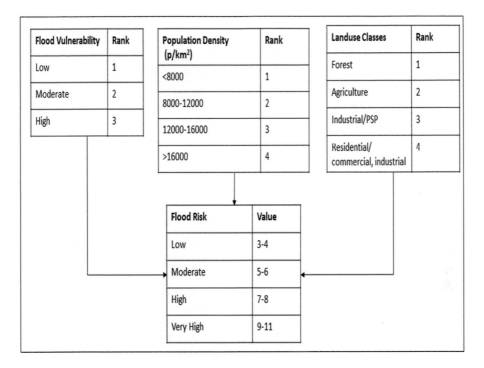

Flood Vulnerability	Rank
Low	1
Moderate	2
High	3

Population Density (p/km²)	Rank
<8000	1
8000–12000	2
12000–16000	3
>16000	4

Landuse Classes	Rank
Forest	1
Agriculture	2
Industrial/PSP	3
Residential/ commercial, industrial	4

Flood Risk	Value
Low	3–4
Moderate	5–6
High	7–8
Very High	9–11

FIGURE 1.3 Flood risk ranking,

property, human life and for a whole community. Vulnerability analysis is a measure of a community's exposure to incur a loss or focus on physical, political, economic, and social parameters. Vulnerability analysis can identify the geographic areas that may be affected, individuals who may be subject to injury or death, and facilities, property, or environment may be susceptible to damage from the event. Risk analysis is described as an assessment of the likelihood (probability) of an accidental release of hazardous material and the consequences that might occur, based on the estimated vulnerable zones. Vulnerability-, land use- and village-wise population density maps were primarily utilized. Overlay analysis in GIS was carried by assigning four ranks to each of the thematic map, and risk classes as low, moderate, high and very high were obtained from the multicriteria method (Figure 1.3).

1.4 RESULTS AND DISCUSSION

An integrated approach by incorporating a hydrodynamic model (HEC-RAS) with GIS technology to predict a flood event in both spatial and temporal contexts for future scenarios has been developed. Probability distributions for the yearly maximum discharges were computed by carrying out flood frequency analysis using six distribution methods viz., (i) extreme value type I distribution (EVI), (ii) generalized extreme value distribution (GEV), (iii) log-normal distribution parameter two (LN2), (iv) log-normal distribution parameter three (LN3), (v) Pearson type III distribution (P3) and (vi) log Pearson type III distribution (LP3).

TABLE 1.1

Flood Magnitude at various return periods

Return Period (Years)	Discharge in m³/s					
	EV	GEV	LN2	LN3	P3	LP3
2	3,799	2,421	2,120	2,099	3,096	1,864
5	9,986	6,348	6,250	6,552	9,478	5,904
10	14,083	10,532	10,998	11,588	14,086	11,619
25	19,259	18,736	20,095	21,115	20,109	25,634
50	23,099	27,994	29,661	31,031	24,662	44,176
100	26,910	41,161	42,101	43,823	29,233	73,860

Maximum instantaneous flow recorded at Tapi River (Ghala Station) for a period of 30 years (1978–2007) was analyzed using the abovementioned six flood distribution methods. The flood magnitude for 2-, 5-, 10-, 25-, 50- and 100-year return periods based on six flood distribution methods are summarized in Table 1.1. The results indicate that the flood discharge till 25-year return period is similar; however, for higher return period, it differs in a large amount. EV and P3 both methods are giving quite similar outputs. GEV, LN2 and LN3 results match each other. The LP3 distribution method provides a very high frequency of water for 100 years. The existing methods of flood frequency analysis produce variable outcomes, particularly at the higher return periods. The result from the distribution method was correlated with observed data, and the highest correlation coefficient of 0.98 is observed for the LP3 distribution method. LP3 outcomes were used as an input in hydrodynamic flood inundation for the study area. The size of the flood with a return period of 100 years or an annual exceedance probability of 1% is 73,860 m³/s from the LP3 distribution method. Probability distribution plays a vital role in designing structures for the proper management of water resources and mitigation of flood disasters.

The purpose of a flood vulnerability assessment is to identify the areas prone to flooding based on various factors. Flood vulnerability is categorized based on the level of susceptible to flooding when the discharge of a river exceeds the bankfull stage. Flood vulnerability assessment is the estimation of the overall adverse effects of flooding. It depends on many parameters such as depth of flooding, duration of flooding, flood wave velocity and rate of rising of water level. In the present study, the depth of flooding and velocity of river water were considered for flood vulnerability assessment. Overlay analysis in GIS was carried out using two parameters, i.e., depth and velocity by assigning four ranks. It led to the mapping of vulnerability classes as low, moderate and high. The intensity of flood vulnerability is always given by a relative scale, which represents the degree of vulnerability, called a vulnerability class. A lower vulnerability class is assigned for a lower depth and lower velocity or low vulnerability, while a higher vulnerability class is assigned to indicate a higher vulnerability. Flood vulnerability maps of 2, 5, 10, 25, 50 and 100 year return periods were generated for

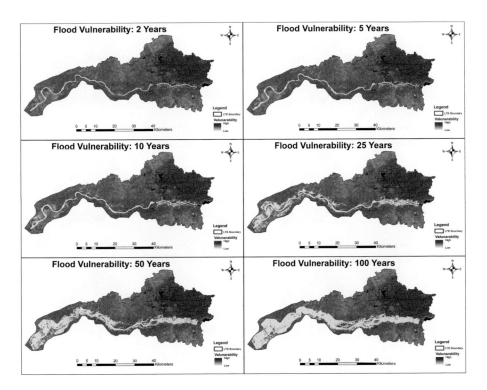

FIGURE 1.4 Flood vulnerability maps of LTB for the different return periods.

TABLE 1.2

Area Under Flood Vulnerability Classes for Different Return Periods

Flood Vulnerability	Area (km²)					
	2 Year	5 Year	10 Year	25 Year	50 Year	100 Year
Low	2	6	11	20	14	9
Moderate	7	12	25	45	40	27
High	48	68	99	201	315	410

LTB from the Ukai Dam to the Arabian Sea (Figure 1.4). The vulnerable classes were classified into three major categories as high, moderate and low.

The results are analyzed based on area, which is vulnerable to flood (Table 1.2)

Flood risk analysis has been done using vulnerability and land use with population density. Flood risk map were generated on various return periods i.e. 2, 5, 10, 25, 50 and 100 years. Spatial analyses were carried out for assessing the area under low, moderate, high and very high-risk zone. Below Figure 1.5 depicts the spatial extent of risk zones, and Table 1.3 shows the quantified result.

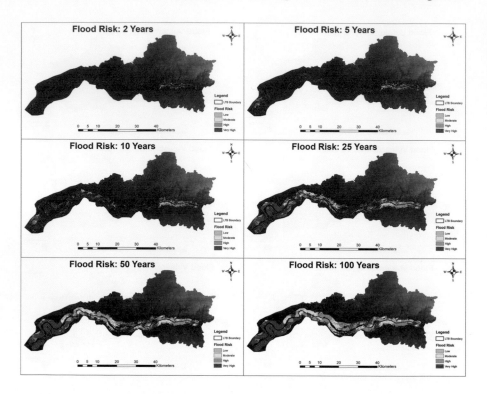

FIGURE 1.5 Flood risk maps.

TABLE 1.3
Total Area Under Flood Risk Classes for Different Return Periods

Flood Risk	2 Year	5 Year	10 Year	25 Year	50 Year	100 Year
Low	4	8	15	33	39	40
Moderate	3	7	16	40	65	79
High	9	9	18	41	65	91
Very high	2	10	32	95	139	172
Total (km^2)	17	35	81	209	308	381

1.4.1 LAND USE UNDER RISK

Flood risk map overlaid on the land use map depicting risk classes for different return periods were analyzed. Flood risk for land use classes for different return periods was computed and is shown in Table 1.4. It is observed that agriculture and residential areas are under the very high-risk category in LTB.

1.4.2 VILLAGES UNDER RISK

There are a total of 388 villages and one urban city (Surat) in the LTB from the Ukai Dam to the Arabian Sea. It is observed that there is no major loss during floods of 2-, 5- and

TABLE 1.4
Flood Risk for Land Use/Land Cover Classes

Land Use	Risk Area (km²) for Different Return Periods					
	2	5	10	25	50	100
Agriculture	8	22.4	50	125.7	194.5	247.5
Commercial	0	0	0.1	0.9	1.7	2.2
Forest	0.5	1.8	3	6.5	9.7	12
Industrial	0.2	0.5	1.4	2.8	3.6	4.2
Open	0	0	0.2	0.5	0.6	0.7
PSP	0.1	0.5	1.5	5.5	8.5	9.9
Transpiration	0.1	0.5	1.2	3.5	4.8	5.5
Recreational	0	0	0.3	0.6	1	1.2
Residential	1.3	6.4	22.9	64.2	85.5	96.8

TABLE 1.5
Villages Under Very High Risk for Different Return Periods

S. No.	Return Period (years)	Very High Risk			
		No. of Villages	Population (lac)	No. of Households (lac)	No. of Workers (lac)
1	2	No major loss			
2	5				
3	10				
4	25	10	44.67	9.46	17.94
5	50	15	48.80	10.35	19.92
6	100	23	49.25	11.22	21.08

10-year return period. However, ten villages are under very high risk with 4,467,000 population for 25-year return period (Table 1.5).

During 50 years of flood event, 15 villages are likely to submerge including Surat city and approximately 48.80 lacs population shall be affected by the flood. There are 23 villages including Surat city under very high risk for a 100-year return period, and an approximate 49.25 lacs population shall be affected. 21.08, 19.92 and 17.94 lac workers may be affected during floods with 100-, 50- and 25-year return period.

1.5 CONCLUSION

The study observed a high correlation between predicted discharge and inundation area as well as the return period and inundation area. Steady flow analysis executed in HEC-RAS indicated that the discharge of 1,863, 5,903, 11,618, 25,633, 44,176, and 73,859 m³/s inundated 57, 88, 139, 270, 374 and 446 km² area for return periods of 2, 5, 1,025, 50 and 100 years, respectively. Validation of simulated results with observed data for five flood events, viz., 1988, 1990, 1994, 1998 and 2006 for 13 locations in LTB shows a good correlation, i.e., $R^2 = 0.87$.

- In this study, 59 river cross-sections were prepared using SRTM DEM data of 30 m resolution for computing channel geometry characteristics using HEC-GeoRAS. In addition to the usage of this outcome in the HEC-RAS hydrodynamic model, river cross-sections were also examined for understanding any anomaly along and across the river. Tapi River has reduced drastically at Surat. Field observations indicate that high siltation and encroachment are the two major factors for this reduction.
- Flood vulnerability of LTB for flood return periods of 2, 5, 10, 25, 50 and 100 years could be assessed based on predicted flood depth and velocity in the GIS environment. GIS-based weighted overlay analysis was performed using the multicriteria method for calculating flood vulnerability. Three classes of vulnerability were ascertained, viz., Low, Moderate and High.

Flood vulnerability of LTB for flood return periods of 2, 5, 10, 25, 50 and 100 years could be assessed based on predicted flood depth and velocity in the GIS environment. An approach of data integration in a GIS environment is observed to be extremely useful in carrying out flood vulnerability and risk assessment for LTB.

REFERENCES

Altaf, F., Meraj, G., and Romshoo, S. A. "Morphometric Analysis to Infer Hydrological Behaviour of Lidder Watershed, Western Himalaya, India." *Geography Journal* 2013(-2013): 1–14.

Altaf, F., Meraj, G., and Romshoo, S. A. "Morphometry and Land Cover Based Multi-Criteria Analysis for Assessing the Soil Erosion Susceptibility of the Western Himalayan Watershed." *Environmental Monitoring and Assessment* 186.12(2014): 8391–8412.

Central Water Commission. NTBO Gandhinagar. Tapi Basin-Year Book, 2009. http://old.cwc.gov.in/regionaloffices/ntbo/Water%20yer%20book/TapiWYB12-13.pdf, page accessed on 13 April 2017.

Gogoi, C., et al. "Flood Risk Zone Mapping of the Subansiri Sub-Basin in Assam, India." *International Journal of Geomatics and Geosciences* 4.1(2013): 75–88.

Guhathakurta, P., Sreejith, O. P., and Menon, P. A. "Impact of Climate Change on Extreme Rainfall Events and Flood Risk in India." *Journal of Earth System Science* 120.3(2011): 359–373.

Gujree, I., Arshad, A., Reshi, A., and Bangroo, Z. "Comprehensive Spatial Planning of Sonamarg Resort in Kashmir Valley, India for Sustainable Tourism." *Pacific International Journal* 3.2(2020): 71–99.

Hassanin, M., Kanga, S., Farooq, M., and Singh, S. K. "Mapping of Trees outside Forest (ToF) From Sentinel-2 MSI Satellite Data Using Object-Based Image Analysis." *Gujarat Agricultural Universities Research Journal* 207(2020): 204–213.

Joy, J., Kanga, S., and Singh, S. K. "Kerala Flood 2018: Flood Mapping by Participatory GIS Approach, Meloor Panchayat." *International Journal on Emerging Technologies* 10.1(2019): 197–205.

Kanga, S., Rather, M. A., Farooq, M., and Singh, S. K. (2021). "GIS Based Forest Fire Vulnerability Assessment and Its Validation Using field and MODIS Data: A Case Study of Bhaderwah Forest Division, Jammu and Kashmir (India)." *Indian Forester* 147.2(2021): 120–136.

Khalid, O., et al. "Flood Risk Mapping for Direct Damage to Residential Buildings in Quebec, Canada." *International Journal of Disaster Risk Reduction* 33(2019):44–54.

Khan, B., et al. "Flood Risk Assessment of River Indus of Pakistan." *Arabian Journal of Geosciences* 4.1–2(2011): 115–122.

Meraj, G., et al. "Assessing the Influence of Watershed Characteristics on the Flood Vulnerability of Jhelum Basin in Kashmir Himalaya." *Natural Hazards* 77.1(2015): 153–175.

Meraj, G., Romshoo, S. A., and Altaf, S. "Inferring Land Surface Processes from Watershed Characterization." In J. Janardhana Raju (ed.), *Geostatistical and Geospatial Approaches for the Characterization of Natural Resources in the Environment.* Springer, Cham, 2016, 741–744.

National Disaster Management Authority, Government of India, New Delhi. National Disaster Management Guidelines – Preparation of State Disaster Management Plan, 2007.

Pandey, A. C., Singh, S. K., and Nathawat, M. S. (2010). "Waterlogging and Flood Hazards Vulnerability and Risk Assessment in Indo Gangetic Plain." *Natural Hazards* 55.2: 273–289.

Pawar, A., Sarup, J., and Mittal, S. K. "Application of GIS for Flood Mapping: A Case Study of Pune City." *International Journal of Modern Trends in Engineering and Research* 3.4(2016): 474–478.

Prasad, A. K. "Potentiality of Multi-Sensor Satellite Data in Mapping Flood Hazard." *Journal of India Society of Remote Sensing* 34.3(2006): 219–231.

Rincón, D., Khan, U. T., and Armenakis, C. "Flood Risk Mapping Using GIS and Multi-Criteria Analysis: A Greater Toronto Area Case Study." *Geosciences* 8.8(2018): 1–27.

Saini, S. S., et al. "Risk and Vulnerability Assessment of Flood Hazard in Part of Ghaggar Basin : A Case Study of Guhla Block, Kaithal, Haryana, India." *International Journal of Geomatics and Geosciences* 3.1(2012): 42–54.

Samantaray, D., et al. "Flood Risk Modelling for Optimal Rice Planning for Delta Region of Mahanadi River Basin in India." *Natural Hazards* 76.1(2015): 347–372.

Scheuer, S., Haase, D., and Meyer, V. "Exploring Multicriteria Flood Vulnerability by Integrating Economic, Social and Ecological Dimensions of Flood Risk and Coping Capacity: From a Starting Point View Towards an End Point View of Vulnerability." *Natural Hazards* 58.2(2011): 731–751.

Scheuren, J.-M., et al. Annual Disaster Statistical Review, 2017. https://www.unisdr.org/-files/2796_DREDAnnualStatisticalReview2007.pdf, page accessed on 1 January 2016.

Sharma, S. S. V, R. G. Srinivasa, and V. Bhanumurthy. "Development of Village-Wise Flood Risk Index Map Using Multi-Temporal Satellite Data: A Study of Nagaon District, Assam, India." *Current Science* 103.6(2012): 705–712.

Sharma, S. S. V., et al. "Flood Risk Assessment Using Multi-Criteria Analysis: A Case Study from Kopili River Basin, Assam, India." *Geomatics, Natural Hazards and Risk* 5705 (2017): 1–15.

Singh, A. K., and Sharma, A. K. "GIS and a Remote Sensing Based Approach for Urban Flood-Plain Mapping for the Tapi Catchment, India." *IAHS Publications* 331 (2009): 389–394.

Sinha, R., et al. "Flood Risk Analysis in the Kosi River Basin, North Bihar Using Multi Parametric Approach of Analytical Hierarchy Process (AHP)." *Journal of the Indian Society of Remote Sensing* 36.4(2008): 335–349.

Sisir, P., and Balan, K. "Flood Risk Mapping of Kadalundi River Basin Using GIS." *International Journal of Scientific & Engineering Research* 5.7(2014): 235–240.

Taubenböck, H., et al. "Flood Risks in Urbanized Areas – Multi-Sensoral Approaches Using Remotely Sensed Data for Risk Assessment." *Natural Hazards and Earth System Science* 11.2(2011): 431–444.

United Nations/International Strategy for Disaster Reduction. Living with Risk: A Global
    Review of Disaster Reduction Initiatives, United Nations International Strategy for
    Disaster Reduction, 2004.
Ward, P. J., et al. "How Are Flood Risk Estimates Affected by the Choice of Return-Periods?"
    Natural Hazards and Earth System Science 11.12(2011): 3181–3195.
Winter, B., et al. "Sources of Uncertainty in a Probabilistic Flood Risk Model." *Natural
    Hazard* 91.2(2018): 431–446.

2 Estimation of Parameters in Ungauged Catchment Using Map-Correlation Method
A Case Study on Krishna-Godavari Basin

Laxmipriya Mohanty, Anindita Swain, and Bharadwaj Nanda
Veer Surendra Sai University of Technology

CONTENTS

2.1 Introduction 19
2.2 MCM 20
2.3 Case Study on Krishna-Godavari River Basin in Southern India 21
2.4 The Effect of Distance on the Correlation of Daily Streamflow Data 23
2.5 Map Correlation Results for Krishna-Godavari Basins 26
2.6 Application Map Correlation Technique in Ungauged Catchments 31
2.7 Conclusions 33
References 34

2.1 INTRODUCTION

The rivers play a crucial role in the civilization of India. It can be evident from the fact that most of the major cities in India are located near one or other river banks. The rivers facilitate irrigation, transportation, and electricity generation apart from their direct consumption. The availability and the quality of the water are directly dependent on the properties of its catchment area; an area that eventually drains the water, which is caught in it, to the river. The amount of water caught by the catchment and the amount of water that is drained out of it are directly dependent on several hydro-morphological and environmental parameters (Hassanin et al., 2020; Tomar et al., 2021; Chandel et al., 2021; Bera et al., 2021). Some of these parameters like catchment area, soil type, catchment slope, etc. are perpetual, while few other parameters like daily streamflow, precipitation, temperature, humidity, wind, etc. are

DOI: 10.1201/9781003147107-3

19

measured in regular intervals for the assessment of flow characteristics at the stream outlet.

Daily streamflow time series data is one of the most important parameters for planning and designing a hydraulic structure, whose estimation is challenging both analytically and conceptually (Sivapalan, 2003; Arsenault and Brissette, 2013). In the meantime, India has more ungauged catchment than any other comparable country. This deficit is sometimes alleviated through the selection of a reference stream gauge for those ungauged catchments. The oldest and most prevalent method used to determine these ungauged parameters is the drainage area ratio method. Apart from that, regional regression and nonlinear spatial interpolation methods are also used in a limited manner. All the techniques are based on the common principle i.e. identification of donor catchment similar to the ungauged one concerning the location or behavior and transformation of information from donor to the ungauged basin (He et al., 2011; Pandey et al., 2013). For the prediction of continuous streamflow data, the entire set of rainfall-runoff data is to be transferred to an ungauged catchment as a whole. Since the prediction of time series data in ungauged catchment depends on donor gauges. the most critical task is to identify the proper reference (donor) gauge for daily streamflow estimation.

Most of the conventional methods preferably use the nearest stream gauge station as the reference one. However, the nearest catchment may not always be the most correlated one. This confuses a probable source of error. The recent map correlation method (MCM), which is another geostatistical method employed to analyze such missing data, can overcome such impediments (Gorsich and Genton, 2000.). The present chapter demonstrates this method through a case study to predict catchment parameters for ungauged sites in Krishna-Godavari river basin in southern India using a single reference stream gauge method (Altaf et al., 2013, 2014; Meraj et al., 2015, 2016).

2.2 MCM

MCM has the advantages of selecting the most correlated stream gauge, which need not be the nearest one, as the donor for the ungauged basin. Two assumptions are made in this method viz. (i) there exists a hydrologic similarity among daily streamflow and the correlation coefficients associated with them and (ii) the correlation coefficients are isotropic in the selected study area. In this method, the reference stream gauge providing the best correlation is selected instead of the nearest one for an ungauged catchment. For this, a variogram used to model the spatial structure of the correlation coefficient with each of the stream gauges resulting from a leave-one-out cross-validation scheme, to calculate spatial covariance, of the respective variograms. Advantageously for any location in the study area, the cross-correlation between the data of a gauged and ungauged site can be figured out using this method. A bare minimum of stream gauges in the network is needed. Additionally, it doesn't assume the study area is limited by regions of homogeneity or other hydrology boundaries. MCM is based on ordinary kriging which is used to interpolate the values for the construction of a continuous map showing the correlation between the pseudo-ungauged catchment (i.e. gauged in actual but

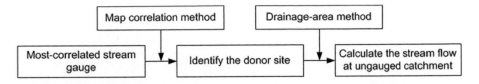

FIGURE 2.1 Flowcharts for streamflow estimation at ungauged catchments.

considered as ungauged to cross-check the method) and rest gauged catchments. For the pseudo-ungauged catchment, correlation maps with all the gauged catchments need to be formed. Then the gauged catchment producing the best correlation map with pseudo-ungauged catchment is considered as donor (reference) catchment for that. The MCM types used in this study was the type suggested by Archfield and Vogel (2010) as shown in the flowchart given in Figure 2.1. For a reliable prediction, a minimum number of catchments in the network is necessitated for using the MCM; however, this study does not focus on that issue.

2.3 CASE STUDY ON KRISHNA-GODAVARI RIVER BASIN IN SOUTHERN INDIA

The MCM is demonstrated through a case study on the Krishna-Godavari river basin in southern India to predict the catchment parameters for different ungauged sites. Krishna and Godavari rivers are two major rivers of peninsular India originating from the Western Ghat mountain range near Mahabaleshwar and Trimbakeshwar, respectively, in the state of Maharashtra, India. Lengthwise, the river Krishna is over 1,400 km, and the Godavari river is over 1,465 km. Both are east-flowing rivers that ultimately discharge into the Bay of Bengal. With a total drainage area of, respectively, 258,948 and 321,813 km², the Krishna and Godavari river system covers approximately 18% of the total geographic area of India.

Nine gauged stations are located in the Krishna river basin for which are Dhannur, Haralahalli, Karad, Madhira, Murvakonda, Mudhol, Shimoga, Takli and Yadgir stream gauge stations. Similarly, eight gauged stations namely Dhalegaon, Jagdalpur, Konta, Mancherial, Pathagudem, Polavaram, Purna, and Yelli are located in the Godavari river basin. The locations of these stream gauge stations are pointed on the India map in Figure 2.2.

The drainage area, the latitude and longitude, and the land use and land cover patterns for associated drainage areas are collected from the Central Water Commission (CWC, India) shown in Table 2.1. The daily streamflow, precipitation, temperature, relative humidity, and wind data at these gauge locations are collected from the Global weather website (https://globalweather.tamu.edu/) for a ten-year duration (from 1st January 1992 to 31st December 2001). The shape of the drainage basin is obtained from the ESRI (Environmental System Research Institute) website. ArcGIS 10.3 is used to capture spatial information.

DEM data was used for 30 m SRTM, 80 numbers of tiles were downloaded from earth explorers. And with the help of earthdata.gov six numbers of LULC tiles were downloaded here to examine the LULC classification of the selected study region.

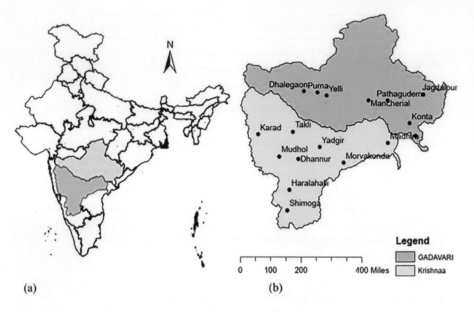

(a)　　　　　　　　　　　　　　　　　(b)

FIGURE 2.2　Location of stream gauge stations in Krishna-Godavari river basin. (a) Location of Krishna-Godavari Basin. (b) Location of 17stream gauges.

TABLE 2.1
Details of 17 Stream Gauges in Southern India

Stream Gauge	River	Drainage area in Sq. Km	Latitude (N) and Longitude (E)	LU/LC
Dhalegaon	Godavari	30,840	19.22 N 76.36 E	Cropland
Jagdalpur		7,380	19.10 N 82.02 E	Water body
Konta		19,550	17.80 N 81.38 E	Barren land
Mancherial		102,900	18.83 N 79.45 E	Open forest
Pathagudem		40,000	18.81 N 80.35 E	Shrub land
Polavaram		307,800	17.24 N 81.65 E	Forest
Purna		15,000	19.17 N 77.01 E	Vegetation
Yelli		53,630	19.04 N 77.45 E	Grassland
Dhannur	Krishna	48,900	16.20 N 76.10 E	Waterbody
Haralahalli		14,582	14.82 N 75.67 E	Barren land
Karad		5,462	17.29 N 74.19 E	Waterbody
Madhira		1,850	16.91 N 80.35 E	Herbaceous
Morvakonda		210,500	16.03 N 78.26 E	Woodland
Mudhol		6,734	16.31 N 75.20 E	Forest
Shimoga		2,831	13.93 N 75.57 E	Barren land
Takli		33,916	17.41 N 75.84 E	Forest
Yadgir		69,863	16.73 N 77.12 E	Waterbody

The arc map geographic coordinate system was WGS-1984 (world geodetic system). WGS-1984 has earth-centered geodetic datum and earth-fixed terrestrial system of reference. The World Geodetic System 1984 explains the earth's shape and size. To know the LU/LC pattern, the UTM-46 was used.

2.4   THE EFFECT OF DISTANCE ON THE CORRELATION OF DAILY STREAMFLOW DATA

The stream gauge nearest to the ungauged catchment can be selected as the reference station, only if there exists a good mutuality among distance and correlation among them. To estimate this mutuality, the distance between each set of the catchment is measured and differentiated to the correlation coefficient Pearson's r, which is calculated from the logarithms of day-to-day discharge data between the sets of stream gauges. Further quantification of the correlation between two streamflow time series can be done using Kendall's r, a nonparametric estimator of correlation to measure monotonic (linear or nonlinear) relation between these data sets (Helsel and Hirsch, 2002; Gujree et al., 2017).

The relation between Pearson's r correlation coefficient (X-axis) computed using logarithms of the daily streamflow values between each pair of stream gauges located in southern India in 10 years (from 1 January 1992 to 31st December 2001) and distance (Y-axis) for 17 study stream gauges is presented in Figure 2.3. The most correlated stream gauges for the different reference gauges are shown in Table 2.2. The correlation between stream-flows at the most correlated reference stream gauge and the study stream gauge is represented through Pearson's r. It can be observed in the table that the nearest stream gauge has not always the highest relationship.

Further, the goodness of fit among observed and the estimated arithmetic and log-transformed streamflows was evaluated using the Nash-Sutcliffe efficiency (NSE) value which equals one minus the ratio of the error variance of the modelled time series divided by the variance of the observed time series (Nash and Sutcliffe, 1970). The NSE is computed as followed:

$$\text{NSE} = \frac{\sum_{t=1}^{T}\left(Q'_m - Q'_0\right)^2}{\sum_{t=1}^{T}\left(Q'_0 - \overline{Q_0}\right)^2}$$

where $\overline{Q_0}$ denotes the most observed discharge, Q_m the model discharge, and Q'_0 the observed discharge at time t. If the model considered is perfect with an estimation error variance equal to zero, then the resulting Nash-Sutcliffe Efficiency equals one. As the variation is in the order of magnitude of streamflow, the magnitude of NSE using logarithmic streamflow values is more trusted over arithmetic computations. The use of E in evaluating the goodness of fit for high and low streamflow values has been discussed by Oudin et al. (2006). Table 2.3 provides a comparison of NSE values for the present 17 stream gauge stations obtained from the drainage area method

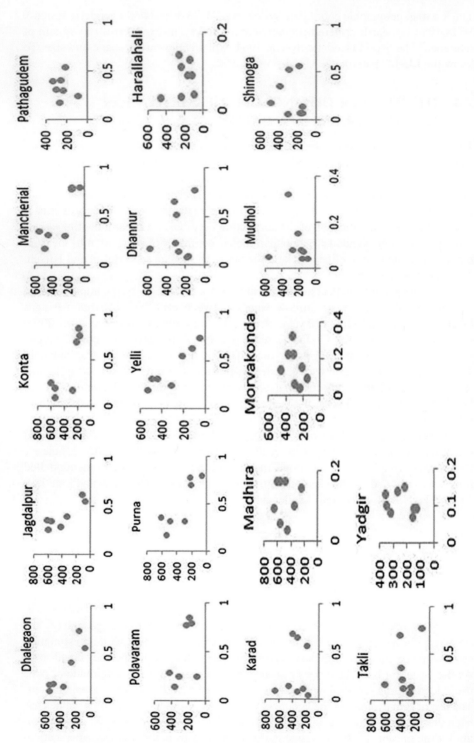

FIGURE 2.3   Correlation coefficient (presented in X-axis) vs. distance in km (presented in Y-axis) for 17 study stream gauges.

TABLE 2.2

Closest Reference Stream Gauge and Reference Stream Gauge Having the Stream-Flows Most Correlated to 17 Stream Gauges Located in Southern New India

Study Stream Gauge	Nearest Stream Gauge	Distance (km) of Nearest Stream Gauge from the Study Stream Gauge	Most Correlated Stream Gauge	Pearson's r Coefficient
Dhalegaon	Purna	68.64	Yelli	0.62
Jagdalpur	Konta	158.93	Pathagudem	0.85
Konta	Polavaram	65.87	Polavaram	0.79
Mancherial	Pathagudem	95.03	Pathagudem	0.56
Pathagudem	Mancherial	95.03	Jagdalpur	0.85
Polavaram	Konta	65.87	Konta	0.79
Purna	Yelli	48.70	Yelli	0.73
Yelli	Purna	48.70	Purna	0.73
Dhannur	Mudhol	96.93	Takli	0.21
Haralahalli	Shimoga	99.23	Karad	0.65
Karad	Mudhol	153.67	Shimoga	0.69
Madhira	Polavaram	142.88	Morvakonda	0.38
Morvakonda	Yadgir	146.40	Yadgir	0.42
Mudhol	Dhannur	126.66	Morvakonda	0.32
Shimoga	Haralahalli	99.23	Haralahalli	0.77
Takli	Mudhol	139.00	Karad	0.56
Yadgir	Dhannur	126.66	Morvakonda	0.42

by adopting the reference station of the nearest stream gauge and the stream gauge having the most correlated streamflow.

The range and median NSE values using the nearest stream gauge as the reference stream gauge and the most correlated stream gauge calculated using arithmetic and log-transformed data sets are presented in Figures 2.4–2.7. It may be observed from Table 2.3 and Figures 4a–7a that the median NSE value computed using both arithmetic and log-transformed data gives a higher magnitude if the most correlated stream gauge is considered as the reference stream gauge. It can be seen from Figures 4b and 6b that except in only two cases selecting the most correlated gauges as the reference yielded improved NSE value using arithmetic-transform data. Similarly, using log-transform data, the most correlated gauges perform better than the nearest stream gauge except only three cases out of 17 stream gauges considered in this study.

Those stream gauges having lower NSE values have been considered to be not having their correlated stream gauge in the selected study area. The correlation between Pearson r and NSE is used to provide information regarding uncertainty estimation in ungagged stream gauges. For the present study, NSE values calculated using log transformation data were compared to the distance and Pearson r-correlation coefficient values of the respective reference stream gauges mentioned in Table 2.3 and

TABLE 2.3

Comparison of NSE Values by Adopting Reference Station of the Nearest Stream Gauge and the Stream Gauge Having the Most Correlated Streamflow

Study Stream Gauge	NSE Value Computed from Arithmetic Transformed Observed and Estimated Values for the Reference Stream Gauge		NSE Value Computed from Log-Transformed Observed and Estimated Values for the Reference Stream Gauge	
	Nearest Stream gauge	Most Correlated Stream Gauge	Nearest Stream Gauge	Most Correlated Stream Gauge
Dhalegaon	0.37	0.37	0.09	0.09
Jagdalpur	−0.67	0.39	−0.21	0.3
Konta	0.56	0.69	0.28	0.81
Mancherial	0.69	0.69	0.47	0.47
Pathagudem	0.69	0.69	0.47	0.47
Polavaram	0.55	0.55	0.27	0.27
Purna	0.93	0.93	0.61	0.61
Yelli	−0.3	0.94	−0.01	0.62
Dhannur	0.09	0.22	−0.01	−0.01
Haralahalli	−0.1	−0.1	−0.5	−0.5
Karad	−0.02	0.2	0.02	0.05
Madhira	−0.03	0.51	−0.06	0.21
Murvakonda	0.28	0.28	−0.18	−0.18
Mudhol	0.29	0.87	0.05	0.54
Shimoga	0.36	0.36	−0.06	−0.06
Takli	0.82	0.82	0.51	0.51
Yadgir	−0.35	0.3	−0.1	0.23

presented in Figure 2.8. It is interesting to note from the figure that the certainty in the estimation of streamflow is more for a higher r-value. This signifies that the most correlated stream gauge as the reference stream gauge has more importance than separation distance for accurate prediction.

2.5   MAP CORRELATION RESULTS FOR KRISHNA-GODAVARI BASINS

To ensure the estimated correlation can represent the whole range of daily discharge time series values for the stream gauges in the study area accurately, it is a primary necessity that the record of each stream gauge should coincide with others for a longer duration. Further, another assumption made in the present case study is that the Pearson "r" is isotropic across the study area. The estimated correlation between a reference stream gauge at Dhalegaon with 16 other stream gauges present in the study region is geographically illustrated in Figure 2.9. The estimated correlation

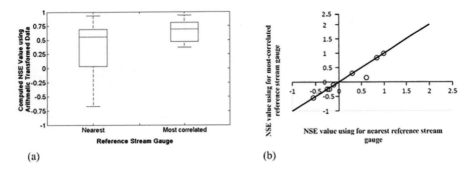

(a)

(b)

FIGURE 2.4 NSE values for stream gauge stations in Godavari basin using arithmetic-transform data. (a) Range and median of NSE values for daily stream flow data. (b) Comparison of NSE values using nearest and most correlated reference stream gauge.

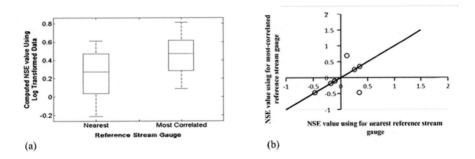

(a)

(b)

FIGURE 2.5 NSE values for stream gauge stations in Godavari basin using log transform data. (a) Range and median of NSE values for daily stream flow data. (b) Comparison of NSE values using nearest and most correlated reference stream gauge.

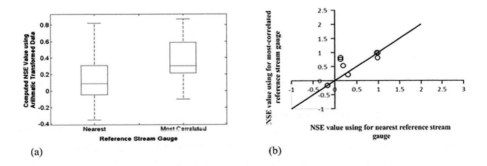

(a)

(b)

FIGURE 2.6 NSE values for stream gauge stations in Krishna basin using arithmetic transform data. (a) Range and median of NSE values for daily stream flow data. (b) Comparison of NSE values using nearest and most correlated reference stream gauge.

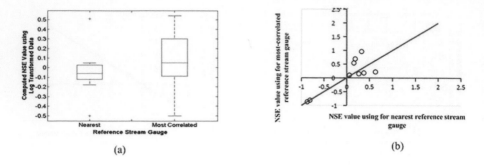

FIGURE 2.7 NSE values for stream gauge stations in Krishna basin using log transform data. (a) Range and median of NSE values for daily stream flow data. (b) Comparison of NSE values using nearest and most correlated reference stream gauge.

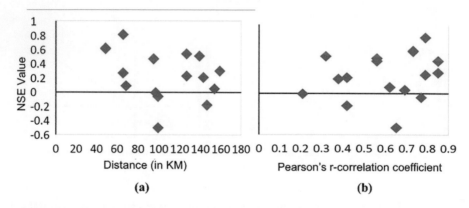

FIGURE 2.8 Relation between NSE value with distance and Pearson's r correlation coefficient. (a) NSE value vs. distance in km. (b) NSE value vs. Pearson's-r.

values for all stream gauges are summarized in Figure 2.10b considering the semi-variogram model in Figure 2.10a.

Between each pair of stream gauges, a relationship between the semivariance, *i.e.*, the squared difference among *r*-values and corresponding separation distance as the Euclidean distance, is obtained for each pair of stream gauges in the study area and plotted as per the suggestions of Isaaks and Srivastava (1989). This plot usually has no discernable pattern containing ($n=2$) points, due to which it is often called the variogram cloud, where n represents the number of stream gauge stations under study. In the present case, the value of $n=17$.

The relationship between semivariance and separation distance is presented by sample variogram models wherein the points constituting the variogram cloud are estimated for the separation distances for each stream gauge station from the others (Isaaks and Srivastava, 1989). Thereafter, a variogram model, the continuous function establishing the relationship between the semivariance and the separation distance, is constituted and fitted to obtain the sample variogram.

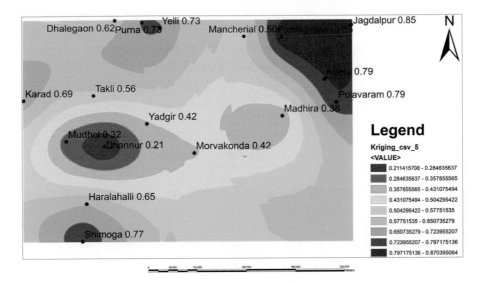

FIGURE 2.9   Pearson's *r* correlation coefficient for Dhalegaon with 16 other stream gauges.

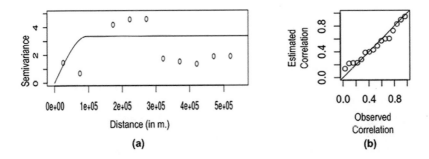

FIGURE 2.10   (a) Variogram model used in this study. (b) Observed vs. estimated correlations for the variogram model.

Then the semivariance of a bin is considered to be the average value for that bin considered in the study. During binning, sample variograms are formed, which can be further used to fit the model variogram. The length and the number of bins are fixed visually through the trial and error method from the plots of sample variograms and fitted models. The bin length thus obtained was used to fit the variogram model for all 17 stream gauges. The weighted least squares method was used to fit the variogram model where the number of points used to estimate each position on the sample variogram is used to calculate the weights (Ribeiro and Diggle, 2001).

Among several types of model variograms here, the spherical type has been selected as it is relatively easier to formulate and visualize for most of the sample variograms (Meraj et al., 2021). Variables of the spherical variogram model should be calculated for every study stream gauge. From the above two equations of spherical variogram, the covariance between two *r* values at any Euclidean distance can be

TABLE 2.4
The Spherical Variogram Model Parameters Calculated from the Pearson's *r* Correlation Coefficient

Study Stream Gauge	Variance Parameter	Range Parameter
Dhalegaon	0.034	129682.1
Jagdalpur	0.061	133708.7
Konta	0.029	142814
Mancherial	0.360	132072.5
Pathagudem	0.350	131256.5
Polavaram	0.257	141216.7
Purna	0.351	129707.2
Yelli	0.357	167346.2
Dhannur	0.362	130814
Haralahalli	0.352	131620.5
Karad	0.319	135648.4
Madhira	0.322	173754.9
Morvakonda	0.294	170449.8
Mudhol	0.622	131918.9
Shimoga	0.336	95735.78
Takli	0.361	134766
Yadgir	0.368	133848.5

determined. Using the geo-R software package, the spatial covariance structure was estimated. For this, a leave-one-out cross-validation experiment was conducted for every study stream gauge. The result of the cross-validation and the variogram model represents a powerful spatial structure between the Pearson's correlation coefficient r values for the selected study stream gauge.

For all 17 variogram models shown in Table 2.4, 16 r-values were removed systematically from the sample variogram followed by reestimation of the variogram model parameters. For each removed stream gauge, the r-value for the removed stream gauge was estimated using a model variogram and compared with the observed value. The E value was calculated from the 16 estimated and observed values of r to provide a measure of goodness of fit for each of the 17 variogram models. For a given study stream gauge, the spatial structure between R-values can be interpreted from the variogram models and cross-validation results. For most of the stream gauges though it is possible to estimate reliability through geostatistics, some stream gauges can have a stronger spatial structure than others.

The covariance describing the spatial relationship among the separation distance and the r-values are obtained along with parameters obtained from the model variogram. The unbiased value of r between any ungauged location and each of the stream gauges is estimated by ordinary kriging using the above-mentioned inputs. The estimated covariance function and parameters are used to obtain a 17×17 matrix of separation distances between each pair of stream gauges known as a covariance matrix. Further, the distance from 17 stream gauges and the outlet of the ungauged catchment is found followed by the execution of covariance function over these distances to obtain a 17×1 matrix. A set of 17 weights are estimated through ordinary kriging by multiplying the transposed 17×1 matrix with the 17×17 matrices obtained above. Finally, the unbiased, minimum variance of the correlation between the ungauged catchment and the given study stream gauge is estimated by multiplying these weights with respective r values. Extended discussion on this method for estimation of krigged value has been described by Isaaks and Srivastava (1989).

With the application of ordinary kriging, a continuous map of correlation between a given stream gauge and the total study area can be obtained. These correlation maps represent the spatial distribution of the correlation between a given stream gauge and any other point in the study area.

The appropriate measure needs to be taken while using Pearson's r correlation coefficient with data sets containing zero data (Koutsoyiannis et al., 2008). To estimate Pearson's correlation coefficient r which was not evaluated, the density of the catchment network and period of records are utilized; however, in the restriction (some catchment with less no coincident period of record), this would assuredly present a critical constraint for the usability of the method.

2.6 APPLICATION MAP CORRELATION TECHNIQUE IN UNGAUGED CATCHMENTS

The spatial relation can be easily represented for the study stream gauge using the ordinary kriging technique. The use of the MCM considering the reference stream gauge as the most correlated basin is demonstrated in this section. In the previous part, the correlation between the reference stream gauge and the study stream gauge was calculated directly from the daily streamflow data. However, in the present section, it is demonstrated whether the selected MCM can execute the selection of the reference stream gauge which is the nearest one when the correlation coefficient can't be calculated directly from the available data.

The entire 17 study stream gauges was supposed to be an ungauged catchment where there was a lack of data due to the absence of a meteorological station. The variogram model was developed at one, and all of the other 16 stream gauges were used to calculate correlation among the ungauged catchment and other 16 possible reference steam gauges. The ungauged catchment was excluded while making the sample variogram. To ascertain that the ungauged basin did not impact the selection of the reference stream gauge, the variogram representation parameters were then refitted. The calculated correlations among one and all of the 16 possible reference stream gauges were estimated, and the stream gauge possessing the greatest calculated r-value was chosen as the reference stream gauge. Further, the drainage area ratio method was applied to the ungauged catchment. In the previous section, NSE values were calculated from both arithmetic and log-transformed calculated and observed daily streamflow values and other parameter's value at every ungauged catchment. The objective of the NSE calculation is comparison only.

A comparison is required for the selection of the nearest stream gauge as a reference stream gauge.

Being roundabout one-third of the entire stream gauge, the ungauged basin in the MCM was accurately chosen and the most correlated stream gauge as the reference stream gauge. The reference stream gauge which was not selected by the MCM, the difference in r values linking the stream gauge selected by MCM, and most correlated stream gauge was standard 0.25.

The selected MCM submits equal outputs as the previous experiment, which applied the perceived correlations to choose the reference stream gauge. Normally the NSE values derived from the selected reference stream gauge using the MCM shown to greater NSE values than the nearest stream gauge was chosen. In reality, most of the NSE values are greater, in some cases significantly greater, than the NSE values deriving from the election of the nearest reference stream gauge; when the nearest stream gauge was chosen as reference stream gauge, the stream gauge having higher NSE value was chosen.

It is essential to point out that, for any cases where the nearest stream gauge was selected as a reference stream gauge exceeding the MCM, distance does not prepare an authentic framework to assess the unreliability of the calculated streamflow values. Based on the entire study, using the above-discussed method, the geostatistical analysis is possible. But the selected MCM is an improved class of all the methods that can be applied for geostatistical analysis at ungauged catchment. The correlation between daily streamflow at an ungauged catchment and a potential reference stream gauge by Kriging which is shown in Figure 2.11. Finally, using the MCM, the reference stream gauge is identified as shown in Table 2.5. The difference in the correlation value for the stream gauge selected by this method and the most correlated stream gauge station identified in Section 2.4 are shown in the table. It can be observed from the difference that the present MCM can identify the most correlated stream gauge in eight of the seventeen cases identified by a 0 (zero) difference in the correlation values. Apart from this, in all other cases, the difference in the correlation values is less. Therefore, the MCM can be adapted to identify the potential reference stream gauge stations.

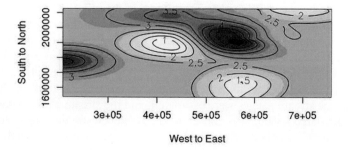

FIGURE 2.11   Ordinary kriging by g stat for the entire stream gauge.

TABLE 2.5

Reference Stream Gauges Identified from MCM for Estimation of Discharge at Different Ungauged Stations

Study Stream Gauge	Most Correlated Stream Gauge	The Difference of r between the Most Correlated Stream Gauge and Stream Gauge Selected by MCM
Dhalegaon	Yadgir	0.2
Jagdalpur	Yadgir	0.43
Konta	Yadgir	0.3
Mancherial	*Yadgir*	*0*
Pathagudem	Yadgir	0.4
Polavaram	Yadgir	0.23
Purna	Jagdalpur	0.21
Yelli	Jagdalpur	0.2
Dhannur	*Jagdalpur*	*0*
Haralahalli	*Jagdalpur*	*0*
Karad	Jagdalpur	0.1
Madhira	*Jagdalpur*	*0*
Morvakonda	*Jagdalpur*	*0*
Mudhol	*Purna*	*0*
Shimoga	Konta	0.21
Takli	*Konta*	*0*
Yadgir	*Jagdalpur*	*0*

2.7 CONCLUSIONS

This study presents the MCM, an improved class of methods to calculate the correlation in all the parameters linking the stream gauges at the ungauged location. As an example, a study to calculate the streamflow value at ungauged catchments in Krishna-Godavari basin is considered. This method has a possible implementation in several hydrological problems such as calibration of hydrologic models, assessment of stream gauge networks, and catchment classification. The correlation among all the parameters for ungauged catchment and a suitable reference stream gauge using kriging the perceived correlation among all the parameter time series at a group of stream gauges were evaluated using the MCM. A strong spatial relationship among the correlations over the study area was identified, and these relations were then modelled by the conventional drainage area ratio method. Spatial models were applied to evaluate the correlation among the ungauged stream gauge and a group of stream gauges and choose the reference stream gauge developed at the maximum correlation value. It was observed from these extensive studies that the MCM delivered an improved estimation of all the parameters over the statistics and could be able to identify the most correlated stream gauge in eight out of the seventeen catchments, and in all other cases, the difference was less.

REFERENCES

Altaf, F., Meraj, G., and Romshoo, S. A. 2013. "Morphometric analysis to infer hydrological behaviour of Lidder watershed, Western Himalaya, India." *Geography Journal* 2013, 1–14.

Altaf, F., Meraj, G., and Romshoo, S. A. 2014. "Morphometry and land cover based multi-criteria analysis for assessing the soil erosion susceptibility of the western Himalayan watershed." *Environmental Monitoring and Assessment* 186(12), 8391–8412.

Archfield, S. A., and Vogel, R. M. 2010. "Map correlation method: selection of a reference stream gauge to estimate daily streamflow at ungauged catchments." *Water Resources Research* 46, W10513. http://dx.doi.org/10.1029/2009WR008481

Arsenault, R., and Brissette, F. P. 2013. "Continuous streamflow prediction in ungauged basins: the effects of equifinality and parameter set selection on uncertainty in regionalization approaches." *Water Resources Research* 50, 6135–6153. http://dx.doi.org/10.1002/2013/WR014898

Bera, A., Taloor, A. K., Meraj, G., Kanga, S., Singh, S. K., Durin, B., and Anand, S. 2021. "Climate vulnerability and economic determinants: Linkages and risk reduction in Sagar Island, India; A geospatial approach." *Quaternary Science Advances* 100038. http://dx.doi.org/10.1016/j.qsa.2021.100038

Chandel, R. S., Kanga, S., & Singhc, S. K. 2021. "Impact of COVID-19 on tourism sector: a case study of Rajasthan, India." *AIMS Geosciences* 7(2), 224–242.

Gorsich, D., and Genton, M. 2000. "Variogram model selection via nonparametric derivative estimation." *Mathematical Geology* 32(3), 249–270.

Gujree, I., et al. (2017). "Evaluating the variability and trends in extreme climate events in the Kashmir Valley using PRECIS RCM simulations." *Modeling Earth Systems and Environment* 3(4), 1647–1662.

Hassanin, M., Kanga, S., Farooq, M., and Singh, S. K. 2020. "Mapping of Trees outside Forest (ToF) from Sentinel-2 MSI satellite data using object-based image analysis." *Gujarat Agricultural Universities Research Journal* 207, 207–212.

He, Y., Bardossy, A., and Zehe, E., 2011. "A review of regionalization for continuous stream-flow simulation." *Hydrology and Earth System Sciences* 15, 3539–3553.

Helsel, D., and Hirsch, R. 2002. *Statistical Methods in Water Resources Techniques of Water Resources Investigations*, Book 4, Chap. A3. U.S. Geological Survey, United States.

Isaaks, E. H., and Srivastava, R. M. 1989. *An Introduction to Applied Geostatistics*, 1st ed. Oxford University Press, New York. ISBN: 9780195050127.

Koutsoyiannis, D., Yao, H., and Georgakakos, A. 2008. "Medium-range flow prediction for the Nile: A comparison of stochastic and deterministic methods." *Hydrological Sciences Journal* 53(1), 142–164.

Meraj, G., et al. 2015. "Assessing the influence of watershed characteristics on the flood vulnerability of Jhelum basin in Kashmir Himalaya." *Natural Hazards* 77(1), 153–175.

Meraj, G., Romshoo, S. A., and Altaf, S. 2016. "Inferring Land Surface Processes from Watershed Characterization." In N. Janardhana Raju (ed.), *Geostatistical and Geospatial Approaches for the Characterization of Natural Resources in the Environment*. Springer, Cham, 741–744.

Meraj, G., Singh, S. K., Kanga, S., and Islam, M. N. 2021. "Modeling on comparison of ecosystem services concepts, tools, methods and their ecological-economic implications: a review." *Modeling Earth Systems and Environment* 1–20. https://doi.org/10.1007/s40808-021-01131-6

Nash, J. E., and Sutcliffe, J. V. 1970. "River flow forecasting through conceptual models part I—A discussion of principles." *Journal of Hydrology* 10(3), 282–290.

Pandey, A. C., Singh, S. K., Nathawat, M. S., and Saha, D. 2013. "Assessment of surface and subsurface waterlogging, water level fluctuations, and lithological variations for evaluating groundwater resources in Ganga Plains." *International Journal of Digital Earth* 6(3), 276–296.

Ribeiro, P., Jr., and Diggle, P. 2001. "geoR: A package for geostatistical analysis." *R-News* 1(2), 15–18. ISSN 1609-3631.

Sivapalan, M. 2003. "Prediction in ungauged basins: A grand challenge for theoretical hydrology." *Hydrological Processes* 17, 3163–3170.

Tomar, J. S., Kranjčić, N., Đurin, B., Kanga, S., and Singh, S. K. 2021. "Forest fire hazards vulnerability and risk assessment in Sirmaur district forest of Himachal Pradesh (India): A geospatial approach." *ISPRS International Journal of Geo-Information* 10(7), 447.

3 Soil and Water Assessment Tool for Simulating the Sediment and Water Yield of Alpine Catchments
A Brief Review

Ishfaq Gujree and Fan Zhang
University of Chinese Academy of Sciences
CAS Center for Excellence in Tibetan Plateau Earth Sciences

Gowhar Meraj and Majid Farooq
Government of Jammu and Kashmir

Muhammad Muslim
University of Kashmir

Arfan Arshad
University of Agriculture Faisalabad

CONTENTS

3.1 Introduction 38
3.2 Studies Using SWAT for Climate and Land-Use Changes Assessment 40
3.3 SWAT Use in Hydrology with Different Sources of Hydro-Metrological
    Input Data 41
    3.3.1 Studies Using SWAT Runoff Assessment 41
    3.3.2 Comparison of SWAT with Other Models with Future SWAT
          Improvements 42
    3.3.3 SWAT Model Using Snow and Glacier Parameter Assessment 43

DOI: 10.1201/9781003147107-4

    3.3.4   The Efficiency of SWAT Model on Soil Erosion and Sediment
            Flux in Streamflow 44
    3.3.5   Studies Using SWAT for Soil Erosion 45
    3.3.6   Studies Using SWAT for Sediment Dynamics 45
    3.3.7   The Efficiency of SWAT in Simulating the Impact of Land Use
            Land Cover Change on Hydrological Processes 46
    3.3.8   Studies Using SWAT for LULC Change in Affecting Sediment
            Dynamics 47
    3.3.9   Studies Using SWAT for Assessing LULC and Climate on
            Watershed Hydrology 47
    3.3.10 Studies Using SWAT for Forest Assessment 48
    3.3.11 Studies Using SWAT for Nutrient Modelling 48
3.4   Conclusions 49
Acknowledgments 52
Competing Interest 52
References 52

3.1   INTRODUCTION

For water resource evaluation and management, the quantification of water balance component variations is essential. The best approach for sustainable watershed management is the knowledge of the efficacy of water management interventions. The combinations of driving forces increase the pressure on the irrigation, energy production, industrial, domestic, and environmental water supplies needed by local, national, and regional governments. Water is a vital part of life. It is an integral part of society's social and economic well-being. Extremes in hydroclimate, like droughts or floods, have probably increased due to climate change and could result in severe socio-economic and environmental impacts (Gujree et al., 2017; Arshad et al., 2019). The absence of hydrological data is the main constraint on the structuring of hydrological models such as SWAT. The detailed knowledge of changing climate and land use/cover change (LUCC) influx promotes the long-term planning, management, and management of water in various alpine regions worldwide (Kanga et al., 2017a, b; Hassanin et al., 2020; Kanga et al., 2021). The working of SWAT model is shown in Figure 3.1.

Khelifa et al. (2021) tried to achieve sustainable watershed management with knowledge about soil and water preservation effectiveness. Their study tests the transferability to the El-Gouazine catchment by the Soil and Water Assessment Tool (SWAT) model of the terraced agricultural catchment experimental site (Sbaihia) of a Bench Terrace parameter set (BTPS). Their results showed potential transfer of BT parameters from the catchment using a SWAT model as a workable option for other associated surface discharge and sediment yield studies, based on descriptive CN/MUSCLE parameters. Hu et al. (2021) have been analyzed further by an extensive ecological restaurant program in the Chinese Loess Plateau (e.g., grain-for-green). They have dramatically modified their use of the Wei River (the largest tributary in the Yellow River Basin, the WRB). Their study was carried out using the soil and water evaluation instrument to simulate the impacts of land use and land cover

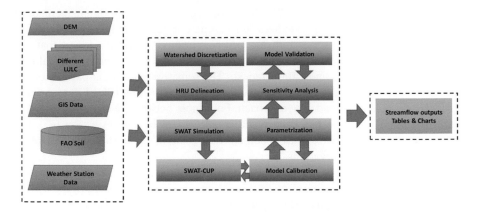

FIGURE 3.1   Working part of soil and water assessment tool (SWAT).

change (LUCC) on major hydrological components (SWAT). Moreover, in areas where crops with slopes above 15° have been transformed into grassland or forests, the change in ET was evident, which suggests that slope is also crucial for hydraulic responses to LUCC.

Singh and Kanga (2020) analyzed the simulated water flow and water balance components of their study using SWAT model in the Ib River Mahanadi River Basin. The daily precipitation and monthly streamflow data for the Model development from 1993 to 2011 were used to measure tropical rainfall. Their study shows that evapotranspiration lost almost 26%–50% of precipitation. Liu et al. (2015) recommended that their study be part of the CEAP to quantify conservation practices' environmental and economic benefits in cultivated cropland across the United States. In their study, SWAT under the United States Hydrologic Modeling (HUMUS) was used. The analysis showed multiple model calibration challenges, such as good quality data availability, numerous reservoir accounting within a sub-sub-water-shaft, inadequate elevation, and slope reporting in mountainous areas. Further, insufficient carriage capacity representation of canals and insufficient capture of irrigation return flows accentuate the problems.

Cuceloglu et al. (2017) suggested their study was first conducted with a high-resolution hydrological model to evaluate the water budget components of aquatic resources in Istanbul. A continuous, half-distributed, process-based model, SWAT. The results show that Istanbul's annual blue-water potential is 3.5 billion m³, with green-water flow and storage of 2.9 million m³, respectively, and 0.7 billion m³. Pontes et al. (2016) discussed the primary limitation on water resource management due to Brazil's lack of hydrological information. In an adjacent drainage basin, the SWAT Hydrological Model was used for the Basin of the Camanducaia River. Their results have concluded that the parametric transfer is a promising technique for modeling unheard catchments that may contribute to water resources management in the Mantiqueira river basins and in other hydrological monitoring regions.

Swain et al. (2016) concluded that to simulate hypothetical, actual, and future scenarios, the SWAT was a valuable method of assessing alternative land-use effects on sediment and pollutant losses. SWAT has been used in watershed management to incorporate future climate parameters from climate models, land use patterns/land cover patterns, and soil series information. Narsimlu et al. (2015) found that the Ton basin has great relevance in India regarding water resources and ecological balance to Madhya Pradesh and Uttar Pradesh. The sensitive hydrologic parameters of the basin were identified using a Sequential Uncertainty Fitting (SUFI-2) technique through a hydrological modeling approach. For fitness between observation and ultimate best simulation use has been made of coefficient of determination (R^2), the efficiency of Nash–Sutcliffe (NSE), percent bias (PBIAS), and Standard Deviation Ratio (RSR) RMSE observations.

The key parameter that controls the groundwater flow regime in broken aquifers has been accessed by Partsinevelou (2017) as a crucial fracture pattern in the hydrogeological study. They analyzed the pattern of fractures which shows a high degree of uniformity in the fragmentation in all lithologies. The model SWAT showed that the hydrological parameters calculated could be linked to a pattern of fractures. High rates of infiltration also occur in areas at which density and degree of interconnection are high. Abbaspour et al. (2015) studied the severe degradation and decreasing water levels in many parts of Europe, leading to adverse environmental effects. They are working to build a comprehensive European hydrological model by using the SWAT, simulating various water resource components, and considering crop yields and water quality at the level of the Hydrological Response Unit (HRU). They have a general approach and methods which can be used in any large region of the world.

3.2 STUDIES USING SWAT FOR CLIMATE AND LAND-USE CHANGES ASSESSMENT

Kibii et al. (2021) claimed that Kenya was the world's 21st largest country regarding water access. Their study utilized the SWAT model to evaluate the effect on the catching yields of land use and climatic variability that lead to high fluvial fluctuations. The use of land has also been changed due to increased catchment settlement, leading to a reduction in forest cover, from around 37% in 1989 to 26% in 2019. The impact of different land-use change scenarios, such as deforestation, forestation, and urbanization, on water balance, simulated by Zhang et al. (2020), using an improved SWAT in eastern Australia. The result shows that change in land use affected all the hydraulic variables and had the most notable annual impact on surface runoff. However, during wet season months (December to May), the scenarios for land usage changes showed absolute change. Samimi et al. (2020) studied the SWAT publications for 20 years in agricultural watersheds arid and semi-arid irrigated. In their review, the model's use to better understand the aspects of water quantity management was strictly dominant. They concluded that the modelers must be encouraged to apply a multi-component (e.g., streamflow, evapotranspiration, groundwater recharging) calibration and report limitations of parameter values and regional relevance to the contextualization of model performance. Abbas et al. (2016) analyzed that the Lesser Zab, a half-arid region of northern Iraq with increased variations and is supposed to contribute to more significant droughts and fluxes due to climate change. For a half centennial lead time to 2,046–2,064 and a centennial lead time to 2,080–2,100, they used outputs of six GCMs in three emission scenarios: A1B, A2, and B1 used the SWAT model for evaluating the impact of climate change on their hydrological components. Its results showed an aggravation of the system of water resources.

Zhang et al. (2016a) analyzed in a similar study the contributions made by the LUCC to hydrologic change in the Heihe River Basin in Northwest China to the effects of climate change and land use. SWAT model showed that the LUCC had the most hydrological effect in the HRB middle stream area, accounting for 92.3%, 79.4%, and 92.8% in the 1980s, 1990s, and 2000s. Guo et al. (2016) proposed promoting a longer-term water planning and management in the arid regions of North-West China through SWAT through a separation approach to better understand the effects of climate change and LUCC. Their results show that during 1990–2008, climate change (i.e., precipitation and temperature change) dominated the streamflow, resulting in a 102.8% increase in the average annual current, while LUCC generated a 2.8% decrease. Pradhan et al. (2015) concluded that an assessment of the potential climate change impacts on the agriculture of Indrawati Basin, Nepal, was necessary to plan adequate adaptation measures. It describes farm owners' perception of climate change, analysis of water availability, and future temperature and precipitation projections. Farmers are already identifying the adaptation strategies and recommendation of new ones based on primary farmer information and an in-depth assessment of climate data.

3.3   SWAT USE IN HYDROLOGY WITH DIFFERENT SOURCES OF HYDRO-METROLOGICAL INPUT DATA

The influence of spatial resolution on the Garonne river watershed by a gridded weather dataset was explored by Grusson et al. (2015) using SWAT. The following data groups are evaluated by comparing: Ground-based weather stations, the SAFRAN 8 km product, and several derived SAFRAN networks with upscale grids of 16, 32, 64, and 128 km. Their results indicate that the difference in climate representation is more important than the spatial resolution of the model, an analysis confirming similar performance achieved in the SWAT model calibrated on SAFRAN networks of 16 and 32 km. Leta et al. (2015) introduced the procedure to evaluate the uncertainty of semi-distributed hydrological models with additional unknown calibration algorithm factors and the autoregressive conditional nature of the neighborhood errors, in addition to a list of model parameters. They used DREAM(zs) to determine the parameter of the retrospective distributions and the resulting uncertainty of a River Senne SWAT model (Belgium). Hetero-consolidation and rainfall insecurity explicitly leads to more reasonable parameter values, better water balance components, and uncertainty intervals.

3.3.1   STUDIES USING SWAT RUNOFF ASSESSMENT

SWAT was used to study mountain hydrology. Over the past half-century, in a mountainous watershed in northwestern Chinese, the different parameters of the water balance approximation on different spatial scales were calculated by Zhang et al.

(2019). During the study period of 1964–2013, there were increasing patterns in total runoff (+30.5%). In addition to the increase in the surface and groundwater runoff, 42.7% and 57.3%, respectively, contributed toward the total runoff. The study concluded that a differing spatial distribution of quantity and trend changes in the water balance components was observed during the study period. A study done by Deb et al. (2015) observed that discharge and surface fluctuations were declining by 54% and by 41% in modest forested watersheds toward the end of this century. In addition, in Laos' Xedone basin south of Laos, Vilaysane et al. (2015) implemented the SWAT model. In the study, the performance and feasible performance of the SWAT model in the river basin were tested.

Wang et al. (2016) conducted another study to investigate the impact of the watershed subdivision on identifying priority management areas (PMAs). The research focuses on confirming the relevant project to simulate nonpoint source PMAs in a watershed division and location of comprehensive and cost-effective watershed management practices, especially in large watersheds. In Puarenga Stream (77 km^2), Me et al. (2015) performed model parameters for the SWAT model in Rotorua, New Zealand, with plantation and forest as the significant land utilization. Unknown parameter values were auto-calibrated in the SWAT model. The study suggests that main channel process parameters should be sensitive to base flow estimates. The study recommended. In contrast, the flow estimates were more responsive to those concerning overland processes. They used a SUFI-2 procedure to study the essential impacts of identifying uncertainties in the sensitivity and performance of parameters of hydrological models.

3.3.2 COMPARISON OF SWAT WITH OTHER MODELS WITH FUTURE SWAT IMPROVEMENTS

Cong et al. (2020) examined how different methods for assessing ecosystem services provide vastly varied simulation outcomes. By comparing the results of three hydrological ecosystem services (HESs) estimated by SWAT and the Integrated Valuation of Ecosystem Services and Tradeoffs (InVEST) model for the Nansihu Lake basin in China, including water supply, soil conservation, and water purification. According to the findings, the HES values in the overall basin simulated by the SWAT model were somewhat greater than those in the InVEST model. However, due to the same simulation methodology and input data, the two models' geographical distribution of soil conservation services was almost identical across the basin. In similar research, Qi et al. (2017), SWAT and GWLF were employed in two catchments in data-scarce China, the Tunxi and Hanjiaying basins, with differing meteorological circumstances (humid and semi-arid, respectively). According to the results, both models were able to reproduce monthly streamflow and sediment loadings over the simulation period. SWAT is better suited to tasks requiring high precision and provides an advantage when measurable data are sparse.

Water and Snow Balance MODeling System (WASMOD) and SWAT with different structures and spatial capabilities were chosen to simulate runoff and actual evapotranspiration (AET) in Yingluoxia watershed, the upper levels of Heihe River basin in northwest China, by Li et al. (2013). The results indicated that a

WASMOD with a basic model structure might produce similar, if not better, results than a semi-distributed SWAT with a complicated structure. Furthermore, Tan et al. (2020) noted that though the SWAT is undoubtedly one of the most widely used eco-hydrological models globally for forecasting the consequences of future hydro-climatic changes, its usage for severe streamflow circumstances is still rare. They discovered that future research needs in this area include:

1. A centralized SWAT extreme program evaluation framework.
2. SWAT improvements that lead to improved reproduction of peak and low flows.
3. Consistency evaluation of global and satellite products for SWAT extreme projections.
4. Bias rectification of CMIP6 and regional climate projections.
5. Comparison of SWAT+ and SWAT simulations.

3.3.3 SWAT Model Using Snow and Glacier Parameter Assessment

Snow, mainly when the watershed contains a hilly area, is a major hydrological reservoir in the water cycle. Snow is a primary source of surface and groundwater recharge in mountainous locations across the world (Singh and Pandey, 2014; Bhatt et al., 2017; Sharma and Kanga, 2020). The interconnections between the climatic and hydrological systems influence the water balance in mountainous areas. The periodic phase of snow storage bridges the winter-dominant rainfall and summer-dominant water demand. The ratios of snowmelt-derived runoff to total runoff have been used to quantify the critical role of snow in the water supply. Water stored in the snow-pack and its impact on water distribution are frequently overlooked by modelers, especially when just a part of the catchment is snow-dominated (Pall et al., 2019; Romshoo et al., 2020). Snow is commonly included in hydrological modeling to enhance statistics but without concern for the realism of its depiction or its impact on the water cycle. A glacier module was included in the SWAT model to mimic glacial-hydrological processes in various mountainous regions.

Moghadam et al. (2021) coupled MODIS remote sensing data with in situ data to validate the snowpack simulated by the SWAT, a semidistributed, physically based model used over a partially snow-dominated watershed. Three key points emerge from the findings: Higher elevation bands across mountainous headwaters enhanced hydrological modeling performance, even far downstream of the snow-dominated region; using MODIS data, SWAT provided an excellent spatial and temporal depiction of the snow cover, and elevation bands resulted in inconsistent changes in water distribution across the hydrological cycle of implemented watersheds, bringing them closer to predicted flow patterns. Liu et al. (2015) examined the possible impacts of mid-twentieth-century changing climate on the hydrological of the Cook Inlet catchment in south-central Alaska in similar research. The North American Regional Climate Change Assessment Program (NARCCAP) library of dynamically downscaled climate products was used to create climate datasets reflecting a range of potential change scenarios for 2041–2070. However, they found that the general hydrology of the basin will continue to be dominated by snowmelt until the mid-twentieth century without a substantial shift in the regime. Zhang et al. (2016) used the glacier-enhanced SWAT model to simulate twenty-four headwater catchments in the East and Central Tian Shan Mountains with changing glacier area ratios (GARs) from 1961 to 2007. They discovered that the GAR is a valuable measure for interpreting the various impacts of glaciers on discharge throughout catchments quantitatively. They concluded: Power functions fit the relationships among RGMC, RIMC, CV, and GAR to high certainty. The CV decreases with the increasing GAR while others increase. Such power functions start changing sensitively with GAR when GAR is less than 10%, implying that a slight change in GAR can cause significant changes in RGMC, RIMC, RC, and CV in time.

According to Tuo et al. (2018), the hydrological calibration process can be a valuable tool for reducing model equifinality and parameter uncertainty. Their research analyses and contrasts three different calibrating methods. The first two methods are single-objective calibration processes, in which all SWAT model parameters were calibrated only based on river discharge. The third technique is a multi-objective calibration, which included information on snow water equivalent (SWE) at two distinct geographical scales (subbasin and elevation band), as well as discharge data. Their findings provide a method for improving SWAT modeling in alpine catchments to the broad community of SWAT users.

Because of the lack of observational data and geodetic observations in basins in the Karakoram Mountains, Western China, acquired precipitation data for the high mountains are fraught with uncertainty and difficulty, according to Wang et al. (2016). However, according to model simulations, the proportions of glacier melt combined ice melt to total runoff were 52% and 31%, respectively. Furthermore, decreased ice melt and slower glacier contraction on the northern slope of the Karakoram during the period 1968–2007 may be due to a substantial precipitation surge and a nonsignificant warming trend during the melt season.

3.3.4 THE EFFICIENCY OF SWAT MODEL ON SOIL EROSION AND SEDIMENT FLUX IN STREAMFLOW

A sediment output estimate is required to analyze reservoir sedimentation, river morphology, and implement any soil and water conservation measures. For the supply of environmental and economic services, sediment management is a top priority in the alpine zones. Climate change has been demonstrated to substantially influence the complete management of complicated big river basins by causing significant fluctuation in water and sediment (Meraj et al., 2018a, b; Kanga et al., 2020a). A better knowledge of anticipated future changes in hydrologic and sediment dynamics is essential for water management in various alpine areas. Soil erosion caused by increasing runoff poses a significant danger to the alpine zones' long-term viability. In hydrological models, good predictions of soil temperature are critical for measuring hydrological and biological processes. In hilly areas, the SWAT model has been utilized to estimate sediment outputs across the basin and estimate sediment retention along distinct watersheds.

3.3.5 Studies Using SWAT for Soil Erosion

Panda et al. (2021) used SWAT to estimate soil loss from the Subarnarekha watershed in Odisha, India. The catchment's mean annual soil loss was 4.84 Mg/ha. The watershed classified as SW18 had the most significant soil loss, ranging from 10 to 15 Mg/ha/year. According to Dutta and Sen (2018), Hirakud, Asia's longest earthen dam on the Mahanadi River, located in the Indian states of Chhattisgarh and Odisha, has been plagued by sedimentation issues that have impacted reservoir storage capacity and reservoir operations. Due to the sensitivity of topographic variables employed in the Modified USLE, the results predicted that the sediment output is susceptible to the sizes of distinct sub-basins. The graphical and statistical data, on the other hand, highlight important erosion-prone regions within the watershed.

Changes in Land-use affects watersheds resulting in environmental deterioration, according to Salsabilla and Kusratmoko (2017). Because of geography, land use, climate, and human activity, estimating soil erosion loss is challenging. Using the SWAT hydrological model, they estimated soil erosion danger and sediment yield for the Komering watershed (806,001 ha) in South Sumatera Province. The SWAT model was used to show that it can forecast runoff water, soil erosion loss, and sediment production. According to Qi et al. (2016), soil temperature forecasts are frequently employed in the SWAT; however, they have high-prediction errors when used to locations with considerable snow cover during the winter. Their findings show that both variants of the soil-temperature module can offer accurate temperature forecasts in different levels of the soil during the summer, spring, and autumn seasons. On the other hand, the original module significantly underestimates winter soil temperatures (within $-10°C$ to $-20°C$), but the new module gives values more in line with field observations (within $-2°C$ to $-2°C$).

3.3.6 Studies Using SWAT for Sediment Dynamics

SWAT has been used to analyze the spatiotemporal responses of sediments and water in the Yellow River Basin to high rainfall events, Xu et al. (2021) constructed basin-scale water and sediment output model. According to their findings, the slope sediment production during the wet season in observing reaches between Toudaoguai to Tongguan is still exceeding 128.57 t/ha, indicating significant erosion. At the Tongguan hydrological station, annual river discharge and carried sediments were 52.5 billion m^3 and 443 million tonnes, respectively.

Sarkar et al. (2017) investigated how to estimate sediment output while studying reservoir sedimentation, river morphology, and implementing any soil and water conservation measures. Their research intended to define the zones of high flow and, as a result, soil degradation in the Chotki Berghi basin in Eastern India, which is dominated by agriculture. Their calibration and validation period findings were excellent.

SWAT was used by Ogbu (2017) to forecast river flow and sediment output from the Ebonyi River watershed, utilizing globally available data acquired from the internet for a large basin (3,765 km^2) in south-eastern Nigeria. For the event of no land-use change, the model found daily mean streamflow and sediment discharge of 24.3 $2m^3/s$ and 341.31 metric tonnes, respectively, at the watershed outlet. Their findings revealed that increasing agricultural land by 19% increased daily mean streamflow and sediment output by 29% and 44%, respectively. When farmland was enlarged to 94% of the watershed area, streamflow and sediment output increased by roughly 72% and 99%, respectively. In the same year, Vigiak et al. (2016) discovered that riparian land provides various ecosystem services critical for the excellent quality of water and aquatic biodiversity, habitats and hydrological connections, and pollution and sediment retention. SWAT model was used to analyze sediment production and estimate sediment retention by vegetative cover across the Danube Drainage basin. At the hillslope scale, the influence of riparian filtering on decreasing sediment flows to the stream network was positive, with a median efficacy of 50%. With Strahler's order, sediment filtration in riparian buffers became more efficient in narrower spans, dropping from around 17% to 5%. According to Vigiak et al. (2016), sediment management is a top priority in the Danube Basin for providing environmental and economic services. Their research intended to use the SWAT model to estimate present Danube Basin sediment flows (1995–2009) and identify sediment budgeting knowledge gaps. Model under-estimations were correlated to Alpine and karst areas. In contrast, underestimations occurred in high seismicity areas of the Lower Danube, indicating that sediment trapping in reservoirs and floodplain deposition were likely underestimated and counterbalanced by high stream deposition. Factor analysis revealed that model under-estimations were strongly linked to Alpine and karst areas, whereas underestimations occurred in high seismicity areas of the Lower Danube.

Water management in the Caatinga-Atlantic forest ecotone of Brazil, according to Santos et al. (2015), is crucially dependent on a better knowledge of anticipated future changes in streamflow and sediment dynamics. They used projected climate data derived from the global circulation model HadGEM2-ES coupled to the regional circulation model ETA-CPTEC/HadCM3 for two typical concentration pathways (RCP 4.5 and 8.5), with bias correction, to evaluate both the future impacts of land use and land cover (LULC) variations and the impacts of climate change on streamflow and sediment yield. Their findings show that the suggested technique can contextualize potentially substantial regional hydrological changes that have consequences for land and biodiversity conservation and water resource sustainability. According to Roth et al. (2016), data-scarce locations such as Ethiopia frequently lack discharge or sediment load data, forcing researchers to depend on data input from global models with more inadequate resolution and accuracy. Their study assessed a model parameter transition from a 100 hectare (ha) large sub-watershed (Minchet) to a 4800 ha basin in the Andean highlands using the SWAT. However, their findings revealed that employing model parameter transfer, validation of the model for streamflow functioned admirably for both the sub-catchment and the whole catchment.

3.3.7 THE EFFICIENCY OF SWAT IN SIMULATING THE IMPACT OF LAND USE LAND COVER CHANGE ON HYDROLOGICAL PROCESSES

The driving forces of hydrological processes are land-use shift and climate change. Changes in land use primarily drive watershed hydrological change. As a result, thorough analyses of hydrological responses to LULC changes must guarantee that water resources and ecosystem services are managed sustainably (Hassanin et al., 2020; Kanga et al., 202). Understanding the flow and distribution of hydrologic processes inside the catchment necessitates modeling techniques to predict the influence of land use/cover and changing climate on hydrologic responses. The impact of LUCC and forest management activities on hydrological processes are increasingly being studied with hydrological models. The SWAT models the effects of LUCC on major hydrological components. The SWAT model can predict how different cropping and management techniques may affect river basin responses.

3.3.8 STUDIES USING SWAT FOR LULC CHANGE IN AFFECTING SEDIMENT DYNAMICS

Using the SWAT model, Moghadam et al. (2021) examined the independent and combined future impacts of climate and land-use modifications on water balance components on different geographical and temporal scales. Their research looked at how water yield and sediment yield changed over time in Iran's diverse Taleghan Catchment. They calculated a WYLD of 295 mm and an SYLD of 17 t/ha. Their innovative method may be used to identify the most susceptible areas and prioritize management methods for better water and soil conservation in watersheds. LULC changes have a significant impact on watershed hydrology and sediment production, according to Aghsaei et al. (2020). Using the SWAT 2012) model, they investigate the hydrological effects of fluid LULC changes in the Anzali wetland catchment in Iran. Their findings indicated that including dynamic LULC variability into the SWAT model may be used as a planning tool in the future to govern LULC change in the Anzali wetland basin. The Hun River Basin (7,919 km^2), a typical mountainous terrain with just four gauge meteorological stations, was chosen as the research location by Zhang et al. (2019). In East Asia, the China Meteorological Assimilation Driving Datasets for the SWAT model are frequently used. Their daily runoff and weekly sediment yield models yielded reasonable findings. However, the authentication of daily sediment load was lacking. During the rainy season, forestland lowered sediment production, higher water percolation, and reduced runoff, whereas, during the dry season, it decreased water penetration and increased runoff.

3.3.9 STUDIES USING SWAT FOR ASSESSING LULC AND CLIMATE ON WATERSHED HYDROLOGY

Puno et al. (2019) used modeling techniques to anticipate the influence of land cover and climate variability on hydrologic responses. Evaluating the movement and distribution of hydrologic processes within the watershed is critical. Using a SWAT model, they looked at the effects of land transformations and climate change predictions on the hydrologic processes of the Muleta watershed. The outcome was statistically acceptable after accuracy assessment using observed streamflow and the same statistical measures ($R^2 = 0.79$, NS $= 0.67$, and RSR $= 0.57$). However, with anticipated climate change, a 13% drop in rainfall directly impacts the decline in hydrologic processes. Local governments and policymakers can utilize their research findings to make decisions about issues and concerns about hydrological processes in the uplands. Zhang et al. (2016) combined two land-use models (the Markov chain model and Dyna-CLUE) with the SWAT model to study the completely separate and combined hydrological implications of land-use change and climate changes in the headwater area of arid inland Heihe River Basin, China, in for periods 1995–2014 and 2015–2024. They concluded that climate change would have a far more significant impact on hydrological regimes in the near future than land-use changes, resulting in considerable changes in all hydrological components. However, the importance of land-use changes should not be underestimated, particularly if the climate gets dry in the future since this would amplify the hydrological effects.

3.3.10   STUDIES USING SWAT FOR FOREST ASSESSMENT

Climate change has damaged British Columbia's wooded watersheds by bringing very extreme temps and beetle infestations, which has coincided with an increase in the number of forest fires, according to Katimbo et al. (2018). Every year, around 10,000 fires are expected to occur in Canada, destroying 2 million ha of the 400 million ha forested area. Water yield simulations were performed using the SWAT model to assess the hydrologic processes of watersheds following forest fires. Because some physical processes, like snowmelt in the watershed, cannot be reproduced in SWAT, the model's findings may be inaccurate. According to their findings, forest cutting and pine beetle infestation cause fast land-use change dynamics before the outbreak of fires.

The Halstead Fire of the Salmon-Challis National Forest damaged about 174,000 acres of forest in 2012, according to Liu et al. (2015). The post-fire logging activity to recover commercial timber necessitates a thorough study of the logging activity's environmental implications on forestry restoration and sediment/water yields. Using Arc SWAT, they investigated these difficulties and examined seasonal variations between winter and summertime logging. Summer logging (SL) can create 0.64 Mg/ha or 7,798 Mg/yr more sediment for the entire watershed than winter logging, and also 6,573 kg nitrogen as well as 1,827 kg phosphorus yearly. In the Sierra Nevada, where wood extraction and controlled fire are famous for forest conservation, hydrological models are infrequently utilized, especially in small watersheds, according to Bailey (2015). The SWAT model was used to simulate streamflow, a small headwater mountain basin in the southern Sierra Nevada, on a daily time step. The daily time-step simulated streamflow findings are valuable for assessing watershed processing and functioning, but not so much for forest and land management. To evaluate forest management activities in P301, SWAT will require additional model changes and monthly and annual water yield estimations.

3.3.11   STUDIES USING SWAT FOR NUTRIENT MODELLING

Veettil et al. (2021) used the fully distributed Agricultural Ecosystems Services (AgES) and the SWAT models to investigate spatial patterns of water balance components and nitrate transport in the Big Dry Creek Watershed, a hybrid agricultural and suburban watershed in semi-arid Colorado, USA. They looked at hydrological flow trends using hourly model outputs from 2010 to 2018. Their research showed the benefits of modeling spatially explicit hydrology, which incorporates connections between fields and other regions in a controlled watershed. Pollution from agricultural production is the dominant contributor of phosphorus and nitrogen in the Corn Belt region's stream systems, according to Kannan et al. (2019). The study's advanced modeling technique yielded accurate forecasts in most areas of the Corn Belt region. It may be used to evaluate pollution mitigation strategies and agri-economic scenarios, providing critical information to policymakers and suggestions on similar regional initiatives.

Thodsen et al. (2015) found that certain agricultural regions lose much more nutrients to waterways than the norm (high-risk areas, HRAs), while others lose significantly less (low-risk areas, HRAs). Whenever river catchment managers want to decrease nutrient inputs to lakes and coastal regions while yet allowing intensive agriculture, these locations are of significant importance. In a heavily farmed lowland watershed in Denmark, they utilized the SWAT model to detect HRAs and LRAs. Two scenarios were performed independently for HRAs (reduced fertilizer intake by 20%) and LRAs (increased fertilizer input by 20%), and two scenarios were conducted with both HRAs and LRAs. The HRA scenario decreased (3.3%) in nitrate load at the watershed scale, whereas the LRA scenario resulted in just a minor increase (0.9%). Overall, the combined scenarios revealed a reduction in nitrate load (2.2%). Some of the significant publications on SWAT model are shown in Tables 3.1 and 3.2. The overall applications of the SWAT model is shown Figure 3.2.

3.4   CONCLUSIONS

The SWAT model's widespread use, both domestically and internationally, has proven a strong and complete hydrological model. However, current local and international studies on this model are primarily concerned with model applicability, model accuracy, and model coupling. Overall, research revealed that models could accurately predict hydrological processes in small to medium-sized watersheds under various climate, land, and forest management scenarios. These findings also revealed that the factors obtained from managing land, groundwater features, soil properties, or plant conditions had the most considerable impact on the hydrological process. Other research (Anand et al., 2018; Boru et al., 2019; Jnior and Montenegro, 2019; Terskii et al., 2019) have also discovered this feature. SWAT for severe applications was calibrated and validated using a time-continuous technique based on daily or monthly streamflow data, similar to other applications. The SWAT model is one of the most widely used hydrological models in environmental, ecological, and hydrological research worldwide. SWAT, a semidistributed physically based model, has seen a surge in hydro-climatic severe research during the last decade, notably after 2015. China and the United States have carried out half of these studies, with 52% of them taking place in river basins with a drainage area of 10,000 km^2. From medium and large size basins, this illustrates the possibility for severe changes in hydrology. The calibration and validation of SWAT are carried out using daily, monthly, and annual streamflow data, with the first two techniques being used chiefly for dynamic simulations. Future research in SWAT could include bias correction of GCMs and regional climate projections, SWAT+ and SWAT development

TABLE 3.1

Summary of the Papers Published in the Special Issue by China and United States in 2017

	Important Processes	Location	Country	Calibration	Water Quality	Crop	Authors
1	Thermal stress	Rhode Island	USA	SWAT-CUP	Temp.		Chambers et al.
2	Fish Habitat	Rhode Island	USA	SWAT-CUP	Temp.		Chambers et al.
3	Leaching	North China Plain	China		NO3	Maize	Chen et al.
4	TN loss	Yangtze River	China		TN		Ding et al.
5	Climate change	Continental US	USA		WT, DO, TN, TP		Fant et al.
6	Nonpoint source pollution	Baoding City	China	manual	TN, TP		Li et al.
7	Nonpoint source pollution	Hunt River Rhode Island	USA	SWAT-CUP	TN		Paul et al.
8	SWAT, GWLF comparison	Tunxi and the Hanjiaying basins	China	SWAT-CUP	Sediment, TN		Qi et al.
9	Optimal design	Clear Creek watershed (TX)	USA	SWAT-CUP			Seo et al.
10	Water quality	Clear Creek watershed (TX)	USA	SWAT-CUP			Seo et al.
11	Climate data	USA	USA				White et al.

TABLE 3.2

Reviewed Publications that Assess Water Resources Dynamics in the Climate Change Context Using the SWAT Model in USA and ASIA

Study	Watershed(s)	Location	Area (km²)	Forest Coverage (%)	Processes Analyzed
				ASIA	
Murari et al. (2007)	Moolehole	India	4.54	a	Hydrological processes
Joh et al. (2011)	Seolma-Cheon	Korea	8.54	88.1	Hydrological components
Qiu et al. (2012)	Zhifanggou	China	8.24	51	Runoff and sedimentation
Ahn et al. (2013)	Seolma-Cheon	Korea	8.54	a	Hydrological components
Park et al. (2014)	Seolma-Cheon	Korea	8.54	88	Hydrological components

(Continued)

TABLE 3.2 (*Continued*)
Reviewed Publications that Assess Water Resources Dynamics in the Climate Change Context Using the SWAT Model in USA and ASIA

Study	Watershed(s)	Location	Area (km²)	Forest Coverage (%)	Processes Analyzed
Tyagi et al. (2014)	Arnigad	India	3.04	80	Water balance and sedimentation
	Bansigad		2.09	83	
Abbas et al. (2016)	Soan	Pakistan	159	50	Runoff and sedimentation
Briones et al. (2016)	Palico	Philippine	236.91	a	Hydrological response after land use and land cover change
Sharannya et al. (2018)	Gurupura (India)	India	712	a	Hydrological components
Marhaento et al. (2018)	Samin	Indonesia	278	a	Hydrological response after land use change
Jung and Kim (2018)	Seolmacheon	Korea	8.48	a	Hydrological processes
		USA			
Amatya and Jha (2011)	Turkey Creek	Carolina/ USA	72.6	87	Testing SWAT applicability
Perazzoli et al. (2013)	Concórdia	Brazil	30.74	58	Water resources
Zhu and Li (2014)	Little River	USA	981	a	Hydrological impact of land use and land cover change
Fu et al. (2014)	Harp Lake	Ontario	5.42	75	Testing SWAT-CS
Blainski et al. (2017)	Camboriú	Brazil	195	63	Simulating land use and soil erosion
Chambers et al. (2017)	Queen	USA	7	a	Streamflow
	Beaver Cork Brook				
Leta et al. (2015)	Kalihi and Nuuanu	USA	7	a	Streamflow

[a] Defined by the authors as predominant or fully forested, without specifying the percent occupied by forest land use.

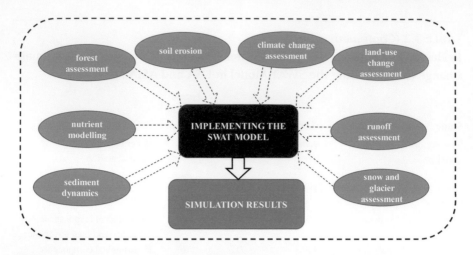

FIGURE 3.2 Applications of SWAT model.

for extreme simulation studies in various arid/nonarid basins, and the integration of machine learning glacial-hydrological processes in different mountainous regions using the SWAT model. Improvements to SWAT for hydro-climatic extremities research will necessitate new drought or flood indicators, improved calibration methods, uncertainty analysis, and SWAT changes, among other things.

ACKNOWLEDGMENTS

The research work was supported by the CAS-TWAS President's PhD Fellowship Programme University of Chinese Academy of Sciences, China.

COMPETING INTEREST

The authors declare that no competing interest exists.

REFERENCES

Abbas, N., Wasimia, S. A., & Al-Ansari, N. (2016). Assessment of climate change impact on water resources of lesser Zab, Kurdistan, Iraq using SWAT model. *Engineering, 8,* 697–715.

Abbaspour, K. C., Rouholahnejad, E., Vaghefi, Srinivasan, R., Yang, H., & Kløve, B. (2015). A continental-scale hydrology and water quality model for Europe: Calibration and uncertainty of a high-resolution large-scale SWAT model. *Journal of Hydrology, 524,* 733–752.

Aghsaei, H., Dinan, N. M., Moridi, A., Asadolahi, Z., Delavar, M., Fohrer, N., & Wagner, P. D. (2020). Effects of dynamic land use/land cover change on water resources and sediment yield in the Anzali wetland catchment, Gilan, Iran. *Science of the Total Environment, 712,* 136449.

Ahn, S. R., Park, G., Jang, C. H., & Kim, S. J. (2013). Assessment of climate change impact on evapotranspiration and soil moisture in a mixed forest catchment using spatially calibrated SWAT model. *Journal of Korea Water Resources Association, 46*(6), 569–583.

Amatya, D. M., & Jha, M. K. (2011). Evaluating the SWAT model for a low-gradient forested watershed in coastal South Carolina. *Transactions of the ASABE, 54*(6), 2151–2163.

Anand, J., Gosain, A. K., & Khosa, R. (2018). Prediction of land use changes based on Land Change Modeler and attribution of changes in the water balance of Ganga basin to land use change using the SWAT model. *Science of the Total Environment, 644,* 503–519.

Arshad, A., Zhang, Z., Zhang, W., & Gujree, I. (2019). Long-term perspective changes in crop irrigation requirement caused by climate and agriculture land use changes in Rechna Doab, Pakistan. *Water,* 11(8), 1567.

Bailey, D. (2015). *Using SWAT (soil water and assessment tool) to evaluate streamflow hydrology in a small mountain watershed in the Sierra Nevada, CA* (Doctoral dissertation, California State University, Northridge).

Briones, R., Ella, V., & Bantayan, N. (2016). Hydrologic impact evaluation of land use and land cover change in Palico Watershed, Batangas, Philippines Using the SWAT model. *Journal of Environmental Science and Management,* 19(1), 96-106.

Bhatt, C. M., et al. (2017). Satellite-based assessment of the catastrophic Jhelum floods of September 2014, Jammu & Kashmir, India. *Geomatics, Natural Hazards and Risk,* 8(2), 309–327.

Boru, G. F., Gonfa, Z. B., & Diga, G. M. (2019). Impacts of climate change on stream flow and water availability in Anger sub-basin, Nile Basin of Ethiopia. *Sustainable Water Resources Management,* 5(4), 1755–1764.

Blainski, É., Porras, A. A., Pospissil, E. H., Garbossa, L., & Pinheiro, A. (2017). Simulation of land use scenarios in the Camboriú River Basin using the SWAT model. *Brazilian Journal of Water Resources* (Simulação de cenários de uso e ocupação das terras na bacia hidrográfica do Rio Camboriú utiliza. *Revista Brasileira de Recursos Hidricos*), 22, 1-12.

Chambers, B. M., Pradhanang, S. M., & Gold, A. J. (2017). Simulating climate change induced thermal stress in coldwater fish habitat using SWAT model. *Water,* 9(10), 732.

Chambers, B., Pradhanang, S. M., & Gold, A. J. (2017). Assessing thermally stressful events in a Rhode Island coldwater fish habitat using the SWAT model. *Water,* 9(9), 667.

Chen, S., Sun, C., Wu, W., Sun, C. (2017). Water leakage and nitrate leaching characteristics in the winter wheat–summer maize rotation system in the North China Plain under different irrigation and fertilization management practices. *Water,* 9, 141.

Cong, W., Sun, X., Guo, H., & Shan, R. (2020). Comparison of the SWAT and InVEST models to determine hydrological ecosystem service spatial patterns, priorities and trade-offs in a complex basin. *Ecological Indicators,* 112, 106089.

Cuceloglu, G., Abbaspour, K. C., & Ozturk, I. (2017). Assessing the water-resources potential of Istanbul by using a soil and water assessment tool (SWAT) hydrological model. *Water,* 9(10), 814.

Deb, D., Butcher, J., & Srinivasan, R. (2015). Projected hydrologic changes under mid-21st century climatic conditions in a sub-arctic watershed. *Water Resources Management,* 29(5), 1467–1487.

Ding, X., Xue, Y., Lin, M., Jiang, G. (2017). Influence mechanisms of rainfall and terrain characteristics on total nitrogen losses from regosol. *Water,* 9, 167.

Dutta, S., & Sen, D. (2018). Application of SWAT model for predicting soil erosion and sediment yield. *Sustainable Water Resources Management,* 4(3), 447–468.

Ezenne, G. I., Ndulue, L. C., Onyeoriri, V. C., & Mbajiorgu, C. C. (2017). Simulation of water quality on Upper Ebonyi watershed using SWAT.

Fant, C., Srinivasan, R., Boehlert, B., Rennels, L., Chapra, S. C., Strzepek, K. M., Corona, J., Allen, A., Martinich, J. (2017). Climate change impacts on us water quality using two models: HAWQS and US basins. *Water,* 9, 118.

Fontes Júnior, R., & Montenegro, A. (2019). Impact of land use change on the water balance in a representative watershed in the semiarid of the state of Pernambuco using the SWAT model. *Engenharia Agrícola,* 39(1), 110–117.

Fu, C., James, A. L., & Yao, H. (2014). SWAT-CS: Revision and testing of SWAT for Canadian Shield catchments. *Journal of Hydrology*, 511, 719–735.

Grusson, Y., Anctil, F., Sauvage, S., & Sánchez Pérez, J. M. (2017). Testing the SWAT model with gridded weather data of different spatial resolutions. *Water*, 9(1), 54.

Gujree, I., et al. (2017). Evaluating the variability and trends in extreme climate events in the Kashmir Valley using PRECIS RCM simulations. *Modeling Earth Systems and Environment*, 3(4), 1647–1662.

Guo, J., Su, X., Singh, V. P., & Jin, J. (2016). Impacts of climate and land use/cover change on streamflow using SWAT and a separation method for the Xiying River Basin in north-western China. *Water*, 8(5), 192.

Hassanin, M., Kanga, S., Farooq, M., & Singh, S. K. (2020). Mapping of Trees outside Forest (ToF) From Sentinel-2 MSI satellite data using object-based image analysis. *Gujarat Agricultural Universities Research Journal*, 207–213.

Hu, J., et al. (2021). Impacts of land-use conversions on the water cycle in a typical watershed in the southern Chinese Loess Plateau. *Journal of Hydrology*, 593, 125741.

Kanga, S., et al. (2020a). Modeling the spatial pattern of sediment flow in lower Hugli Estuary, West Bengal, India by quantifying suspended sediment concentration (SSC) and depth conditions using geoinformatics. *Applied Computing and Geosciences*, 8, 100043.

Kanga, S., Kumar, S., and Singh, S. K. (2017b). Climate induced variation in forest fire using Remote Sensing and GIS in Bilaspur District of Himachal Pradesh. *International Journal of Engineering and Computer Science*, 6(6), 21695–21702.

Kanga, S., Rather, M. A., Farooq, M., & Singh, S. K. (2021). GIS based forest fire vulner-ability assessment and its validation using field and MODIS data: A case study of bhad-erwah forest division, Jammu and Kashmir (India). *Indian Forester*, 147(2), 120–136.

Kanga, S., Singh, S. K., & Sudhanshu. (2017a). Delineation of urban built-up and change detection analysis using multi-temporal satellite images. *International Journal of Recent Research Aspects* 4(3), 1–9.

Kannan, N., Santhi, C., White, M. J., Mehan, S., Arnold, J. G., & Gassman, P. W. (2019). Some challenges in hydrologic model calibration for large-scale studies: A case study of SWAT model application to Mississippi-Atchafalaya River Basin. *Hydrology*, 6(1), 17.

Katimbo, A., Lavkulich, L., & Schreier, H. (2018). Using SWAT to simulate the effects of forest fires on water yield in forested watershed: A case study of Bonaparte Watershed, Central Interior of British Columbia, Canada.

Khelifa, W. B., Strohmeier, S., Benabdallah, S., & Habaieb, H. (2021). Evaluation of bench terracing model parameters transferability for runoff and sediment yield on catchment modelling. *Journal of African Earth Sciences*, 178, 104177.

Kibii, J. K., Kipkorir, E. C., & Kosgei, J. R. (2021). Application of soil and water assessment tool (SWAT) to evaluate the impact of land use and climate variability on the Kaptagat catchment river discharge. *Sustainability*, 13(4), 1802.

Joh, H. K., Lee, J. W., Park, M. J., Shin, H. J., Yi, J. E., Kim, G. S., ... & Kim, S. J. (2011). Assessing climate change impact on hydrological components of a small forest water-shed through SWAT calibration of evapotranspiration and soil moisture. *Transactions of the ASABE*, 54(5), 1773–1781.

Jung, C. G., & Kim, S. J. (2018). Assessment of the water cycle impact by the Budyko curve on watershed hydrology using SWAT and CO_2 concentrations derived from Terra MODIS GPP. *Ecological Engineering*, 118, 179–190.

Leta, O. T., Nossent, J., Velez, C., Shrestha, N. K., van Griensven, A., & Bauwens, W. (2015). Assessment of the different sources of uncertainty in a SWAT model of the River Senne (Belgium). *Environmental Modelling & Software*, 68, 129–146.

Leta, O. T., El-Kadi, A. I., & Dulai, H. (2018). Impact of climate change on daily streamflow and its extreme values in pacific island watersheds. *Sustainability*, 10(6), 2057.

Li, Z., Shao, Q., Xu, Z., & Xu, C. Y. (2013). Uncertainty issues of a conceptual water balance model for a semi-arid watershed in north-west of China. *Hydrological Processes*, 27(2), 304–312.

Li, C., Zheng, X., Zhao, F., Wang, X., Cai, Y., & Zhang, N. (2017). Effects of urban non-point source pollution from Baoding City on Baiyangdian Lake, China. *Water*, 9, 249.

Liu, J., Paul, S., & Manguerra, H. (2015). ArcSWAT modeling analysis for post-wildfire logging impacts on sediment and water yields at Salmon-Challis National Forest, Idaho, USA. *Watershed Management*, 240–250. doi:10.1061/9780784479322.021

Marhaento, H., Booij, M. J., & Hoekstra, A. Y. (2018). Hydrological response to future land-use change and climate change in a tropical catchment. *Hydrological Sciences Journal*, 63(9), 1368–1385.

Me, W., Abell, J. M., & Hamilton, D. P. (2015). Effects of hydrologic conditions on SWAT model performance and parameter sensitivity for a small, mixed land use catchment in New Zealand. *Hydrology and Earth System Sciences*, 19(10), 4127–4147.

Meraj, G., et al. (2018a). Geoinformatics based approach for estimating the sediment yield of the mountainous watersheds in Kashmir Himalaya, India. *Geocarto International*, 33(10), 1114–1138.

Meraj, G., et al. (2018b). An integrated geoinformatics and hydrological modelling-based approach for effective flood management in the Jhelum Basin, NW Himalaya. *Multidisciplinary Digital Publishing Institute Proceedings*, 7(1), 8.

Moghadam, N. T., Abbaspour, K. C., Malekmohammadi, B., Schirmer, M., & Yavari, A. R. (2021). Spatiotemporal modelling of water balance components in response to climate and landuse changes in a heterogeneous mountainous catchment. *Water Resources Management*, 35(3), 793–810.

Murari, R. R. V., Ruiz, L., Sandhya, C., Braun, J. J., & Kumar, M. M. (2007). Study of hydrological processes in a small forested watershed in South Karnataka (India), *National Seminar on Forest, Water and People at NIH*, 29th–30th July 2004.

Narsimlu, B., Gosain, A. K., Chahar, B. R., Singh, S. K., & Srivastava, P. K. (2015). SWAT model calibration and uncertainty analysis for streamflow prediction in the Kunwari River Basin, India, using sequential uncertainty fitting. *Environmental Processes*, 2(1), 79–95.

Ogbu, K. N. (2017). Use of the SWAT model to evaluate the impact of land-use change on Ebonyi River Watershed (Doctoral dissertation).

Park, G. A., Park, J. Y., Joh, H. K., Lee, J. W., Ahn, S. R., & Kim, S. J. (2014). Evaluation of mixed forest evapotranspiration and soil moisture using measured and swat simulated results in a hillslope watershed. *KSCE Journal of Civil Engineering*, 18(1), 315–322.

Pall, I. A., Meraj, G., and Romshoo, S. A. (2019). Applying integrated remote sensing and field-based approach to map glacial landform features of the Machoi Glacier valley, NW Himalaya. *SN Applied Sciences*, 1(5), 488.

Panda, C., Das, D. M., Raul, S. K., & Sahoo, B. C. (2021). Sediment yield prediction and prioritization of sub-watersheds in the Upper Subarnarekha basin (India) using SWAT. *Arabian Journal of Geosciences*, 14(9), 1–19.

Partsinevelou, A. S. (2017). Using the SWAT model in analyzing hard rock hydrogeological environments. Application in Naxos Island, Greece. *Bulletin of the Geological Society of Greece*, 51, 18–37.

Paul, S., Cashman, M.A., Szura, K., Pradhanang, S.M. (2017). Assessment of Nitrogen Inputs into Hunt River by OnsiteWastewater Treatment Systems via SWAT Simulation. *Water*, 9, 610.

Perazzoli, M., Pinheiro, A., & Kaufmann, V. (2013). Assessing the impact of climate change scenarios on water resources in southern Brazil. *Hydrological Sciences Journal*, 58(1), 77–87.

Pontes, L. M., Viola, M. R., Silva, M. L. N., Bispo, D. F., & Curi, N. (2016). Hydrological modeling of tributaries of cantareira system, Southeast Brazil, with the Swat model. *Engenharia Agrícola*, 36(6), 1037–1049.

Pradhan, N. S., Sijapati, S., & Bajracharya, S. R. (2015). Farmers' responses to climate change impact on water availability: Insights from the Indrawati Basin in Nepal. *International Journal of Water Resources Development*, 31(2), 269–283.

Puno, R. C. C., Puno, G. R., & Talisay, B. A. M. (2019). Hydrologic responses of watershed assessment to land cover and climate change using soil and water assessment tool model. *Global Journal of Environmental Science and Management*, 5(1), 71–82.

Qi, J., Li, S., Li, Q., Xing, Z., Bourque, C. P. A., & Meng, F. R. (2016). A new soil-temperature module for SWAT application in regions with seasonal snow cover. *Journal of Hydrology*, 538, 863–877.

Qi, Z., Kang, G., Chu, C., Qiu, Y., Xu, Z., & Wang, Y. (2017). Comparison of SWAT and GWLF model simulation performance in humid south and semi-arid north of China. *Water*, 9(8), 567.

Qiu, L. J., Zheng, F. L., & Yin, R. S. (2012). SWAT-based runoff and sediment simulation in a small watershed, the loessial hilly-gullied region of China: capabilities and challenges. *International Journal of Sediment Research*, 27(2), 226–234.

Romshoo, S. A., et al. (2020). Satellite-observed glacier recession in the Kashmir Himalaya, India, from 1980 to 2018. *Environmental Monitoring and Assessment*, 192(9), 1–17.

Roth, V., Nigussie, T. K., & Lemann, T. (2016). Model parameter transfer for streamflow and sediment loss prediction with SWAT in a tropical watershed. *Environmental Earth Sciences*, 75(19), 1–13.

Salsabilla, A., & Kusratmoko, E. (2017, July). Assessment of soil erosion risk in Komering watershed, South Sumatera, using SWAT model. In *AIP Conference Proceedings* (Vol. 1862, No. 1, p. 030192). AIP Publishing LLC Depok, Jawa Barat, Indonesia.

Samimi, M., Mirchi, A., Moriasi, D., Ahn, S., Alian, S., Taghvaeian, S., & Sheng, Z. (2020). Modeling arid/semi-arid irrigated agricultural watersheds with SWAT: Applications, challenges, and solution strategies. *Journal of Hydrology*, 590, 125418.

Santos, J. D., Silva, R. M., Carvalho Neto, J. G., Montenegro, S. M. G. L., Santos, C. A. G., & Silva, A. M. (2015). Land cover and climate change effects on streamflow and sediment yield: a case study of Tapacurá River basin, Brazil. *Proceedings of the International Association of Hydrological Sciences*, 371, 189–193.

Sarkar, S., Vaibhav, V., & Singh, A. (2017). Estimation of sediment yield by using soil and water assessment tool for an agricultural watershed in Eastern India. *Indian Journal of Soil Conservation*, 45(1), 52–59.

Seo, M., Jaber, F., Srinivasan, R., & Jeong, J. (2017). Evaluating the impact of low impact development (LID) practices on water quantity and quality under different development designs using SWAT. *Water*, 9, 193.

Seo, M., Jaber, F., & Srinivasan, R. (2017). Evaluating various low-impact development scenarios for optimal design criteria development. *Water*, 9, 270.

Sharannya, T. M., Mudbhatkal, A., & Mahesha, A. (2018). Assessing climate change impacts on river hydrology–A case study in the Western Ghats of India. *Journal of Earth System Science*, 127(6), 1–11.

Sharma, A., & Kanga, S. (2020). Surface runoff estimation of Sind River Basin using SCS-CN method and GIS technology.

Singh, L., & Saravanan, S. (2020). Simulation of monthly streamflow using the SWAT model of the Ib River watershed, India. *HydroResearch*, 3, 95–105.

Singh, S. K., & Pandey, A. C. (2014). Geomorphology and the controls of geohydrology on waterlogging in Gangetic Plains, North Bihar, India. *Environmental Earth Sciences*, 71(4), 1561–1579.

Swain, K. C., Singha, C., & Rusia, D. K. (2016). SWAT simulation model for climate change impact on runoff. *International Journal Environmental & Agricultural Science*, 2(2), 322–332.

Tan, M. L., Gassman, P., Yang, X., & Haywood, J. (2020). A review of SWAT applications, performance and future needs for simulation of hydro-climatic extremes. *Advances in Water Resources*, 143, 103662.

Terskii, P., Kuleshov, A., Chalov, S., Terskaia, A., Belyakova, P., Karthe, D., & Pluntke, T. (2019). Assessment of water balance for Russian subcatchment of Western Dvina River using SWAT model. *Frontiers in Earth Science*, 7, 241.

Thodsen, H., Andersen, H. E., Blicher-Mathiesen, G., & Trolle, D. (2015). The combined effects of fertilizer reduction on high risk areas and increased fertilization on low risk areas, investigated using the SWAT model for a Danish catchment. *Acta Agriculturae Scandinavica, Section B—Soil & Plant Science*, 65(sup2), 217–227.

Tuo, Y., Marcolini, G., Disse, M., & Chiogna, G. (2018). A multi-objective approach to improve SWAT model calibration in alpine catchments. *Journal of Hydrology*, 559, 347–360.

Tyagi, J. V., Rai, S. P., Qazi, N., & Singh, M. P. (2014). Assessment of discharge and sediment transport from different forest cover types in lower Himalaya using Soil and Water Assessment Tool (SWAT). *International Journal of Water Resources and Environmental Engineering*, 6(1), 49–66.

Veettil, A. V., Green, T. R., Kipka, H., Arabi, M., Lighthart, N., Mankin, K., & Clary, J. (2021). Fully distributed versus semi-distributed process simulation of a highly managed watershed with mixed land use and irrigation return flow. *Environmental Modelling & Software*, 140, 105000.

Vigiak, O., Malagó, A., Bouraoui, F., Grizzetti, B., Weissteiner, C. J., & Pastori, M. (2016). Impact of current riparian land on sediment retention in the Danube River Basin. *Sustainability of Water Quality and Ecology*, 8, 30–49.

Vilaysane, B., Takara, K., Luo, P., Akkharath, I., & Duan, W. (2015). Hydrological stream flow modelling for calibration and uncertainty analysis using SWAT model in the Xedone river basin, Lao PDR. *Procedia Environmental Sciences*, 28, 380–390.

Wang, G., Chen, L., Huang, Q., Xiao, Y., & Shen, Z. (2016). The influence of watershed subdivision level on model assessment and identification of non-point source priority management areas. *Ecological Engineering*, 87, 110–119.

Wang, X., Sun, L., Zhang, Y., & Luo, Y. (2016). Rationalization of altitudinal precipitation profiles in a data-scarce glacierized watershed simulation in the Karakoram. *Water*, 8(5), 186.

White, M.J., Gambone, M., Haney, E., Arnold, J., & Gao, J. (2017). Development of a station based climate database for SWAT and APEX Assessments in the US. *Water*, 9, 437.

Xu, Z., Zhang, S., & Yang, X. (2021). Water and sediment yield response to extreme rainfall events in a complex large river basin: A case study of the Yellow River Basin, China. *Journal of Hydrology*, 597, 126183.

Zhang, H., Wang, B., Li Liu, D., Zhang, M., Leslie, L. M., & Yu, Q. (2020). Using an improved SWAT model to simulate hydrological responses to land use change: A case study of a catchment in tropical Australia. *Journal of Hydrology*, 585, 124822.

Zhang, L., Meng, X., Wang, H., & Yang, M. (2019). Simulated runoff and sediment yield responses to land-use change using the SWAT model in northeast China. *Water*, 11(5), 915.

Zhang, L., Nan, Z., Xu, Y., & Li, S. (2016a). Hydrological impacts of land use change and climate variability in the headwater region of the Heihe River Basin, Northwest China. *PloS One*, 11(6), e0158394.

Zhang, Y., Luo, Y., Sun, L., Liu, S., Chen, X., & Wang, X. (2016b). Using glacier area ratio to quantify effects of melt water on runoff. *Journal of Hydrology*, 538, 269–277.

Zhu, C., & Li, Y. (2014). Long-term hydrological impacts of land use/land cover change from 1984 to 2010 in the Little River Watershed, Tennessee. *International Soil and Water Conservation Research*, 2(2), 11–21.

4 Temporal Assessment of Sedimentation in Siruvani Reservoir Using Remote Sensing and GIS

J. Brema and A. Tamilarasan
Karunya Institute of Technology and Sciences

CONTENTS

4.1   Introduction 59
4.2   Study Area 61
4.3   Methodology 61
    4.3.1   Morphometric Analysis 62
    4.3.2   USLE 63
    4.3.3   NDWI 66
4.4   Results and Discussion 66
4.5   Conclusion 68
References 70

4.1   INTRODUCTION

Sedimentation in rivers and reservoirs are due to the physical processes: various forms of erosion, sediment movement and deposition. In the case of matured river basins, sediment processes are relatively balanced. Construction of infrastructure such as dams, aqueducts etc. causes change in flow velocity, resulting in sediment deposition. There are three stages in a reservoir's life; they are [Native Advertisement] continuous and rapidly occurring sediment accumulation, partial sediment balance, where often fine sediments reach a balance but coarse sediments continue to accumulate, and full sediment balance, with sediment inflow and outflow equal for all particle sizes. About 0.5%–1% of the total volume of 6,800 km³ of water stored in reservoirs around the world is lost annually because of sedimentation. Loss of reservoir storage reduces flexibility in hydro-power generation and affects the reliability of water supply. Most of the world's reservoirs are in continuous accumulation stage. Many were designed by estimating sedimentation rates to provide a pool with sufficient volume to achieve a specified design life. In most of the occasions, the probability of achieving the design life for a reservoir is less. Therefore, managing reservoirs to achieve a full sedimentation control is inevitable.

DOI: 10.1201/9781003147107-5

Remote sensing techniques are widely used to estimate the turbidity in reservoirs, inland water bodies and estuaries (Hassanin et al., 2020; Tomar et al., 2021; Meraj et al., 2021; Chandel et al., 2021; Bera et al., 2021). There are few studies that focus on estimating turbidity in river waters using MODIS satellite images but are challenging due to the coarse spatial resolution of the data comparing the size of the river section. The latest Landsat sensor, the Operational Land Imager on board Landsat-8, launched in 2013, has a high potential for monitoring aquatic environments. To map and monitor suspended sediment concentration, several empirical regression-based models have been developed. These include single-band, two-band ratio, and three-band combinations. Results from these studies have demonstrated Landsat8 as a suitable satellite asset for the sedimentation study.

In this research, a study is carried out to estimate the soil erosion and further to prioritize the conservation of subwatershed of Siruvani reservoir, Kerala, India. The reservoir receives water from five subwatersheds. These subwatersheds are ranked based on their soil loss due to erosion and morphometric analysis. The Universal Soil Loss Equation (USLE) used in this study and has proved to be very popular, and it estimates long-term average annual gross soil erosion. Morphometric analysis describes the shape, dimension and configuration of a watershed in a mathematical form.

Morphometric analysis is important in any hydrological investigation, and it is inevitable in the development and management of drainage basin (Rekha et al., 2011; Ali and Khan, 2013; Altaf et al., 2013, 2014; Meraj et al., 2015, 2016). The values obtained from morphometric analysis for various parameters like bifurcation ratio, drainage density, circularity ratio, elongation ratio, form factor, stream frequency and drainage intensity indicate that the basin produces a flatter peak of direct runoff for a longer duration, and such flood flows emerging from elongated basins are easier to manage than those from circular basins (Withanage et al., 2014). Analysis of morphometric parameters of a river drainage network will help in determining flooding potential in a river basin. Linear and areal morphometric parameters such as form factor with low values, stream frequency closer to 1, drainage density of less than 1 km/sq. km, mean bifurcation closer to 2 and relief ratio closer to 20 reveal that the watershed is elongated with permeable strata, and this also indicate less potential for flooding in the rivers. Development of synthetic unit hydrograph using microwatershed morphometric parameters extracted by using GIS techniques will lead to drainage characteristics, investigation of distributed curve number technique for runoff estimation and Sediment Yield Index,. Digital terrain models provide elevation data in the form of triangulated irregular network grids or digital elevation models (DEMs), and they are more widely used in watershed modeling.

The sedimentation results for Tuli-Makwe reservoir, Mzingwane catchment, South province of Zimbabwe using the remote sensing approach for 2013 are comparable with the sedimentation results from the 2012 hydrographic survey. The results further confirm the applicability of remote sensing for sedimentation analysis for small reservoirs in semi-arid regions (Mupfiga et al., 2016). Satellite remote sensing techniques can be employed for estimating water spread area of reservoirs. Any reduction in water spread area at specified elevation over a time period is indicative of sediment deposition at that level (Nathawat et al., 2010; Bhavsar et al., 2015; ;

Farooq and Muslim, 2014; Singh et al., 2017; Kanga et al. 2020a, b; Gujree et al., 2020). The geostatistical approach creates digital surfaces that represent relatively accurate reservoir bottom conditions and support automated reservoir volume and surface area calculations (Takal et al. 2010). Assessment of water spread area during different periods of a year serves as a valid input for the estimation of reservoir capacity (Jeyakanthan and Sanjeevi, 2011). The bands corresponding to near-infrared and visible green wavelength can be used to derive Normalized Difference Water Index (NDWI), which discriminates the presence of water bodies from terrestrial and vegetation features (McFeeters, 1996). NDWI enhances the spectral signature from water, uses modified histogram for redefining spectral information in order to separate out water bodies and enhances segmentation of water bodies based on training datasets and iterative classification procedures (Qiao et al., 2012). The USLE has proved to be very popular, and it estimates long-term average annual gross soil erosion. This equation multiplies six parameters: rainfall erosivity factor, soil erodibility factor, slope length factor, slope gradient factor, crop cover or crop management factor and conservation practice factor (Brema and Hauzinger, 2016; Fistikoglu and Harmancioglu, 2002). When compared with existing models like Nayak and Khosla's method, USLE along with GIS gives better results (Ahmad and Verma, 2013).

4.2 STUDY AREA

Siruvani reservoir is in Palakkad District, Kerala, located 46 km away from Palakkad town. This dam was constructed across the Siruvani river, for supplying drinking water to the Coimbatore city in Tamil Nadu. The dam is surrounded by reserve forests. The length and height of the dam are 224 and 57 m, respectively. The reservoir has a water spread area of 1.52 sq. km and perimeter 21 km with a storage capacity of 650 Mcft. Though the reservoir is situated in Kerala state, it is utilized by Tamil Nadu state for the purpose of drinking water supply and irrigation. For this purpose, an agreement was framed between the two state governments during 1973.

The location of the Siruvani reservoir is shown in Figures 4.1, and 4.2 shows the subwatersheds that drain into Siruvani reservoir.

4.3 METHODOLOGY

The watershed was delineated from the DEM (Figure 4.3) using ArcGIS, and the drainage lines were obtained. Based on drainage lines of the watersheds, the morphometric analysis of the study area was carried out. The NDWI was generated from the satellite imageries, and based on this, the changes in water spread area during the period 2011–2019 were obtained. Using the USLE method, the soil loss in each of the subwatersheds was determined.

The DEM, Shuttle Radar Topography Mission (SRTM) with a resolution of 90 m of the study area has been downloaded from USGS EarthExplorer website (Figure 4.3). After delineating the watershed, the stream network and stream orders were identified. The drainage lines and the stream orders of the subwatersheds are shown in Figure 4.5.

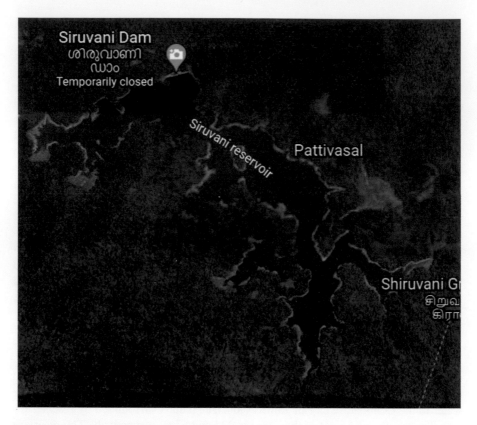

FIGURE 4.1   Location of Siruvani reservoir.

4.3.1 MORPHOMETRIC ANALYSIS

Morphometric analysis considers three aspects, namely, linear representing one dimensional, areal representing two dimensional and relief representing three-dimensional aspects (Figure 4.4). The behavior of the watershed is generally defined from this analysis, and also the results are further used for assessment of soil erosion, etc.

The various parameters studied are stream order (Figure 4.5), stream number, stream length, texture ratio, bifurcation ratio, shape factor, stream frequency, form factor, relief ratio, etc. This study was first introduced by Horton, and further developments were carried out in the later years by Coates and Strahler. It has importance in any hydrological investigation like watershed management, groundwater potential assessment, management of groundwater, environmental assessment and physical changes in a drainage system over time to time. Many areas which still remain inaccessible to humans can be studied with the help of geospatial techniques such as geographic information system and remote sensing. These new developments in this field have attracted many scientists, and now this stands as one of the best management practice for a watershed.

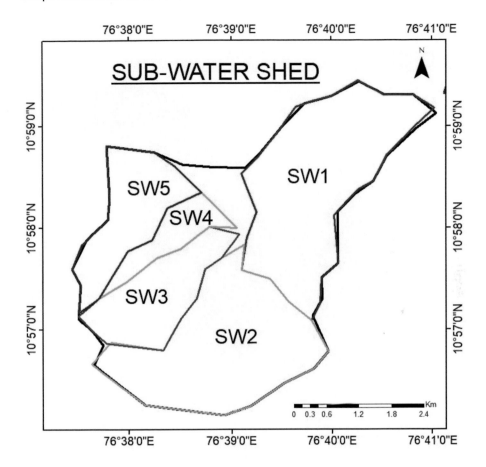

FIGURE 4.2   Siruvani reservoir.

4.3.2   USLE

The USLE proposed by Wischmeier and Smith has been widely accepted and is utilized for soil loss estimation in different parts of the world. The method has been utilized for predicting average annual soil loss in a watershed due to sheet or rill erosion. While it can estimate long-term annual soil loss and guide conservationists on proper cropping, management and conservation practices, it cannot be applied to a specific year or a specific storm. The USLE equation used for estimating average annual soil erosion in a watershed is as follows:

$$A = R \times K \times LS \times C \times P$$

- A = Average annual soil loss in t/ha/year
- R = Rainfall erosivity index
- K = Soil erodibility factor
- LS = Topographic factor – L is for slope length, and S is for slope

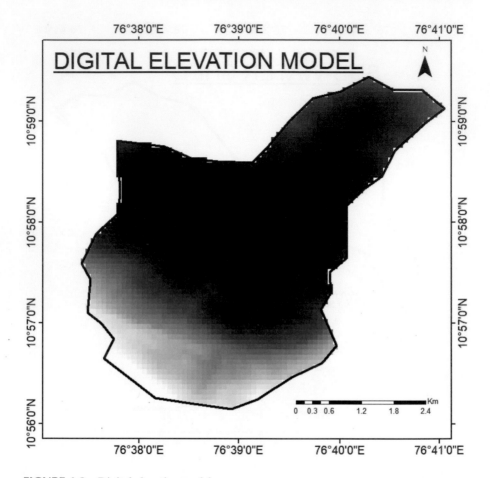

FIGURE 4.3   Digital elevation model.

- C = Cropping factor
- P = Conservation practice factor

Evaluating the factors in USLE:

1. R – the rainfall erosivity index

   Most appropriately called the erosivity index, it is a statistical parameter calculated from the annual summation of rainfall energy in every storm (correlates with raindrop size) times its maximum 30-minute intensity. As expected, it varies geographically.

2. K – the soil erodibility factor

   This factor represents the cohesiveness or bonding nature of a particular soil type, and its behavior toward the raindrop impact and overland flow in the watersheds.

3. LS – the topographic factor

   Steeper slopes produce higher overland flow velocities, whereas, runoff is accumulated in longer slopes resulting in higher flow velocities. Finally,

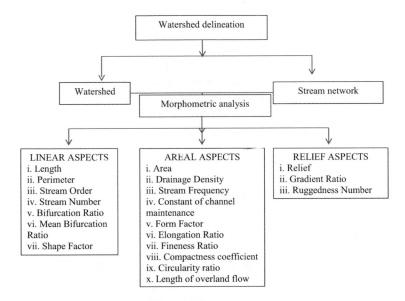

FIGURE 4.4   Methodology for morphometric analysis.

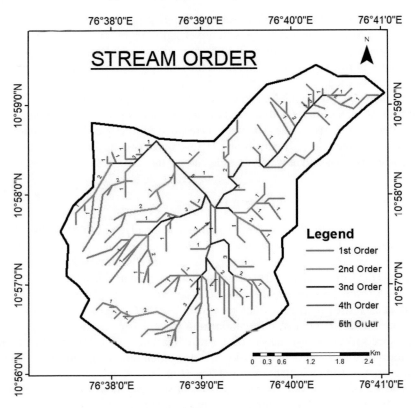

FIGURE 4.5   Stream order.

both the slopes result in increased erosion potential in a nonlinear fashion. For convenience, L and S are frequently lumped into a single term.

4. C – the crop management factor

This factor is the ratio of soil loss from land cropped under specified conditions to corresponding loss under tilled, continuous fallow conditions. The most computationally complicated of USLE factors, it incorporates effects of tillage management (dates and types), crops, seasonal erosivity index distribution, cropping history (rotation) and crop yield level (organic matter production potential).

5. P – the conservation practice factor

Practices included in this term are contouring, strip cropping (alternate crops on a given slope established on the contour) and terracing.

4.3.3 NDWI

NDWI reflects moisture content in plants and soil and is estimated in a similar manner to NDVI:

$$NDWI = (NIR - SWIR)/(NIR + SWIR)$$

NIR – near-infrared range with wavelength ranging between 0.841 and 0.876 nm; SWIR – short-wave infrared range with wavelength ranging from 1.628 to 1.652 nm. Both spatial and temporal distribution of water stress in vegetation can be presented in the form of maps and graphs, based on the results generated using a NDWI index for a given location.

The index is dimensionless, and it varies from −1 to +1, depending on the moisture content, which varies with respect to meteorological parameters and vegetation. Normally, high NDWI values (in blue) correspond to high plant water content, and low NDWI values (in red) correspond to low water content in vegetation. During water stress periods and at water stress locations, the index value decreases.

4.4   RESULTS AND DISCUSSION

Figures 4.6 and 4.7 show the flow direction and accumulation of flow in the Siruvani watersheds, which are obtained from DEM. Flow direction map shows the flow direction from each cell to its down slope neighbor cell or cells, and the flow accumulation represents the number of cells that contributes to a particular cell (Figures 4.6 and 4.7).

Table 4.1 shows the value of linear aspects and areal aspects like stream order (U), length of each stream order in km, total stream length (L_u), mean stream length (L_{sm}), bifurcation ratio (R_b), mean bifurcation ratio (R_{bm}), basin length (L_b) and also the areal aspects like drainage density (D_d), stream frequency (F_s), drainage texture ratio (D_t), form factor (R_f), circulatory ratio (R_c), elongation ratio (R_e), constant of channel maintenance (C) and length of overland flow (L_{of}).

Table 4.2 shows the value of relief aspects of subwatersheds like basin relief (B_h), relief ratio (R_h), ruggedness number (R_n) and relative relief (R_r).

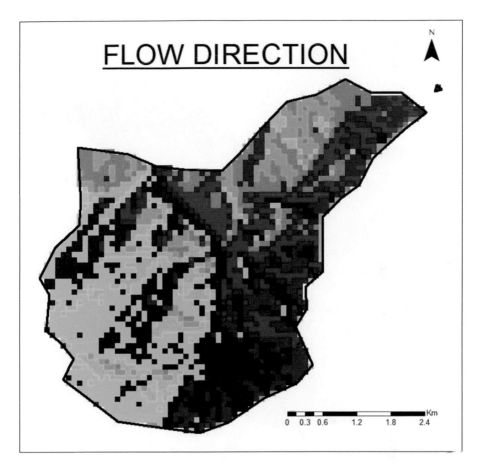

FIGURE 4.6 Flow direction.

In order to identify the watershed which is highly susceptible to erosion, a prioritization analysis has been carried out. The compound parameter values of the five subwatersheds of the catchment area of the reservoir are calculated, and the prioritization rating is shown in Table 4.3. Subwatershed 2 with a compound parameter value of 2.62 received the highest priority (one) with the next in priority being subwatersheds 1 and 3, having a compound parameter value of 2.75. The highest priority shows that the particular watershed is highly susceptible to erosion. The soil conservation measures have to be carried out in this watershed.

Figures 4.8–4.11 show the NDWI of Siruvani watershed corresponding to the years 2013, 2014, 2015 and 2016. Similarly, the NDWI has been generated for the remaining years also during the period 2011–2019.

Figures 4.12–4.15 show the water spread area of the Siruvani reservoir corresponding to the years 2011, 2012, 2013 and 2014, as derived from NDWI maps. Water spread areas corresponding to various years during the period 2011–2019 are shown

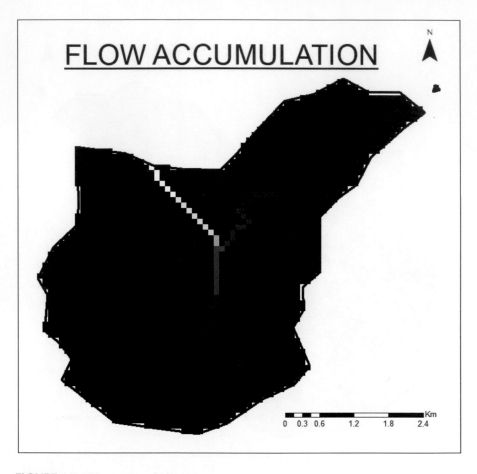

FIGURE 4.7   Flow accumulation.

in Figure 4.16. The gradual increase in water spread area denotes the increase in sediment level in the Siruvani reservoir (Table 4.4).

Figures 4.17 and 4.18 show the annual soil loss during the years 2011 and 2019, estimated using USLE soil loss equation in the subwatersheds of Siruvani reservoir. Table 4.5 shows the annual soil loss corresponding to the years 2011 and 2019 and increase in the soil erosion value during this period in the respective subwatersheds.

4.5   CONCLUSION

Morphometric analysis was carried out using ArcGIS software for the chosen study area. The study shows that morphometric analysis can be carried out accurately and conveniently with help of SRTM data and GIS technology. A procedure has been adapted to extract the water spread area from the NDWI. The water spread area extracted for the study period 2011–2019 shows that there is an increase in water spread area due to sedimentation in the reservoir. In the present study, the higher

TABLE 4.1
Linear and Areal Aspects of the Studied Watersheds

Linear Aspects

Parameter	SW-1			SW-2				SW-3			SW-4		SW-5		
Stream order (U)	1st	2nd	3rd	1st	2nd	3rd	4th	1st	2nd	3rd	1st	2nd	1st	2nd	3rd
No. of streams	19	4	1	20	4	2	1	7	2	1	5	1	9	2	1
Length of each stream Order in km	12.6	2.53	4.2	13.5	3.7	2.9	1.12	6.79	1.46	1.55	1.48	1.78	5.1	1.73	1.1
Total stream length (L_u)	19.38			21.26				9.83			3.26		7.93		
Mean stream length (L_{sm})	0.8075			0.7874				0.983			0.5433		0.6608		

Bifurcation ratio (R_b)

Parameter	SW-1	SW-2	SW-3	SW-4	SW-5
1st/2nd	4.75	5	3.5	5	4.5
2nd/3rd	4	2	2	–	2
3rd/4th	–	2	–	–	-
Mean bifurcation ratio(R_{bm})	4.38	3	2.75	5	3.25
Basin length (L_b)	3.39	3.76	2.38	1.81	2.39

Areal Aspects

Parameter	SW-1	SW-2	SW-3	SW-4	SW-5
Drainage density (D_d)	2.79	3.34	3.43	1.84	2.75
Stream frequency (F_s)	3.46	4.24	3.49	3.39	4.17
Form factor (R_f)	1.73	2.59	1.38	0.61	1.41
Circulatory ratio (R_c)	0.603	0.45	0.5	0.54	0.5
Elongation ratio (R_e)	0.45	0.74	0.68	0.23	0.5
Constant of channel maintenance (C)	0.76	1.96	0.8	0.83	0.8
Length of overland flow (L_{of})	0.358	0.299	0.29	0.54	0.36
Drainage density (D_d)	0.179	0.15	0.15	0.27	0.18

TABLE 4.2
Relief Aspects of Subwatershed

Parameter	SW-1	SW-2	SW-3	SW-4	SW-5
Basin relief	782	1198	1045	894	930
Relief ratio (R_h)	230.7	318.9	438.5	492.6	388.6
Relative relief (R_r)	56.4	115.2	143.8	90.3	108.9
Ruggedness number (R_n)	2181.7	3998.3	3591.7	1646.5	2560.7

TABLE 4.3
Prioritization of Subwatershed

Parameter	SW-1	SW-2	SW-3	SW-4	SW-5
Mean bifurcation ratio (R_{bm})	2	4	5	1	3
Drainage density (D_d)	3	2	1	5	4
Stream frequency (F_s)	4	1	3	5	2
Drainage texture ratio (D_t)	2	1	4	5	3
Form factor (R_f)	5	1	3	4	2
Circulatory ratio (R_c)	2	5	4	1	3
Elongation ratio (R_e)	1	5	3	4	2
Constant of channel maintenance (C)	3	2	1	5	4
Compound parameter	2.75	2.62	2.75	3.75	2.87
Priority	2	1	2	4	3

values of bifurcation ratio indicate strong structural control on the drainage pattern. The bifurcation ratio 5 indicates high overland flow in the catchment area of the reservoir and the discharge due to hilly terrain and high steep slopes in the catchment area. Based on the prioritization and morphometric analysis, the subwatersheds 2 and 5 have been identified as the watershed with high erosive nature. A prioritization study based on two methods gives more accurate spatial and temporal information for the implementation of conservation measures in a watershed.

REFERENCES

Ahmad, I., Verma, M.K. (2013). Application of USLE model & GIS in estimation of soil erosion for Tandula reservoir. *International Journal of Emerging Technology and Advanced Engineering*, 3(4):570–576.

Ahmed, F., Rao, S.K. (2016). Hypsometric analysis of the Tuirini drainage basin: A geographic information system approach. *International Journal of Geomatics and Geosciences*, 6(3):1685–95.

Ali, S.A., Khan, N. (2013). Evaluation of morphometric parameters – A remote sensing and GIS based approach. *Open Journal of Modern Hydrology*, 3:20–27.

Altaf, F., Meraj, G., Romshoo, S.A. (2013). Morphometric analysis to infer hydrological behaviour of Lidder watershed, Western Himalaya, India. *Geography Journal*, 2013: 1–4.

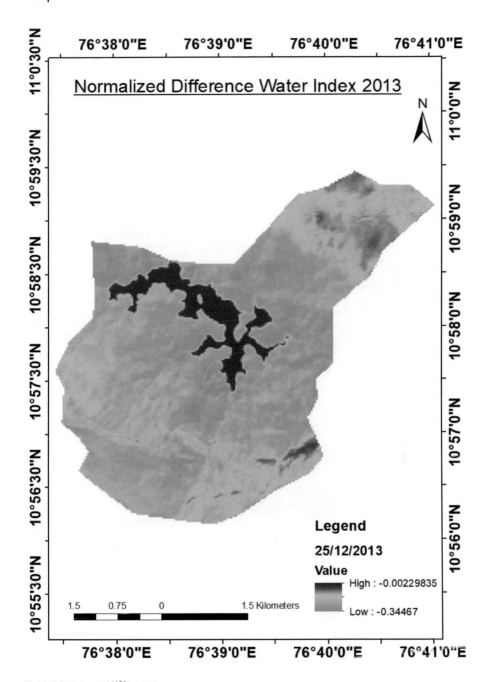

FIGURE 4.8 NDWI 2013.

FIGURE 4.9   NDWI 2014.

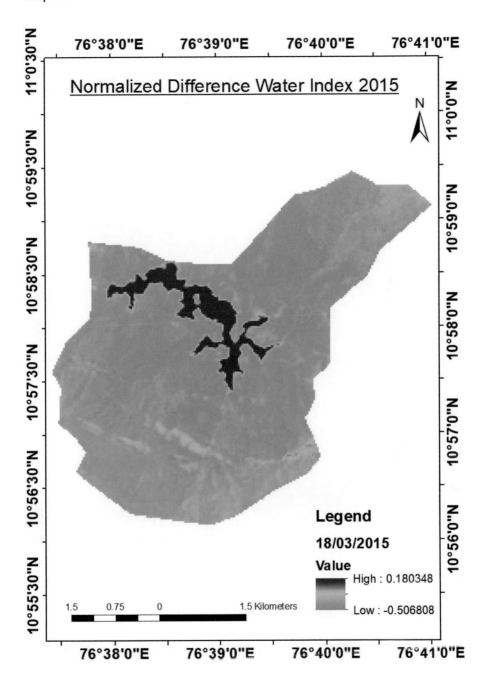

FIGURE 4.10   NDWI 2015.

FIGURE 4.11 NDWI 2016.

Water Spread Area 2011

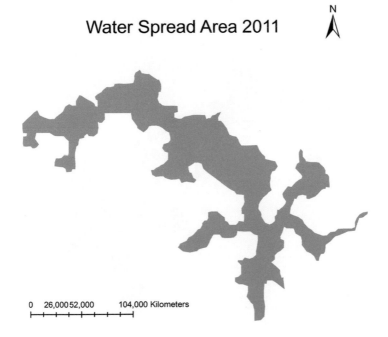

FIGURE 4.12 Water spread area for 2011.

Water Spread Area 2012

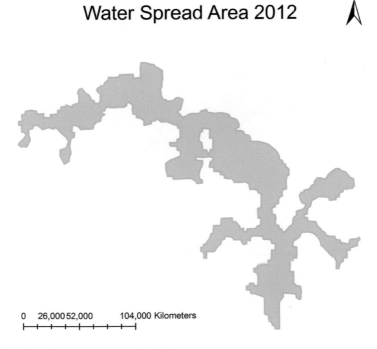

FIGURE 4.13 Water spread area for 2012.

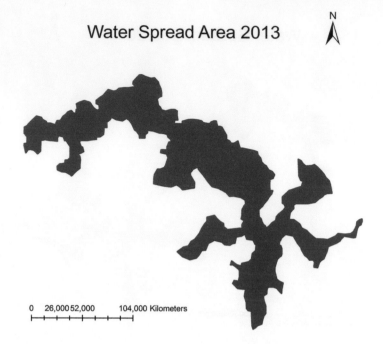

FIGURE 4.14   Water spread area for 2013.

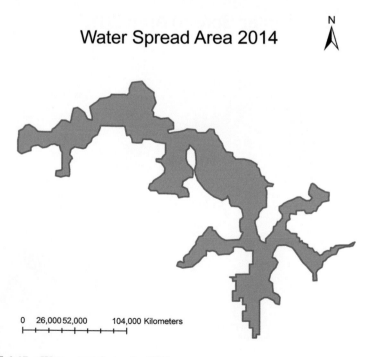

FIGURE 4.15   Water spread area for 2014.

Water Spread Area 2011-19

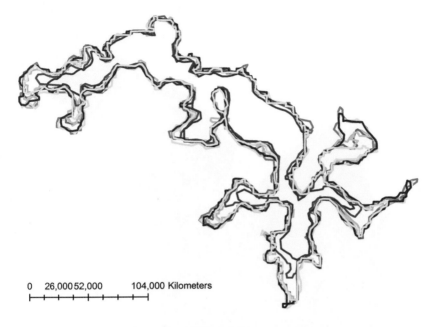

0   26,000 52,000   104,000 Kilometers

FIGURE 4.16   Water spread area from 2011 to 2019.

TABLE 4.4
Water Spread Area Derived from NDWI

S.No.	Year	Water Spread Area (km²)
1	2011	1.19
2	2012	1.42
3	2013	0.8
4	2014	0.89
5	2015	1.26
6	2016	1.27
7	2017	1.51
8	2018	1.14
9	2019	1.45

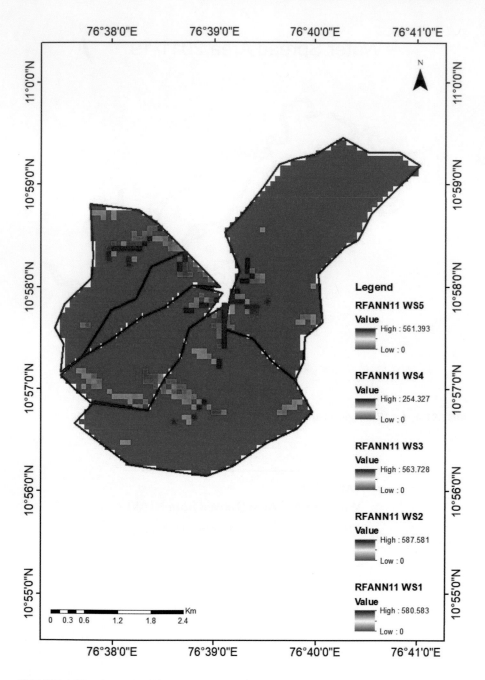

FIGURE 4.17   Annual soil loss 2011.

TABLE 4.5
Annual Soil Loss

Subwatershed	Soil Loss for 2011 (tons/ha/year)	Soil Loss for 2019 (tons/ha/year)	Increase in Soil Loss between 2011 and 2019 (tons/ha/year)
SW1	10,415	10,964	549
SW2	10,341	10,886	545
SW3	8,272	8,708	436
SW4	1,942	2,045	103
SW5	11,340	11,937	597

Altaf, S., Meraj, G., Romshoo, S.A. (2014). Morphometry and land cover based multi-criteria analysis for assessing the soil erosion susceptibility of the western Himalayan watershed. *Environmental Monitoring and Assessment*, 186(12):8391–8412.

Bera, A., Taloor, A.K., Meraj, G., Kanga, S., Singh, S.K., Durin, B., Anand, S. (2021). Climate vulnerability and economic determinants: Linkages and risk reduction in Sagar Island, India; A geospatial approach. *Quaternary Science Advances*: 100038. doi:10.1016/j.qsa.2021.100038.

Bhavsar, M., Gohil, K.B., Shrimali, N.J. (2015). Estimation of reservoir capacity loss of Sukhi reservoir by remote sensing. *International Journal of Advanced Research in Engineering & Management*, 1(8):1766–2394.

Brema, J., Hauzinger, J. (2016). Estimation of the soil erosion in Cauvery watershed (Tamil Nadu and Karnataka) using USLE. *IOSR Journal of Environmental Science, Toxicology and Food Technology*, 10:11.

Symposium on Integrated Water Resources Management (IWRM-2014), CWRDM, Kozhikode, India, pp. 105–111.

Chandel, R.S., Kanga, S., Singh, S.K. (2021). Impact of COVID-19 on tourism sector: A case study of Rajasthan, India. *AIMS Geosciences*, 7(2), 224–242.

Farooq, M., Muslim, M. (2014). Dynamics and forecasting of population growth and urban expansion in Srinagar City – A geospatial approach. *The International Archives of Photogrammetry, Remote Sensing and Spatial Information Sciences*, 40(8):709.

Fistikoglu, O., Harmancioglu, N.B. (2002). Integration of GIS with USLE in assessment of soil erosion. *Water Resources Management*, 16(6):447–467.

Gujree, I., Arshad, A., Reshi, A., Bangroo, Z. (2020). Comprehensive spatial planning of Sonamarg resort in Kashmir Valley, India for sustainable tourism. *Pacific International Journal*, 3(2):71–99.

Hassanin, M., Kanga, S., Farooq, M., Singh, S.K. (2020). Mapping of Trees outside Forest (ToF) from Sentinel-2 MSI satellite data using object-based image analysis. *Gujarat Agricultural Universities Research Journal*, 207–213.

Jeyakanthan, V.S., Sanjeevi, S. (2011). Assessment of reservoir sedimentation using Indian satellite image data. *IUP Journal of Soil & Water Sciences*, 4(1):9

Kanga, S., et al. (2020a). Modeling the spatial pattern of sediment flow in Lower Hugli Estuary, West Bengal, India by quantifying suspended sediment concentration (SSC) and depth conditions using geoinformatics. *Applied Computing and Geosciences*, 8:100043.

Kanga, S., Sheikh, A.J., Godara, U. (2020b). GIscience for groundwater prospect zonation. *Journal of Critical Reviews*, 7(18), 697–709.

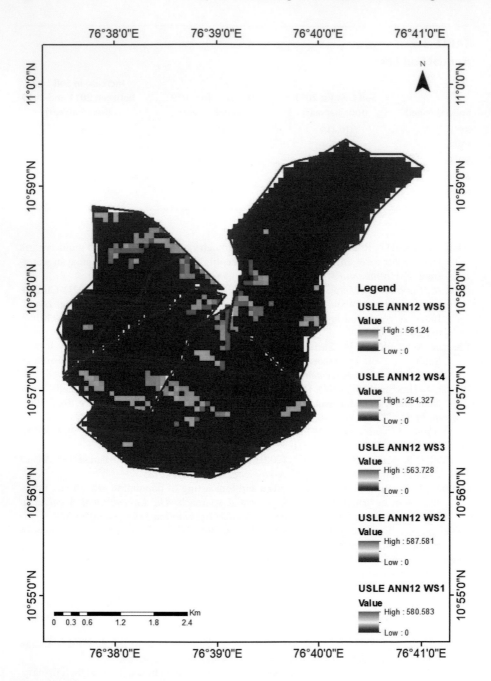

FIGURE 4.18　Annual soil loss 2019.

Kansal, M.L., Chandniha, S.K. (2014). Prioritization of sub-watersheds based on morphometric analysis using geospatial technique in Piperiya watershed, India. *Applied Water Science*. 329–338.

McFeeters, S.K. (1996). The use of the Normalized Difference Water Index (NDWI) in the delineation of open water features. *International Journal of Remote Sensing*, 17(7):1425–1432.

Meraj, G., et al. (2015). Assessing the influence of watershed characteristics on the flood vulnerability of Jhelum basin in Kashmir Himalaya. *Natural Hazards*, 77(1):153–175.

Meraj, G., Romshoo, S.A., Altaf, S. (2016). Inferring land surface processes from watershed characterization. In J Janardhana Raju, J. (Ed.), *Geostatistical and Geospatial Approaches for the Characterization of Natural Resources in the Environment*. Springer, Cham, 741–744.

Meraj, G., Singh, S.K., Kanga, S., Islam, M.N. (2021). Modeling on comparison of ecosystem services concepts, tools, methods and their ecological-economic implications: A review. *Modeling Earth Systems and Environment*: 1–20. https://doi.org/10.1007/s40808-021-01131-6

Mupfiga, E.T., Munkwakwata, R., Mudereri, B., Nyatondo, U.N. (2016). Assessment of sedimentation in Tuli Makwe Dam using remotely sensed data. *Journal of Soil Science and Environmental Management*, 7(12):230–238.

Nathawat, M.S., et al. (2010). Monitoring & analysis of wastelands and its dynamics using multiresolution and temporal satellite data in part of Indian state of Bihar. *International Journal of Geomatics and Geosciences*, 1(3):297–307.

Qiao, C., Luo, J., Sheng, Y., Shen, Z., Zhu, Z., Ming, D. (2012). An adaptive water extraction method from remote sensing image based on NDWI. *Journal of the Indian Society of Remote Sensing*, 40(3):421–433.

Rekha, B.V., George, A.V., Rita, M. (2011). Morphometric analysis and micro-watershed prioritization of Peruvanthanam sub-watershed, the Manimala River Basin, Kerala, South India. *Aplinkos tyrimai, inžinerija ir vadyba*, 3(57):6–14.

Singh, S.K., Mishra, S.K., Kanga, S. (2017). Delineation of groundwater potential zone using geospatial techniques for Shimla city, Himachal Pradesh (India). *International Journal for Scientific Research and Development*, 5(4):225–234.

Takal, K.M., Mittal, S.K., Sarup, J. (2010). Estimation of soil erosion and net sediment trapped of Upper-Helmand catchment in Kajaki reservoir using USLE model and remote sensing GIS technique. *International Journal of Advanced Engineering Research and Science*, 4(2), 23056.

Tigray, E. Small-scale reservoir sedimentation rate analysis for a reliable estimation of irrigation schemes economic lifetime (A case study of Adigudom area, Tigray, northern Ethiopia).

Tomar, J.S., Kranjčić, N., Đurin, B., Kanga, S., Singh, S.K. (2021). Forest fire hazards vulnerability and risk assessment in Sirmaur district forest of Himachal Pradesh (India): A geospatial approach. *ISPRS International Journal of Geo-Information*, 10(7):447.

Withanage, N.S., Dayawansa, N.D.K., De Silva, R.P. (2014). Morphometric analysis of the Gal Oya River Basin using spatial data derived from GIS. *Tropical Agricultural Research*, 26(1):175–188.

5 Review of Conceptual Models of Estimating the Spatio-Temporal Variations of Water Depth Using Remote Sensing and GIS for the Management of Dams and Reservoirs

Abdulkadir Isah Funtua
Abubakar Tafawa Balewa University

CONTENTS

5.1 Introduction 84
    5.1.1 Dams and Reservoirs...85
5.2 Concept of Space, Time, and Spatio-Temporal Analysis 86
    5.2.1 Space...86
    5.2.2 Time..87
    5.2.3 Spatio-Temporal Analysis..87
5.3 The Technology and Approaches 87
    5.3.1 Remote Sensing 87
    5.3.2 Satellite-Derived Bathymetry (SDB)..88
    5.3.3 LiDAR 91
5.4 Conceptual Model Overview...93
    5.4.1 Model Development..94
    5.4.2 Model Application 95
    5.4.3 Satellite Imageries Based Models..96

DOI: 10.1201/9781003147107-6

        5.4.4   LiDAR-Based Models...100
5.5   Conclusion 103
References...105

5.1   INTRODUCTION

Water depth measurement (Bathymetry) is required whenever a detailed survey of waterbed level is to be carried out; it is a method of quantifying depths to study the topography of water bodies, including oceans, seas, rivers, streams, and lakes (Jawak et al., 2015). It is also very important for the studies of the riverbed and seabed morphology, environmental research, and resource management of dams and reservoirs as well as coastal zones.

Bathymetric surveys are undertaken either directly or with remote sensing in several ways (Hell, 2011) to allow professionals to measure water depth, map the underwater terrain, classify submerged vegetation and habitat as well as study marine ecology, water quality, contaminant spills, and hydrodynamics (Webster and Lockhart, 2020, Vasszi, 2019).

Knowledge of the status of water depth is crucial in dam and reservoir capacity estimating to meet the supply-demand for human survival and development. Consequently, several methods of measuring water depth have evolved over time and have been in use for centuries by man through different civilizations and using varying technologies from ship-borne to airborne and satellite-borne platforms.

The need for a precise and detailed outlook of the seafloor or riverbed morphology and the recent technological advancement in the science of remote sensing, particularly the use of space-based satellites and other manned and unmanned remote sensing systems, have brought on new approaches for measuring and mapping water depth. This has necessitated the development of conceptual models to help better understand the dynamics of these natural systems, especially the changes in their underwater profiles. Of particular interest in this chapter are some of the developed conceptual models of remote sensing techniques, of airborne light detection and ranging (ALB) and satellite-derived bathymetry (SDB), applicable in coastal, water, river, and dam environments.

The physical basis for modeling water depth from remote sensing spectral data lies in the ability of light to penetrate the water body (Deng and Zhang, 2008; Lillesand and Kiefer, 1994; Ojinnaka, 2007; Setiawan, 2013; Elsahabi et al. 2017; Meraj et al., 2018a, b).

The use of these new approaches in evaluating water features, though at the infant stage of development and of low accuracy, is fast becoming popular and prefers alternative mode of measuring and analysis of data in water resource management, against the classical empirical methods. The commonly used satellite-based methods are classified under three categories, namely (i) single-band method, (ii) multiband spectral threshold method, and water index method (Yang et al., 2015; Kanga et al., 2017a, b; Rather et al., 2018; Hassanin et al., 2020; Kanga et al., 2021).

The attraction to these new remote sensing methods lies in their abilities to detect and measure surface characteristics of extensive areas on a consistent, timely, and repetitive basis. The data are captured in a suitable format, for computer-based visualization, analysis, and presentation. This is important for understanding the dynamics of water feature which is regarded as the most complex spatial feature on the Earth's surface.

Natural systems such as rivers and dams are very complex. Thus, to understand and manage them effectively, it is often necessary to make simple assumptions by representing them as a conceptual model. A conceptual model comprises a composition of concepts that represent a system (Wood, 2002). Therefore, models can only be understood fully within the perspective that they are made. They are used to help people know, simulate, and have a better understanding of the system or subject of interest. Models succeed and become more effective if they represent a comparative representation of the system to which they are dedicated and have combined an understanding of both the mathematics of the model and the system's behavior.

Effective management of dams and reservoirs requires updated information about the status of the available water to meet the supply demands for which they are designed. However, the dynamic nature of sedimentation and siltation processes affects their volumetric capacity, and over time, results in shortening their designed lifespan (Singh et al., 2017; Kanga et al. 2020a, b). This often poses management challenges, particularly during critical periods requiring timely intervention.

Even though there are significant research efforts in water depth extraction from satellite imageries and LIDAR data since the 1970s, these were largely centered on the ocean and tidal water environments. There are, however, scanty research efforts that have so far been made which utilized these technologies in the study of water depth dynamics of nontidal water environments particularly in rivers and natural and man-made lakes, respectively (Abdulkadir, 2019).

This chapter examines some remote sensing-based conceptual models of extracting and estimating the spatio-temporal variations of water depth. It focuses on the approaches and roles of satellite remote sensing and LIDAR technologies in facilitating the development of the digital presentation of these variations.

The reviewed models are, in broader terms and context, relating to the physical riverbed profile dynamics.

5.1.1 DAMS AND RESERVOIRS

Dam building has a very long history. Dams have been built by diverting water for domestic water supply, power generation, navigation, and flood control, to improve the quality of life of humans. When built, the dam develops a reservoir that has a capacity to store water and builds up ahead and creates a potential for the river water. They have been used for thousands of years to provide a store of water for agriculture, industrial uses, and household uses (Dandekar and Sharma, 2007), but they also resulted in environmental problems as well as human health worries.

The dynamic nature of the water depth profile of a dam due to overtime siltation and sedimentation process at the upstream results in a decrease in its volumetric capacity. The cumulative impacts of these variations have therefore been of concern to the operators and management of dams, especially in times of emergency when a critical decision has to be taken. Effective management of dams and reservoirs requires updated information about the status of the available water to meet the supply demands for which they are designed. Thus, these changes must be monitored and evaluated.

The conventional approaches to mapping bathymetry are labor-intensive and require major capital investment. This is a setback, particularly in low-budget and low-accuracy projects covering large areas as well as for research purpose undertakings. Aside from the conventional or the classic hydrography survey method of sounding, water depth can now be determined using the emerging satellite remote sensing techniques (Abdulkadir, 2019).

The remote sensing technique provides us directly with the surface water area of a reservoir at a particular elevation on the date of the pass of the satellite. This helps us to estimate the available volume of water and also estimate the sedimentation process over a period of time.

Most of the current developed remote sensing data models of water depth extraction are based on coastal environment studies, but with some modifications, these models could be adopted for dams and reservoir studies.

5.2   CONCEPT OF SPACE, TIME, AND SPATIO-TEMPORAL ANALYSIS

5.2.1   SPACE

Philosophically, everything in the world is spatial and temporal, and the concept of space and time is absolute (Spirkin, 1983). It is an accepted disposition that no matter how much time passes before some event, time will fallow it. Also, it does not matter how long ago a certain event occurred, it was preceded by countless other events (Spirkin, 1983).

Space is viewed as a form of harmonization of the concurrent objects and states of matter. It consists in the fact that objects are extra-posed to one another topologically (alongside, beside, beneath, above, within, behind, in front, etc.) and have certain quantitative relationships. Consequently, space can be conceptualized in three ways: (i) absolute space, (ii) relative or topological space, and (iii), a cognitive term.

Absolute space is the mathematical space, the determination of which involves the precise measurement of location and space, such as an X, Y, and Z coordinate. These coordinates provide an unambiguous description of the space.

Absolute space is the mathematical space, the determination of which involves the precise measurement of location and space, such as an X, Y, and Z coordinate. These coordinates provide an unambiguous description of the space.

Topological space is a relative space by which the definition of one location is based on the location of another object, and these topological relationships represent connectivity between considered features. In this case, the relative description of spatial features is more important than the precise measurement of space.

In contrast to the absolute space and topological space, the cognitive space reveals people's perceptions, beliefs and, experiences, about places. Therefore, it is subjective. In conceptual modeling of a phenomenon or feature, the modeler has to take into considerations these space concepts to develop a fit for purpose conceptual models.

5.2.2 Time

Time tells or measures the interval between the occurrences of events or the duration of an event. From Einstein's perspective, time is simply what the clock measures. A clock is an object natural or artificial that moves with a regular cyclic rhythm. Two intervals of time may be described as equal if they contain an equal number of clock cycles.

According to Egenhofer and Golledge (1998), world time can be visualized as a set of the timeline along which our consciences move. Therefore, time is more than a mere concept but something which is conscientiously being felt.

The conceptualization of time was originally assumed to be independent of the spatial conceptualizations, but later, strong links were established. These links include the use of spatial metaphors and processes. The process is to be described reciprocally when a specific process defines the temporal conceptualization as it does the spatial conceptualization. Each focused process provides its unique context and conceptual frame for the cognition of space and time (Egenhofer and Golledge, 1998). Depending upon the conceptual model used, there are several types of time, and a single model of time does not fit all situations. Examples of the model of time include ordinal time, interval time, continuous time, container time, cyclic time, branching time, etc.

Dam and reservoir bathymetry variations are principally due to the natural phenomena of river bed change due to mainly siltation/sedimentation. The bathymetry variations can also be linked to the concept that all reservoirs have a finite lifetime due to the accumulation of sediment behind their dam, and these can evaluate using any or several time models.

5.2.3 Spatio-Temporal Analysis

A spatio-temporal analysis is a concept of spatial analysis. It is used in data analysis for data collected across both space and time. It describes a phenomenon in a specific location (space) and period of time (temporal) having both spatial extension and temporal duration. It is also viewed as a form of geographic reasoning which according to Egenhofer and Golledge (1998) is probably the most common and basic form of human intelligence that when applied to computerized information system can help solve human and environmental problems (Gujree et al., 2020; Farooq and Muslim, 2014).

Water depth data of dam and reservoir is a multidimensional data collected across both space and time that varies over time and space and hence subject of spatio-temporal variation analysis.

5.3 THE TECHNOLOGY AND APPROACHES

5.3.1 Remote Sensing

The principles of remote sensing involve the measurement of radiant energy in many parts of the electromagnetic spectrum (EMS), ranging from gamma rays to radio waves, including visible light. This enables the collection and interpretation of information about an object, area, or event without physical contact (Ndukwe, 1997).

Aerial photography using the visible portion of EMS was one of the earliest form of remote sensing techniques; however, the recent technological advancement, especially concerning the use of satellites and other sensors, has enabled the usage of other wavelengths comprising the near-infrared, thermal infrared, and microwave for the acquisition of information. This has enhanced the data collection process over a large area in a less expensive way.

Different materials reflect and absorbed incident solar radiation at a different wavelength, and each has unique spectral signatures that can be used to identify them in the remotely sensed images provided the sensing system has sufficient spectral resolution to distinguishes them. This capacity of remote sensing is used in the identification and monitoring of land surfaces and environmental conditions. It serves as a useful tool for planning and management.

There are basically two distinctive groups of methods developed for water depth mapping, namely, those that use active and those that are based on passive remote sensing data, respectively. In passive remote sensing, the sensor detects incoming solar radiation reflected or scattered from the surface of the Earth, while active remote sensing uses artificially generated energy sources to receive signals reflected from objects.

The arrival of satellites with high and very higher spatial resolution images have made it a model tool for monitoring certain environmental process of great socio-economic importance. In hydrological modeling, the satellite remote sensing data are used in deriving information on land use, elevation, Metrology, topographic, and roughness to parameterize models (Muthuwatta, 2014).

The physical basis for modeling water depth from remote sensing spectral data is the ability of light to penetrate the water body. The technique is based on the empirical, semi-analytical, or analytical modeling of light transmission through the atmosphere and the water column. The focus here is on the satellite-derived bathymetry (SDB) and the airborne remote sensing techniques of light detection and ranging (LiDAR) for water depth measurement.

5.3.2 Satellite-Derived Bathymetry (SDB)

The satellite-derived bathymetry (SDB) is centered on the remote sensing theory, which states that up to a certain depth, part of the signal recorded by the sensor is reflected by the bottom. The technique based on the empirical, semi-analytical, or analytical modeling of light transmission through the atmosphere and the water column is regarded as an efficient and complementary method for deriving bathymetry in shallow waters (Mavraeidopoulos et al., 2017).

The feasibility of deriving water depth estimates from remote sensing imagery was first demonstrated by Lyzenga (1978) using aerial photographs over clear shallow water. There are however different methods for retrieving water depth from remote sensing data that have evolved. The linear methods and ratio methods are considered as the most basic. Some of the linear methods include those of the Lyzenga (1978) Models, Benny and Dawson (1983) Method, and Jupp (1988). While nonlinear inversion model of Stumpf et al. (2003) is referred to as the ratio method (Abdulkadir, 2019).

Loomis, Jr. (2009), Beer-Lambert Law which describes the absorption effects as light passes through transparent media is the physical principle underlying the flow depth from brightness levels in imagery. Therefore, the fundamentals to retrieving water depth information from remotely sensed radiance values are accurate correction for attenuation and backscattering within the atmosphere and water column.

The mathematical expression of the Beer-Lambert Law is stated as follows:

If a beam of light with intensity I_{in} passes through a transparent media of thickness x, the remaining outgoing intensity I_{out} can be written as in Equation (5.1)

$$I_{out} = I_{in}e^{-cx} \tag{5.1}$$

where

$x=$ is the thickness of the media

$c=$ is the rate of absorption of the medium, which varies according to the properties of the medium such as turbidly, and frequency of the incident light

$I_{out}=$ is the final observed brightness levels in the image.

$I_{in}=$ is the initial brightness of light passing through the water column.

The water depth mapping relies on the determination of the constant c and the initial brightness I_{in} (Carbonneau et al., 2006).

Figure 5.1 shows the path of light from the sun to the satellite. When light passes through water, it becomes attenuated by interaction with the water column. The attenuation light in water varies spectrally and with an increase in depth.

The Lyzenga Model, developed by Lyzenga (1978) was based on Beer–Lambert's law. The method is of historical importance with continued widespread use. The model showed that the relation of observed reflectance or radiance to depth and bottom albedo could be described by Equation (5.2),

$$R_w = (Ad - R\infty)\exp(-gz) + R\infty \tag{5.2}$$

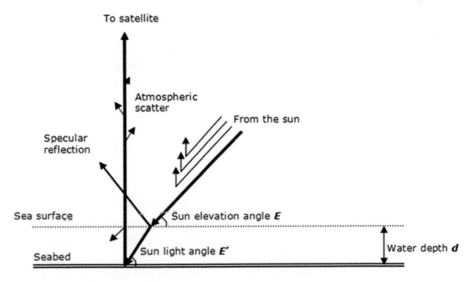

FIGURE 5.1 Path of light from the sun to satellite. (Modified from Nga and Nga, 2007.)

where

   $R\infty$ is the water column reflectance if the water were optically deep
   Ad is the irradiance of the bottom (bottom albedo or bottom reflectance)
   Z is the depth
   g is a function of diffuse attenuation coefficient for both downwelling and
      upwelling light.

The ratio method developed by Stumpf et al. (2003) is based on the ratio of two or
more wavelength and offers more accurate depth estimation over variable bottom
types (reflectance) in the deep and shallow environments by using the ratio of reflectance. While, Benny and Dawson (1983), and Jupp (1988) methods assumed that

   (i) Light attenuation is an exponential function of depth
   (ii) Water quality (attenuation coefficient) does not vary within an image.

Another approach is that of Van Hengel and Spitzer (1991), in short, V-S (1991)
model. This model is in a form of a rotation transformation matrix, for the transformation of satellite imagery to produce relative water depth or depth index, Y_1 value
from the image. The model provides an elaboration on the idea that the reality of
computed water depth depends to a considerable extent on the accuracy of known
water depth data used to calibrate the algorithm.

Van Hengel and Spitzer (1991) utilize three multitemporal images as input data to
get relative water depths from LANDSAT ETM+. The use of this model enables the
relative water depth map to be acquired without prior knowledge of the actual value
of the water depth (Setiawan, 2013). However, in situ measurements are necessary to
calibrate and validate results.

In summary, there are multiple approaches and models available for water depth
extraction, and these can be grouped based on the adopted modeling approach. It is
in this regards that, Deng identified three main groups of models namely;

Theoretical models based on the transmission equation of electromagnetic radiation in water, e.g., Lyzenga and Polycn (1979), and Philpot (1989).

Empirical models are based on the statistical relationship between pixel values
and field-measured water depth (Lyzenga, 1978, 1981).

Models integrate the merits of both theoretical and statistical models by simplify-
ing the former through the use of statistical regression to estimate the photochemical
parameters (Melsheimer and Chin, 2001; Stumpf et al., 2003).

Mavraeidopoulos et al. (2017) stated that the horizontal accuracy of SDB is a
function of the spatial resolution of the satellite sensor used. The uncertainty con-
tained in bathymetric data is usually of the order of 1-pixel size. For the satellite
sensor that has a spatial resolution of 5 m, then for a 5 m resolution product, the horizontal accuracy is <5 m.

The vertical accuracy is a function of an offset parameter equal to 0.5 m and a
depth-dependent factor of 10%–20%. Therefore, for 10m water depths, the accuracy
lies between ±1.5 and ±2.5 m (Mavraeidopoulos et al., 2017).

For dam and reservoir model, the required horizontal and vertical accuracy must
be taken into consideration in the selection of a suitable model.

5.3.3 LiDAR

Light detection and ranging (LiDAR) is an airborne remote sensing technique for measuring the exact distances of objects on the Earth's surfaces. The technique is being used in some countries for seafloor mapping within shallow-water coastal environments. The idea of using LiDAR originates from the problem of measuring water depth without direct contacting with the water body or without any instrument mounted on the water surface in the shallow region (Mohammadzadeh and Zoej, 2018).

LiDAR systems are active remote sensing systems as they emit pulses of light (i.e., the laser beams) and detect the reflected light (NOAA, 2012). LiDAR uses either pulse (time of flight) or continuous wave (phase difference). When the former is used, a high energy laser pulse is emitted, and the reflected pulse from the target is detected by a receiver. The range is determined by multiplying the speed of light with half of the total travel time of the pulse to and back from the target. When the latter is used, a long-wavelength sinuous waveform is modulated onto a continuously emitted low energy short-wavelength carrier Laser signal (Monroy and Khan, 2018).

The principle of LiDAR (Figure 5.2) is actually rather simple: simply shine a small light at a surface and measure the time it takes to return to its source. The principle is similar to the principle of electronic distance measuring instrument (EDMI), where a laser (pulse or continuous wave) is fired from a transmitter, and the reflected energy is captured.

The LiDAR systems are typically made up of several components which comprise a range-finding system and scanning optics (central inertial, D-GPS, etc.) to direct laser pulse and position and orientation system that records the pulse origin.

Specifically, the LiDAR system uses two different wavelengths: an infrared pulse reflected from the water surface and a green pulse which penetrates the water surface and is reflected from the bottom surface.

When used for airborne LiDAR, it is often referred to as "linear mode LiDAR". These linear mode LiDAR systems consume relatively high energy per emitted laser pulse. Each of these pulses travels from the aircraft to the ground and is reflected from there to the scanner (Figure 5.3).

The more energy per pulse, the stronger the reflection that can be recorded as more photons are reflected by the terrain below the aircraft. The output from linear mode systems is impressive, and these systems provide data with high spatial and radiometric precision aircraft (Roth and Sirota, 2017). Therefore, the accuracy of

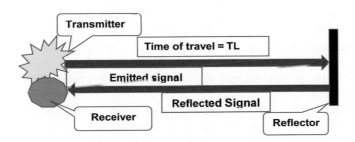

FIGURE 5.2   Principle of measurement using LiDAR.

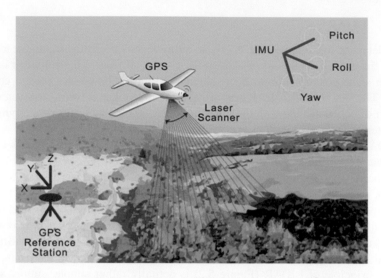

FIGURE 5.3   Schematic diagram of airborne LiDAR data collection. (Modified from MKS, 2019).

both the horizontal and vertical measurements by LiDAR sensor is dependent on the flying height.

The LiDAR technique is adjudged complementary to Sonar and one of the important technology providing accurate 3D point clouds of Earth surfaces and water depth. According to Allouis et al. (2013), literature references on accuracy and limits of bathymetric LiDAR exist in coastal areas. However, only a few exist on rivers and inland waters. However, it is generally accepted that the technology provides low cost, high energy efficiency, and a small measurement footprint.

LiDAR bathymetric surveys involve the use of two different laser beams mounted on a flying aircraft above water surface which provide quickly an accurate and precise geospatial information, especially from shallow water bodies.

According to Saylam et al. (2018), airborne lidar bathymetry (ALB) is an advanced and effective technology for mapping and measuring water depth in relatively shallow inland water bodies and coastal areas. It is an ideal tool to study underwater features in a clear water environment. In particular, ALB technology is used for depicting the depths of relatively clear shallow water bodies using a scanner and pulsed light beam from an airborne platform (Pe'eri et al., 2013). The output from these systems is impressive with high spatial and radiometric precision.

The principle of ALB data collection is shown in Figure 5.3. As shown in Figure 5.4, a laser pulse is transmitted to the water surface where, through Fresnel reflection, a portion of the energy is reflected to the airborne receiver, while the remainder of the pulse continues through the water column to the bottom and is subsequently reflected.

LiDAR sounding works with emitted laser pulses (emissions-receptions) at a steady frequency. The interval between the emission and reception is transformed into the distance. Therefore, the measurements are computed by using and accounting for the speed of light in water and measuring the time interval between them.

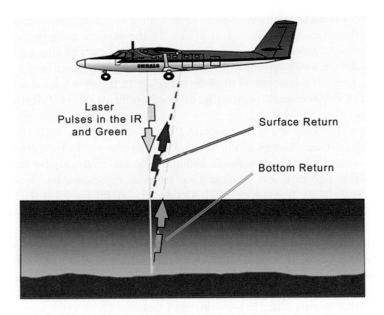

FIGURE 5.4   Principles of ALB. (Modified from MKS, 2019).

LiDAR sensors in clear waters not only have the ability to measure the deepness down to approximately 50 m but also have a higher spatial resolution, below surface object detection larger than $1 \times 1 \, m^2$, high data acquisition rate per m^2, acquires direct 3D position, and needless data postprocessing and rapidly is available on the critical period.

LiDAR bathymetry system can measure depths; subject to the clarity of the water from 0.9 to 40 m, it gives a vertical accuracy of +15 cm and a horizontal accuracy of +2.5 m, respectively. Therefore, presently LiDAR data accuracy satisfies the bathymetric standards.

Among the common applications of bathymetry LiDAR are shoreline and elevation change analysis, submerged sediment characterization, bluff edge detection, and invasive species identification Gao (2015). Detection of the zero-depth which is river banks and its displacement is a very time-consuming and expensive task by traditional hydrographic methods and can be detected by Lidar sensors more rapidly with lower cost.

5.4   CONCEPTUAL MODEL OVERVIEW

Models refer to the simplification of reality developed from the current personal understanding of a system by the modeler. Several types of models exist within the scientific realms. The major types are the conceptual models and empirical models. Generally, models are central to the way to explain and develop scientific theories about real-world phenomena (Brimicombe, 2003). They perform several functions and require clear identification of the type of device they are intended to be employed for.

Empirical models are derived from observation and data from which conclusions are drawn on the outcomes or effects of the process under consideration, while conceptual models are statements or simple formulae and also a mapping of the main elements of a system with the links showing how these are interrelated. The focus here is on conceptual models that relate to the dynamics and processes of a physical environment, specifically, the measurements of water depth of rivers and dams.

Conceptual models of a system contain a high level of abstractions. They display relationships between factors and the flow of data or processes of the observed real world by employing several methods, including the use of descriptive text, charts, tables, box and block arrow, diagrams, or pictorial representation (Wood, 2002; Hassanin et al., 2020; Tomar et al., 2021; Meraj et al., 2021; Chandel et al., 2021; Bera et al., 2021) These types of models can be divided into two core classes: sequential and structured, nonsequential, and unstructured.

To adequately serve their purposes, conceptual models are composed of certain specific elements that increase the understanding of the system or subject being modeled. Therefore, the basis of conceptual models is to make it easier for human understanding of a complex system.

Conceptual models differ in the degrees of freedom given to the user in accomplishing tasks and achieving goals (Parush, 2015). In a given hydrological system, a conceptual model represents, to the extent possible, all physical mechanisms that govern the process within the system (Abdulkadir et al., 2014).

5.4.1 MODEL DEVELOPMENT

Model development is a multistage approach to the development and application of a workable model which demonstrates an understanding of phenomena and how it might be managed effectively. There are several approaches to modeling peculiar to the field of GIS, environmental modeling, and engineering. Full discussions on these are available in the literature (e.g. Brimicombe, 2003; Sanders, 2007).

A typical approach to model development requires the development of an initial conceptual model. According to Wiki (2017), the composition of conceptual models includes the Problem, Plan, Data, Analysis, and Conclusion (PPDAC) model (Figure 5.5), and Analytical Process model (Figure 5.6)

The PPDAC model (Figure 5.5) involves
a. the identification and statement of the study problem: Problem (P),
b. planning and the setting of relevant procedures to achieve the objective: Plan (P)
c. specification of the type of required data and its process of acquisition: Data (D)
d. processing, examination, and interpretation of the acquired data (Analysis: A)
e. drawing conclusion on the outcomes of the study (Conclusion: C).

The Analytical process model (Figure 5.6) involves
(i) objective setting and subject identification for problem specification
(ii) setting and subjects for planning and data acquisition process

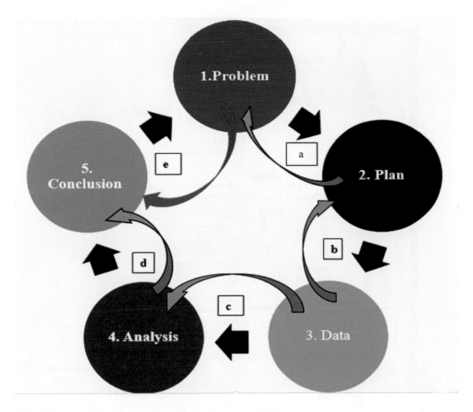

FIGURE 5.5 PPDAC model. (Abdulkadir et al., 2014.)

FIGURE 5.6 Analytical process model. (Abdulkadir et al., 2014.)

(iii) design and measures for analysis
(iv) results and conclusion from the outcomes.

Figure 5.7, adopted from Brimicombe (2003) is a schematic diagram of a model of a model development. It comprises positive and negative feedback loops.

5.4.2 MODEL APPLICATION

The ultimate intent of building a model lies in its application. The first step of applying the model is the identification of the specific problem at hand. The modeler is required to understand the implications of not only the theoretical approaches to the problem but also how these have been incorporated into a model (Brimicombe, 2003). In this chapter, we review some of the developed conceptual models for water

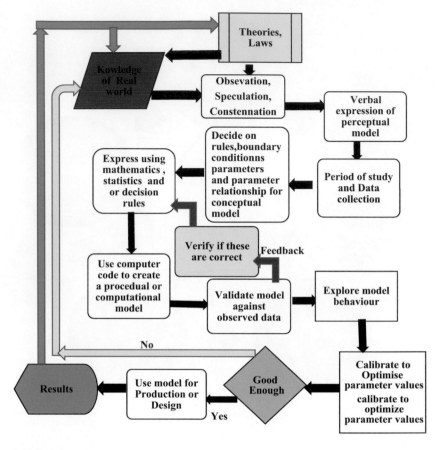

FIGURE 5.7 Schematic outline of the model development process. (Modified from Brimicombe, 2003.)

depth measurements in respect of the shallow depth of rivers and dams. Dams and reservoirs are constructed for multiple purposes, comprising water resource control, irrigation, transportation, and hydroelectricity generation.

5.4.3 SATELLITE IMAGERIES BASED MODELS

Generally, in hydrological modeling satellite, imagery-based model parameters are employed in the conceptual model served to represent not only the catchment's properties but also its characteristics including water depth. Few such developed models exist for dam and reservoir's water depth extraction from satellite imageries as major application. The area of research interest has largely been on coastal waters. According to Jagalingam et al. (2015), several empirical and analytical algorithms are available in the literature for depth extraction from remote sensing data. These include Gholamalifard et al. (2013), Stumpf et al. (2003), Gao (2015), Lyzenga (1978, 1981, 1985, 2003), Van Hengel and Spitzer, Setiawan (2013), Benny and Dawson (1983), Jupp (1988), Setiawan et al. (2017), Zhang et al. (2015), Loomis, Jr (2009). Su et al., Stumpf et al. (2003), and Lyzenga et al. (2006), Phipot (1989), etc. These studies relate to ocean water environments.

The empirical models for bathymetric data estimation that have proved to be effective in modeling virtually any nonlinear function with acceptable accuracy were based on artificial neural network (ANN), support vector machine (SVM), and regression tree. The extraction of water bodies from high-resolution satellite imageries is, however, the starting point for the water depth extraction.

Different extraction techniques such as image analysis for digital number values, radiances and top of atmosphere (TOA) spectral reflectance values, indices analysis using normalized difference water index (NDWI), modified normalized difference water index (MNDWI), automated water extraction index (AWEI), etc. are involved to provide information about water body's pixels and criteria setting. It is to be noted, however, that each of those algorithms has its own strengths and weaknesses, or environments for which it may provide better or poorer retrievals than other algorithms.

According to Loomis, Jr. (2009), Benny and Dawson (1983), and Jupp (1988),

(i) light attenuation is an exponential function of depth (Equation 5.1);
(ii) water quality (attenuation coefficient) does not vary within an image;
(iii) the color of the substrate/seabed (reflective properties) is constant.

The model presented by Benny and Dawson establishes that the clear water in shallow areas allows the reflection of the light, and it reaches the sensor. However, the amount of light that returns depends on the attenuation coefficient for that wavelength and the reflection coefficient of the bottom (Castillo-López et al., 2017).

Jupp's (1988) model is a model that finds large usage in the literature to reconstruct the bathymetry in coastal zones from MS imagery. It is based on the fundamental principle that radiation is attenuated at different rates as it penetrates the water column. Different wavelengths will be attenuated until they became extinct at a certain depth. There are two major parts to Jupp's method:

(i) the calculation of depth of penetration zones or DOP zones,
(ii) the interpolation of depths within DOP zones.

Jupp's model (Equation 5.3) can be mathematically expressed as

$$L_e = \left(e^{-2kz}\right)L_b + \left(1 - e^{-2kz}\right)L_w \tag{3}$$

where
L_e is measured at-sensor radiance,
L_b is the emergent radiance from the seabed,
L_w is the emergent radiance from different layers of water, z is depth,
k is the coefficient of absorption.

If the term L_w is hypothesized as negligible and is directly related to the quality of the water (suspended sediments) and small changes in the seabed, and the depth of the water column, considering the logarithm of the measured at-sensor radiance, then there will be a linear relationship (Jawak et al., 2015).

According to Setiawan (2013), both Benny and Dawnson (1983) and Jupp (1988) methods assumed that the depth of penetration of light is dependent on the water turbidity, hence the accuracy to which optical data may be used to estimate depth is limited as a result of increased attenuation. Thus, clarity of the water, as well as its shallowness, is an important requirement to obtain the best results from satellite-based bathymetry.

Lyzenga's (1978) method, like the two above methods, accepted the first two assumptions but did not the third one (Green et al., 2000). The second and third assumptions are a weakness of these methods because a satellite image normally covers a very large area, and the water and sea-bed properties vary, in some cases, very much.

Similarly, Loomis, Jr. (2009) stated that Stumpf et al. (2003) was an advancement to the Lyzenga (1978). The study proposed a nonlinear inversion model commonly refers to as the "Ratio model". This model offers a more accurate depth estimation over variable bottom types in deep and shallow environments by using the ratio of reflectance.

In the ratio model (Equation 5.4), only two parameters (i.e., m_o and m_i) need to be estimated.

$$Z = m_i \frac{\ln\left(nR_w\left(\lambda_i\right)\right)}{\ln\left(nR_w\left(\lambda_j\right)\right)} - m_o \qquad (5.4)$$

where

$R_w\left(\lambda_i\right)$ and $R_w\left(\lambda_j\right)$ = the atmospherically corrected pixel value for bands i and
m_i = a tunable constant to scale the ratio to depth
n = a fixed constant, mainly for ensuring a positive value after the log transform and a linear response between the ratio and the depth
m_o = an offset value when $Z = 0$.

The water reflectance R_w of particular spectral band is defined by Equation (5.4).

A single band and band ratios were used by Zhang et al. (2015) to derive bathymetry from a linear model and demonstrated that the power exponential model based on Landsat band 2 is the best among other alternatives of a logarithmic model, a power exponential model, and an exponential model. Similarly, the log-linear inversion model by Raj et al. used data from multiple band images and bathymetry measure from echo sounder. Consequently, Gao (2015) opined that shallow water depth inversion from multispectral remote sensing images could provide a reliable measure of water depth and bottom bathymetry.

Van Hengel and Spitzer's (VS) model is an algorithm transformation of satellite imagery to produce value relative depth seawater. The V-S model introduced an elaboration on the idea that water depth is proportional to the radiation algorithm and relative water depth value that can be obtained by transforming specific to the particular value of each band radiation.

This algorithm requires three input band, defined as follows: (Equation 5.5)

$$\begin{bmatrix} Y_1 \\ Y_2 \\ Y_3 \end{bmatrix} = \begin{bmatrix} \cos(r)\,Sin(s) & \sin(r)\cos(s) & \sin(s) \\ -\sin(r) & \cos(r) & 0 \\ -\cos(r)\sin(s) & -\sin(r)\sin(s) & \cos(s) \end{bmatrix} \times \begin{bmatrix} X_1 \\ X_2 \\ X_3 \end{bmatrix} \qquad (5.5)$$

This can be expressed in a linear equation format as (Equation 5.6)

$$Y_1 = \left(\cos(r) \times \sin(s) \times X_1\right) + \left(\sin(r) \times \cos(s) X_2\right) + \left(\sin(s) \times X_3\right) \qquad (5.6)$$

where
   Y: Relative water depth (Depth Index)
   X_i: Reflectance in spectral band i ($i = 1, 2, 3$),
   r and s: Rotation Parameters

Value of rotation angle (r and s) in the algorithm is a constant obtained from the calculation with the following respective the formulas (Equations 5.7 and 5.8):

$$r = \arctan\left(U_r + \sqrt{U_r^{\,2} + 1}\right) \qquad (5.7)$$

$$S = \arctan\left(U_s + \sqrt{U_s^{\,2} + 1}\right) \qquad (5.8)$$

while U_r and U_s are obtained respectively from (Equations 5.9 and 5.10)

$$U_r = \frac{\text{var}\,X_2 + \text{var}\,X_1}{2\,\text{cov}\,X_1 X_2} \qquad (5.9)$$

$$U_s = \frac{\text{var}\,X_3 + \text{var}\,X_1}{2\,\text{cov}\,X_1 X_2} \qquad (5.10)$$

where
   Var X_i: is variant value of band i and
   Cov $X_i\,X_j$: are covariant bands i and j

Depth value is the result of rotation on the relative depth values shown by the variable Y_1. While the result of the rotation Y_2 and Y_3 is the effect produced by this matrix.

The resulting depth of Y_1 is a relative depth from the image. Therefore, the relative depth should be changed to the absolute depths to obtain the actual depth value. This is achieved by regression analysis.

The regression equation obtained is used to convert water depths from Equation 5.11

$$a\,Y_{1,\,i} + b = S_{ti} \qquad (5.11)$$

where
   a and b: Regression constants
   i: Pixel number
   S_{ti}: Water depth

The discrepancy between the in situ measurement and those obtained from the remote sensing can be determined from Equation (5.12) below

$$\varepsilon a = \frac{A_1 - A_0}{A_0} \times 100\% \tag{5.12}$$

where
  εa: Error value
  A_1: Depth measurement using remote sensing data
  A_0: In situ data

Gholamalifard et al. (2013) stated that it was due to the dynamic nature of ocean environments that bathymetric algorithms attempt to separate water attenuation effects and hence depth from other factors by using different combinations of spectral bands. In this respect, the inversion method proposed by Lyzenga (1978, 1981, 1985) was adjudged the most widely used (Gao, 2015).

The basic linear and analytical models of water depth extraction from satellite data provide depth data at the time of the image acquisition, and therefore, different images acquired at different times provide a window for the development of a conceptual model for spatio-temporal analysis.

Khattab and Al-Ani (2012) estimated the water depth of Mosul Dam's reservoir located in North of Iraq by using multispectral image processing techniques. The results of the study show the suitability of the application of a linear regression model using bands 1 and 5, to show a map of depth distributions that range from 0 to 80 m.

Similarly, Mohamed et al. (2016) used ensemble learning algorithms for the bathymetry determination from high-resolution satellite imagery of shallow lakes at El-Burullus Lake.

Environmental conditions such as cloud cover, haze, waves, and water quality can reduce the quality of the image and hence results. Deng identified the several steps involved in the preparation of the spectral and bathymetric data for water depth mapping. These comprise atmospheric correction, geometric correction, tidal correction, and the extraction of bathymetric values from the nautical chart or field measurements (for calibration or validations) and the corresponding pixel values from images metadata. In the case of nontidal inland waters, tidal corrections are not applicable.

Woodlard and Colby (2002) argued that although satellite imagery is extremely valuable for studying phenomena over large spatial extents, it is limited in its ability to provide detailed information about coastal features where the size of the feature is smaller than or barely exceeds the resolution of the data. Therefore, it is important to consider the spatial extents of the dam or reservoir under consideration and the type of remote sensing satellite data in developing or adopting a suitable model.

5.4.4 LiDAR-Based Models

Hydrographic LiDAR approximates the water depth using the time difference of blue-green channel and infrared channel reflected from the sea bottom and water surface, respectively (Mohammadzadeh and Zoej, 2018). The water depth profiles can be derived by suitable algorithms from LiDAR signals of detection wavelengths corresponding to the respective scattering and fluorescence bands of the features or substances being measured Diebel-Langohr et al. (1986). Figure 5.8 adapted from NOAA (2013) shows the various interactions for a bathymetric LiDAR laser pulse.

The intensity of a laser beam propagating in a medium is described by Beer's (or Bouguer's) law (Equation 5.13)

$$\frac{I(\lambda)}{I_0} = T(\lambda, R)\exp\left[-\int_0^R \alpha(r\lambda)dr\right] \tag{5.13}$$

where

I_0 = the intensity at $r = 0$,

I = the intensity at $r = R$,

α = the atmospheric extinction coefficient.

$T(\lambda, R)$ = the transmissivity in $(0, R)$.

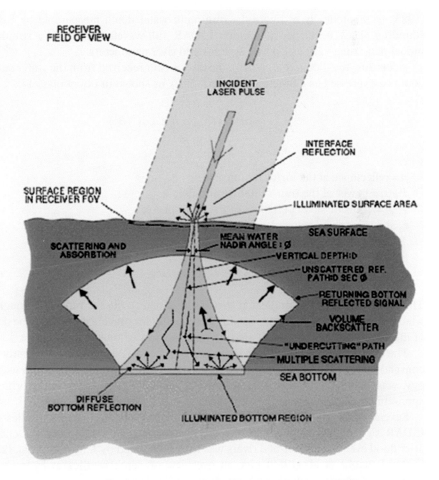

FIGURE 5.8 The interactions for a bathymetric LiDAR laser pulse. (USACE Adopted from NOAA, 2013.)

Atmospheric transmission windows for LiDAR operation lies between the following ranges of visible spectrum (VIS), (0.4–0.7 μm), near-infrared (NIR), (0.7–1.5 μm), and infrared (IR), (3–5 μm and 9–13 μm), respectively, and the "eye-safe" range is $\lambda > 1.4\,\mu m$ (100 mW/cm^2, 1 J/cm^2) (Rocadenbosch, 2007).

According to Wasser (2020), waveform or distribution of light energy that returns to the LiDAR sensor could be recorded in two different ways, namely a discrete return LiDAR system and a full waveform LiDAR system. The discrete return LiDAR systems identify the respective pulse peaks, and record a point at each peak location in the waveform curve. A full waveform LiDAR system similarly records a distribution of the returned light energy. They can often capture more information compared to discrete return LiDAR systems (Wasser, 2020).

To develop a suitable LiDAR model for rivers, Allouis et al. (20013) advanced a hypothesis and introduced additional elements to the equation developed for coastal areas. Allouis et al. (2013) used equations developed for coastal waters by Guenther (1985) to theoretically estimate the minimum water depth measurable by LiDAR sounding. They developed a specific LiDAR full-waveform (GLFW) simulation model data using various LiDAR systems and river parameters.

According to Allouis et al. (2013), the laser power received from the water surface could be expressed as a function of time t, $P_s(t)$, as shown in Equation (5.14):

$$P_s(t) = \frac{\rho P_r(t) T^2{}_{atm} \eta_r \eta_l A_r \cos^2(\theta_0)}{\pi t^2} \qquad (5.14)$$

where:
ρ = reflectance at the air/water interface
$P_T(t)$ = power of the transmitted laser pulse;
$T^2{}_{atm}$ = transmission coefficient of the atmosphere;
η_t and η_r = optical transmission and reception efficiencies, respectively;
A_r = area of the receptor;
θ_0 = incidence angle of the sensor; and
L = distance of the sensor from the water surface

The results of the study indicated that 0.4 m is detectable for moderate rough flat water surfaces, but validation is required especially for gravel-bed rivers. The details of the application of this model are fully discussed in Allouis et al. (2013)

The experimental use of LiDAR for surveying a shallow river concurrently with conventional survey grade, real-time kinematic, global positioning system technology, was evaluated by Kinzel et al. (2007). The study developed an algorithm to approximate the position of the riverbed.

Several other algorithms/models exist for the derivation of depth profiles from LiDAR signals. The application of these in assessing fluvial geomorphic changes after flood events in gravel-bed rivers was fully elaborated by J. Moretto et al. (2013). Diebel-Langohr et al. (1986) give an overview of depth profiles of hydrographic parameter measurement and interpretation of LiDAR signals.

While giving an overview of the application of LiDAR in hydrography and ocean-ography, Mohammadzadeh and Zoej (2018) contended that water surface and bottom reconstructions are the most important concerns in bathymetric LiDAR.

The collection of elevation data using LiDAR has several advantages due to the recent advancement in GPS and IMU technology. Similarly, the combination of LiDAR with imaging sensors and better processing algorithms provides more realis-tic and accurate 3D models of the geospatial objects.

The accuracy of LiDAR measurements is determined by the pulse width and the accuracy of the method that measures the time of flight. The narrower the pulse width and the higher pulse power, the more the accuracy of the measurements.

To detect possible systematic errors or fusion of the data, depth derived from LiDAR data is validated by comparing it with more accurate data such as echo sounder data and also calibration during flight using ground control points (GCP).

5.5 CONCLUSION

For the evaluation, planning, and effective management of water resources for eco-nomic and sustainable development, precise information about water depth and the exact location of underwater features is required. A brief review of techniques and models for generating remote sensing-based water depth (Bathymetric) data for spatio-temporal analysis has been provided in this chapter. Since the early 1970s, the possibility of obtaining water depths using satellite imagery data has been investigated. Imaging and nonimaging bathymetric measurements are the two most common remote sensing methods. Based on the recorded pixel values or digital numbers (DN) (representing reflectance or backscatter) of an image, the imaging methods estimate the water depth. The visible and/or near-infrared (NIR) as well as microwave radiation are used in imaging methods. Analytical modelling, empir-ical modelling, or a combination of both are used to implement imaging methods. A nonimaging method is laser scanners, or LIDAR, which uses a single wave pulse or double waves to measure the distance between the sensor and the water surface or ocean floor (Jawak et al., 2015). These methods for acquiring water depth data have now progressed to the point where useful underwater feature profiles can be extracted from images with sufficient accuracy for a variety of applications. In order to prepare spectral and bathymetric data for water depth mapping, the procedures entail subjecting the acquired relevant images to several steps of image processing techniques. The concepts of space, time, spatio-temporal analysis, and a concep-tual outline of model development and application in relation to satellite-derived bathymetry (SDB) and bathymetric LiDAR are primarily covered in this chapter. SDB surveys, which are considered vital tools in mapping large areas to the vis-ible water depth, have been developed with various data extraction techniques and models. Each of these algorithms has its own set of strengths and weaknesses, as well as environments in which it may perform better or worse than others. The most widely used inversion method was proposed by Lyzenga (1978, 1981, 1985) and GAO (2015). The linear and analytical models of water depth extraction from satellites provide depth data at the time of image acquisition, so different images acquired at different times provide a window for the development of a conceptual model for spatio-temporal analysis. SDB has a lot of advantages when it comes to planning and executing hydrographic activities. It is a cost-effective method for obtaining high-resolution, up-to-date bathymetric and production data quickly. It is used to collect data from remote locations and areas that are physically inaccessible. The main factors influencing satellite-derived bathymetry are turbidity, abrupt topography, and bottom signal. For large-scale water depth mapping at higher resolutions, airborne bathymetric LiDAR (ABL) is regarded as an ideal tool for acquiring data for the study of underwater features in relatively clear shallow waters. To facilitate mapping of water surfaces and underwater topography, the approach involves using the time difference of blue-green channel and infrared channel reflected from the sea bottom and water surface, respectively, from two LiDAR. Green light penetrates the water column and bounces off underwater objects/surfaces, whereas infrared light bounces off the water surface and adjacent topography. The most important factor that determines the acceptable flying height and speed of the aeroplane during data acquisition is the LiDAR source pulse repetition rate. The consumption of electric power rises in tandem with the pulse rate. LiDAR with a pulse rate of 284 kHz achieves better sounding depth and spatial resolution than LiDAR with a pulse rate of 522 kHz. Bathymetry acquisition systems differ in terms of spatial resolution, coverage, temporal resolution, and data type (Jawak et al., 2015). Stumpf and Lyzenga developed empirical approach models that provide practical approaches to derive bathymetric information from remote sensing data. However, it was discovered that the ratio transform (Stumpf model) is more robust than the linear transform (Lyzenga model). In general, for depths less than 30 m, Jupp's, Stumpf, and Lyzenga methods were used. All of the techniques examined are reliant on water clarity and focal distance, as well as the measurement sensor and platform (airborne or space-borne platform), and are thus subject to error due to murky waters and geometric and atmospheric interference. Dams are typically built in freshwater rivers where there are no waves or tidal influences. They are characterized by shallow depth waters that are relatively clear and small geographical areas. All models developed for the ocean and coastal environment can be used to extract water depth data from dams and reservoirs for effective monitoring and management with appropriate modifications. As a result, it's critical to pick the right sensor and bathymetric algorithm. The SDB method has a lot of advantages when it comes to mapping large areas to the visible water depth, whereas the ALB method is better for bathymetric mapping over small geographical areas. Regardless of their merits, the validation of these RS-based bathymetry models requires the use of in-situ acoustic depth soundings. Spatio-temporal analysis is an attempt to connect the dynamics of phenomena operating at various levels in space and time, and the conceptual models for it make excellent decision support tools. Their utility, on the other hand, is contingent on what and how they are used. The variety of existing models and methods for extracting water depth in ocean environments demonstrates how broad the opportunities for their use in the spatio-temporal analysis of underwater profile dynamics of dams and reservoirs are. More research and development for the applications of these approaches in the future could be very interesting.

REFERENCES

Abdulkadir, I. F. (2019). "Evaluation of the hydrological change impacts on a river system." Unpublished Ph.D. Thesis, Department of Geoinformatics and Surveying, Faculty of Environmental Science, University of Nigeria Enugu Campus, Nigeria.

Abdulkadir, I. F., Ojinnaka, O. C., and Okeke, F.I. (2014). "Conceptual analytical framework for the upstream geospatial impact analysis of Zobe Dam Northwest Nigeria." *Book of Proceedings and Abstract 2014 conference of the Nigerian Union of Planetary and Radio Science (NUPRS)*, University of Nigeria Enugu Campus, 16th–18th September 2014, Nigeria (Book of Abstract).

Allouis, T., Bailly, J. S., and Feurer, D. (2013). "Assessing water surface effects on LiDAR bathymetry measurements in very shallow rivers: a theoretical study." http://earth.esa.int/hydrospace07/participants/07_14/07_14_Allouis.pdf.

Benny, A. and Dawson, G. (1983). "Satellite imagery as an aid to bathymetric charting in the Red Sea." *The Cartographic Journal*, 20(1), 5–16. https://www.tandfonline.com/doi/abs/10.1179/caj.1983.20.1.5 [Accessed 23/9/2017].

Bera, A., Taloor, A. K., Meraj, G., Kanga, S., Singh, S. K., Durin, B., and Anand, S. (2021). "Climate vulnerability and economic determinants: linkages and risk reduction in Sagar Island, India; a geospatial approach." *Quaternary Science Advances*, 100038. https://doi.org/10.1016/j.qsa.2021.100038.

Brimicombe, A. (2003). *GIS, Environmental Modelling and Engineering*. Taylor & Francis, London.

Caldwell, J. (2013). "LiDAR best practices." https://eijournal.com/print/articles/lidar-best-practice.

Carbonneau, P. E. et al. (2006). "Feature-based image processing methods applied to bathymetric measurements from airborne remote sensing in fluvial environments." *Earth Surface Processes and Landforms*, 3, 1413–1423. Published online in Wiley InterScience (www.interscience.wiley.com). https://doi.org/10.1002/esp.1341.

Castillo-López, E., Dominguez, J. A., Pereda, R., de Luis, J. M., Pérez, R., and Piña, F. (2017). "The importance of atmospheric correction for airborne hyperspectral remote sensing of shallow waters: application to depth estimation." *Atmospheric Measurement Techniques*, 10, 3919–3929. https://doi.org/10.5194/amt-10-3919-2017. Available from https://www.researchgate.net/publication/320595624_The_importance_of_atmospheric_correction_for_airborne_hyperspectral_remote_sensing_of_shallow_waters_Application_to_depth_estimation [Accessed 20/11/2020].

Chandel, R. S., Kanga, S., and Singh, S. K. (2021). "Impact of COVID-19 on tourism sector: a case study of Rajasthan, India." *AIMS Geosciences*, 7(2), 224–242.

Dandekar, M. M. and Sharma, K. N. (2007). *Water Power Engineering*. Vikas Publishing House PVT Ltd, New-Delhi.

Deng, Z., Ji, M., and Zhang, Z. (2008). "Mapping bathymetry from – source remote sensing image: a case study in the Beilum Estuary Guangxi, China." *The International Archives of the Photogrammetry, Remote sensing and Spatial Information Science*, XXXVII (Part 138), Beijing.

Diebel-Langohr, D., Hengstermann, T., and Reuter, R. (1986). "Depth profiles of hydrographic parameters – measurement and interpretation of Lidar signals." In: Waidelich, W. (ed) *Laser/Optoelektronik in der Technik/Laser/Optoelectronics in Engineering*. Springer, Berlin, Heidelberg. https://doi.org/10.1007/978-3-642-82638-2_120; https://link.springer.com/chapter/10.1007/978-3-642-82638-2_120#citeas.

Egenhofer, M. J. and Golleedge, R. G. (1998). *Spatial and Temporal Reasoning in Geography Information*. Oxford University Press, New York.

Elsahabi, M. A., Makboul, O., and Negm, M. (2017). "Lake Nubia bathymetry detection by satellite remote sensing." *Tiventieth International Water Technology Conference, IWTC 20*. Hurghada, Egyptt.

Farooq, M. and Muslim, M. (2014). "Dynamics and forecasting of population growth and urban expansion in Srinagar City – a geospatial approach." *The International Archives of Photogrammetry, Remote Sensing and Spatial Information Sciences*, 40(8), 709.

Gao, S. (2015). "Shallow water depth inversion based on data mining models." A thesis submitted to the Graduate Faculty of the Louisiana State University and Agricultural and Mechanical College.

Gholamalifard, M., Kutser, T., Esmaili-Sari, A., Abkar, A. A., and Naimi, B. (2013). "Remotely sensed empirical modelling of bathymetry in the Southerneasthen Caspian Sea." *Remote Sensing*, 5, 2746–2762. ISSN 2072-4292. https://www.mdpi.com/2072-4292/5/6/2746.

Green, E. P., Mumby, P. J., Edwards, A. J., and Clark, C. D. (2000). *Remote Sensing Handbook for Tropical Coastal Management*. UNESCO, Paris.

Guenther, G. C. (1985). "Airborne laser hydrography system design and performance factors." NOAA professional paper series NOSI, Rockville, MD.

Gujree, I., Arshad, A., Reshi, A., and Bangroo, Z. (2020). "Comprehensive Spatial planning of Sonamarg resort in Kashmir Valley, India for sustainable tourism." *Pacific International Journal*, 3(2), 71–99.

Hassanin, M., Kanga, S., Farooq, M., & Singh, S. K. (2020). "Mapping of Trees outside Forest (ToF) from Sentinel-2 MSI satellite data using object-based image analysis." *Gujarat Agricultural Universities Research Journal*, 207.

Hell, B. (2011). "Mapping bathymetry from measurement to applications." Doctoral thesis in Marine Geoscience, Department of Geological Science Stockholm University Stockholm Sweden.

Jagalingam, P., Akshaye, B. J., and Hegde, A. V. (2015). "Bathymetry mapping using Landsat 8 satellite imagery." *8th International Conference on Asian and Pacific Coast (APAC)*. Available online at https://www.sciencedirect.com.

Jawak, S. D., Vadlamani, S. S., and Luis, A. J. (2015). "A synoptic review on deriving bathymetry information using remote sensing technologies: models, methods and comparisons". *Advances in Remote Sensing*, 4, 147–162. http://dx.doi.org/10.4236/ars.2015.42013.

Jupp, D. L. B. (1988). "Background and extractions to depth of penetration (DOP) mapping in shallow coastal waters." *Proceedings of the Symposium on Remote Sensing of the Coastal Zone*, Gold Coast Queensland, pp. IV.2.1–IV.2.19.

Kanga, S., Kumar, S., & Singh, S. K. (2017b). "Climate induced variation in forest fire using remote sensing and GIS in Bilaspur district of Himachal Pradesh." *International Journal of Engineering and Computer Science*, 6(6), 21695–21702.

Kanga, S., Rather, M. A., Farooq, M., and Singh, S. K. (2021). "GIS based forest fire vulnerability assessment and its validation using field and MODIS data: a case study of Bhaderwah forest division, Jammu and Kashmir (India)." *Indian Forester*, 147(2), 120–136.

Kanga, S. et al. (2020a). "Modeling the spatial pattern of sediment flow in Lower Hugli Estuary, West Bengal, India by quantifying suspended sediment concentration (SSC) and depth conditions using geoinformatics." *Applied Computing and Geosciences*, 8, 100043.

Kanga, S., Sheikh, A. J., & Godara, U. (2020b). "GIscience for groundwater prospect zonation." *Journal of Critical Reviews*, 7(18), 697–709.

Kanga, S., Singh, S. K., and Sudhanshu. (2017a). "Delineation of urban built-up and change detection analysis using multi-temporal satellite images." *International Journal of Recent Research Aspects*, 4(3), 1–9.

Khattab, M. F. O. and Al-Ani, F. A. M. (2012). "Detection depth of Mosul Dam Reservoir by using image processing techniques." *Tikrit Journal of Pure Science*, 17(2), 190–194.

Kinzel, P. J., Wright, C. W., Nelson, J. M., and Burman, A. R. (2007). "Evaluation of an experimental LiDAR for surveying a shallow, braided, sand-bedded river." *Journal of Hydraulic Engineering*. https://wwwbrr.cr.usgs.gov/gstl/kinzel_papers/kinzel_ASCE_2007.pdf.

Lillesand, T. M. and Kiefer, R. W. 1994. *Remote Sensing and Image Interpretation*. John Wiley & Sons, Chichester.

Loomis, Jr. M. J. (2009). *Depth Derivation from the Worldview-2 Satellite Using Hyperspectral Imagery.* Master of Science in Meteorology and Physical Oceanography, Naval Postgraduate School, Monterey, CA.

Lyzenga, D. R. (1978). Passive remote sensing techniques for mapping water depth and bottom features. *Applied Optics.* 17(3), 379–383.

Lyzenga, D. R. (1981). "Remote sensing of bottom reflectance and water attenuation parameters in shallow water using aircraft and Landsat data." *International Journal of Remote Sensing,* 2(1), 71–82. Available at https://www.researchgate.net/profi [Accessed 02/04/2017].

Lyzenga, D. R. (1985). "Shallow-water bathymetry using combined lidar and passive multispectral scanner data." *International Journal of Remote Sensing,* 6(1), 115–125. Available at https://www.researchgate.net/publication/232980165_ [Accessed 02/04/2017].

Lyzenga, D. R. and Polycn, F. C. (1979). Techniques for the extraction of water depth information from Landsat digital data (Final Report Environmental Research Institute of Michigan-Michigan 48109). Available online. https://apps.dtic.mil/sti/citations/ADA210177

Lyzenga, D. R., Malinas, N. P., and Taris, F. J. (2006). Multispectral bathymetry using a simple physically-based algorithm. *IEEE Transaction on Geoscience and Remote Sensing,* 52(2), 1205–1212.

Mavraeidopoulos, A. K., Pallikaris, A., & Oikonomou, E. (2018). Satellite Derived Bathymetry (SDB) and Safety of Navigation. *The International Hydrographic Review,* (17). Available online. https://journals.lib.unb.ca/index.php/ihr/article/view/26290

Melsheimer, C. and Chin, L. S. (2001). "Extracting bathymetry from multi-temporal sport images." Paper presented at the *22nd Asian Conference and Remote Sensing,* 5–9 November 2001, Singapore.

Meraj, G. et al. (2018a). "Geoinformatics based approach for estimating the sediment yield of the mountainous watersheds in Kashmir Himalaya, India." *Geocarto International,* 33(10), 1114–1138.

Meraj, G. et al. (2018b). "An integrated geoinformatics and hydrological modelling-based approach for effective flood management in the Jhelum Basin, NW Himalaya." *Multidisciplinary Digital Publishing Institute Proceedings,* 7(1): 8.

Meraj, G., Singh, S. K., Kanga, S., & Islam, M. N. (2021). Modeling on comparison of ecosystem services concepts, tools, methods and their ecological-economic implications: a review. *Modeling Earth Systems and Environment,* 1–20. https://doi.org/10.1007/s40808-021-01131-6

MKS Instruments Handbook (2019). *Principles & Applications in Photonics Technologies.* The Office of the CTO. https://www.newport.com/photonics-handbook [Accessed 21/10/2020].

Mohamed, H., Negm, A., Zahran, M., and Saavedra, O. C. (2016). "Bathymetry determination from high resolution satellite imagery using ensemble learning algorithms in shallow lakes: case study El-Burullus Lake." *International Journal of Environmental Science and Development,* 7(4), 295.

Mohammadzadeh, A. and Valadan Zoej, M. J. (2008). "A state of art on airborne Lidar application in hydrology and oceanography: a comprehensive overview." *The International Archives of the Photogrammetry, Remote Sensing and Spatial Information Sciences,* XXXVII(Part B1). Beijing 2008. https://www.isprs.org/proceedings/XXXVII/congress/1_pdf/52.pdf.

Monroy, J. D. and Khan, S. (2018). "Remote sensing techniques make high-resolution terrain mapping easier and quicker." *Civil and Structural Engineer,* September 1, p. 2978. https://csenginermag.com/principles-of-photogrammetry-and-lidar/.

Moretto, J., Delai, F., and Lenzi, M. A. (2013). "Hybrid DTMS derived by LiDAR and colour bathymetry for assessing fluvial geomorphic changes after flood events in gravel-bed rivers (Tagliamento, Piave and Brenta Rivers, Italy)." *International Journal of Safety and Security Engineering,* 3(2), 128–140. http://www.sedalp.eu/download/dwd/publication/Moretto.pdf.

Muthuwatta, L. P. (2014). "Integrating water resources modelling and remote sensing in Karkheh River Basin, Iran." PhD thesis – ITC. https://www.itc.nl/library/papers_2014/phd/muthuwatta.pdf.

National Oceanic and Atmospheric Administration (NOAA) Coastal Services Center. (2012). *Lidar 101: An Introduction to Lidar Technology, Data, and Applications.* Revised. NOAA Coastal Services Center, Charleston, SC.

Ndukwe, N. K. (1997). *Principles of Environmental Remote Sensing and Photo Interpretation.* New Concept Publishers, Enugu.

NOAA (2013). Bathy Lidar: harder than it looks; Digital Coast GeoZone Tech talk for the Digital Coast. https://geozoneblog.wordpress.com/2013/09/25/bathy-lidar-harder-than-it-looks/ [Accessed 20/10/2020].

Ojinnaka, O. C. (2007). *Principles of Hydrographic Surveying From Sextant to Satellite.* El Demak, New Layout Enugu.

Parush, A. (2015). A typology of conceptual model in conceptual design for interactive systems. Available online 13 March 2015: https://doi.org/10.1016/B978-0-12-419969-9.00010-3 [Visited 10/10/2020].

Philpot, W. D. ed. (2019). *Airborne Laser Hydrography II (Blue Book II) eCommons.* https://doi.org/10.7298/jxm9-g971

Philpot, W. D. (1989). "Bathymetry mapping with passive multi-spectral imagery." *Applied Optics*, 28(8), 1569–1578.

Rather, M. A. et al. (2018). "Remote sensing and GIS based forest fire vulnerability assessment in Dachigam National park, North Western Himalaya." *Asian Journal of Applied Sciences*, 11(2), 98–114.

Rocadenbosch, F. (2007). "Introduction to LIDAR (LIDAR (laser radar) (Remote radar) remote sensing sensing systems." Remote Sensing Lab (RSLAB), Universitat Politècnica de Catalunya. http://www.tsc.upc.edu.

Sanders, L. eds. (2007). *Models in Spatial Analysis.* Geographic Information Series. ISTE Publication, ISTE Ltd, London.

Saylam, K., Hupp, J. R., Andrews, J. R., Averett, A. R., and Knudby, A. J. (2018). "Quantifying airborne LiDAR bathymetry quality-control measures: a case study in Frio River, Texas." Published online 2018 November 27. https://doi.org/10.3390/s18124153; https://www.ncbi.nlm.nih.gov/pmc/articles/PMC6308523/.

Setiawan, K. T. (2013). "Study of bathymetry map using Landsat ETM+ data: a case study of Menjanga Island Bali." Unpublished MSc. Thesis, Graduate Study of Environmental Science, Postgraduate UDayana University Denpasar.

Setiawan, K. T., Osawa, T., and Nuarsa, I. W. (2017). "Study of bathymetry map using Landsat ETM+ data: a case study of Menjanga Island Bali." *International Journal of Geoscience.* Available at http://ojs.unud.ac.id/index.php/ijeg/article/view/28832.

Singh, S. K., Mishra, S. K., and Kanga, S. (2017). "Delineation of groundwater potential zone using geospatial techniques for Shimla City, Himachal Pradesh (India)." *International Journal for Scientific Research and Development*, 5(4), 225–234.

Spirkin, A. (1983). *Dialectical Materialism.* Progress Publishers. Available online: https://www.marxists.org/reference/archive/spirkin/works/dialectical-materialism/index.html [Accessed 5/11/2020].

Stumpf, R. P., Holderied, K., and Sinclair, M. (2003). "Determination of water depth with high-resolution satellite imagery over variable bottom types." *Limnology and Oceanography*, 48(1-part 2), 547–556. Available online at https://aslopubs.onlinelibrary.wiley.com/doi/abs/10.4319/lo.2003. [Accessed 12/02/2014].

Tomar, J. S., Kranjčić, N., Đurin, B., Kanga, S., & Singh, S. K. (2021). "Forest fire hazards vulnerability and risk assessment in Sirmaur district forest of Himachal Pradesh (India): a geospatial approach." *ISPRS International Journal of Geo-Information*, 10(7), 447.

Van Hengel, W. and Spitzer, D. (1991). "Multi-temporal water depth mapping by means of Landsat TM." *International Journal of Remote Sensing*, 12, 703–712. Available online at https://www.tandfonline.com/ [Accessed 21/09/2013].

Vasszi, R. (2019). Mapping underwater terrain with bathymetric LiDAR case study. https://leica-geosystems.com/case-studies/natural-resources/mapping-underwater-terrain-with-bathymetric-LiDAR [Accessed 10/10/2020].

Wasser, L. A. (2020). The basics of LiDAR – Light Detection and Ranging – remote sensing. https://www.neonscience.org/resources/learning-hub/tutorials/lidar-basics [Accessed 21/10/2020].

Webster, T. and Lockhart, C. (2020). Mapping underwater terrain with bathymetric LiDAR two similar projects in different parts of the world. https://www.hydro-international..com/content/article/mapping-underwater-terrain-with-bathymetric-LiDAR-2 accessed 20/10/2020

Wiki (2017). PPDAC model. http://wiki.gis.com/wiki/index.php/PPDAC_Model. Retrieved 12 October 2017.

Wood, D. (2002). Conceptual models: definition & characteristics related study materials. https://study.com/academy/lesson/conceptual-models-definition-characteristics.html [Accessed 10/10/2020].

Woolard, J. W. and Colby, J. D. (2002). "Spatial characterization, resolution, and volumetric change of coastal dunes using airborne LIDAR": Cape Hatteras, North Carolina. *Geomorphology*, 48, 269–287. https://www.elsevier.com/locate/geomorph.

Yang, L., Tian, S., Yu, L., Ye, F., Qian, J., and Qian, Y. (2015). "Deep learning for extracting water body from Landsat imagery." *International Journal of Innovative Computing Information and Control*, 11(6), 1913–1929.

Zhang, D., Chen, X., and Yao, H. (2015). "Development of prototype web-base decision support system for watershed management water 2015." *Open Access Journal*. ISSN 2073-4441. https://www.mdpi.com /journal/water.

Part B

Geospatial Modeling in Landslide Studies

6 Geospatial Modeling in Landslide Hazard Assessment

A Case Study along Bandipora-Srinagar Highway, N-W Himalaya, J&K, India

Ahsan Afzal Wani, Bikram Singh Bali, Sareer Ahmad, and Umar Nazir
University of Kashmir

Gowhar Meraj
Government of Jammu and Kashmir

CONTENTS

6.1 Introduction 114
6.2 Materials and Methods 115
    6.2.1 Study Area 115
    6.2.2 Geology of Study Area 116
6.3 Results and Discussion 117
    6.3.1 Slope 117
    6.3.2 Aspect 118
    6.3.3 Geological and Geotechnical Investigations along
           Bandipora-Srinagar Highway 119
    6.3.4 Land Cover 120
    6.3.5 Contour Map 120
    6.3.6 Landslide Hazard Assessment and Inventory Map 121
    6.3.7 Management Plan 122
6.4 Conclusion 123
References 123

DOI: 10.1201/9781003147107-8

113

6.1 INTRODUCTION

Landslides are considered to be the most damaging and devastating natural geological hazard in mountainous regions (Mengistu et al. 2019; Hamza and Raghuvanshi 2017; Girma et al. 2015; Raghuvanshi et al. 2014a; Pan et al. 2008; Kanungo et al. 2006; Crozier and Glade 2005; Dai and Lee 2002; Parise and Jibson 2000; Varnes 1984). The term landslide hazard refers to the possibility of an event within a specified period and within a given area, of a landslide, of a given magnitude (Varnes 1984; Guzzetti et al. 2012). Like most other definitions of hazards, landslide hazard definition incorporates three basic elements: (i) magnitude, (ii) geographical location, and (iii) time. Landslides primarily result from geological, hydrological, and geomorphologic factors. Globally, it is estimated that between the years 2004 and 2010, fatal landslides numbering around 2,620 killed a total of 32,322 people, and this figure excludes landslides triggered by earthquakes (Petley 2012). Furthermore, Martha and Kerle (2008) point out that the effects of landslides on people and property have been mounting in recent decades due to increasing slope instability as a result of deforestation, road cutting, and urbanization. Likewise, Muhammad et al. (2010) have identified that landslides have posed serious threats to settlements and structures that support transportation, natural resource management, and tourism. According to the official figures of the United Nations International Strategy for Disaster Reduction and the Centre for Research on the Epidemiology of Disasters for 2006, landslides ranked third in terms of number of deaths among the top 10 natural disasters (Ramakrishnan et al. 2002; Farooq and Muslim 2014; Nathawat et al. 2010; Kumar et al. 2018; Joy et al. 2019). In Kashmir, a huge landslide recently occurred near Kangan along the Srinagar-Leh highway, on May 30, 2020 that caused a huge loss to the habitation with a number of houses and shops gutted inside the slide (Figure 6.1a). The landslide is believed to have caused due to the heavy vehicular movement. Landslides have also been reported in the steep mountains of Bandipora Srinagar highway near Saderkoot with no loss of life and property (Figure 6.1b).

The limitation with field-based approaches is that they are costly and require a lot of time for the assessment to be completed (Hassanin et al. 2020; Tomar et al. 2021;

(a)　　　　　　　　　　　　　　(b)

FIGURE 6.1 Latest landslide episodes near Kangan (a) and Churteng (b).

Meraj and Kumar 2021; Chandel et al. 2021; Bera et al. 2021). Secondly, the results of the assessments are not easy to explain to the beneficiaries, who include local leaders and the general population. It is also clear from operations of the ministry that most of the effort in landslide hazard management as well as other disasters is mainly focusing on postdisaster management rather than early warning. This paper explores the possibility of adopting geospatial techniques to improve landslide hazard assessment in Bandipora-Srinagar highway based on available and accessible spatial data. GIS and related geo-spatial technologies have proved to be useful landslide assessment tools primarily because they combine mapping, field inventory spatial analysis and other tools (Pall et al. 2019; Romshoo et al. 2020; Gujree et al. 2020; Arshad et al. 2019).

6.2 MATERIALS AND METHODS

Slope stability analysis and delineating the hazard-prone areas in northwest Himalaya are the most important exercise. As the roads are the backbone of the economy of the region, the demarcation and mitigation of these hazards is an important step. In the present study, the geospatial technology has been put forward to demarcate the landslide-prone areas.

The chief objectives of present research work is to formulate reliable methodology to delineate the landslide hazard areas and to predict the landslide susceptibility zones using the information value method and weighted overlay method. The present study is primarily based on satellite data products and its derivatives and field observations. Most of the layers have been generated from ASTER DEM and Google Earth Pro. By using these data sets, primary layers and secondary layers have been generated. Primary layers were directly generated from the satellite data sets by using GIS tools, and the secondary layers represent the layer generated with the help of some other collateral data like existing lithology map, google images, historical records, and other government sources (Kanga et al. 2017a; Rather et al. 2018). To calculate the information value, each raster layer is combined with the landslide map using combined function in ArcGIS spatial analyst tool extension. After combining the landslide map, pixel counts are derived from each layer. Finally, information value calculation has been carried out for each layer, and these values were integrated with the respective layer for weighted overlay analysis. Finally, landslide hazard zones were delineated through classified data ranges.

6.2.1 STUDY AREA

Srinagar-Bandipora Gilgit road was one of the oldest roads of Kashmir. Located on the lap of the Great Himalayan Range, this highway serves the people of Bandipora in a number of ways. While comparing Bandipora-Srinagar highway of the 1890s and 2018, we did not see any improvement; instead, it is even worse. The Bandipora-Srinagar road has become a headache for people, and in return, it has hindered the economy of the district. As it is well said, better roads are the beginning toward real development. The Bandipora-Srinagar highway is a sole highway for the

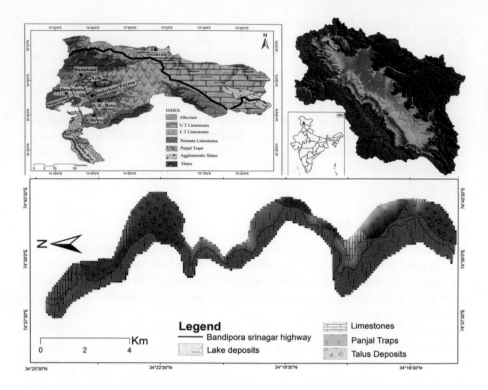

FIGURE 6.2   Location map of the study area.

people of Bandipora and Gurez to connect with Srinagar. The high elevation road
with the steep slopes have made the road vulnerable to the landslides and rock fall.
There are some areas that are highly vulnerable to landslides. However, the anthropogenic activities without any proper guidance have made rods more vulnerable. The
uneven road cuts for road widening presently going on along the Bandipora-Srinagar
highway have made the roads very vulnerable and under high risk. Thus, the present
study has been undertaken to highlight some issues and delineate the landslide-prone
areas along the Bandipora-Srinagar highway (Figure 6.2).

6.2.2   GEOLOGY OF STUDY AREA

The zigzag pattern of the Bandipora-Srinagar highway mostly carved on the ele-
vated terrain. Most of the area along the Bandipora-Srinagar highway is under hard
rock terrain. The Panjal Traps cover the majority of the slopes of the highway, followed by Limestones and quartzites. Good exposures of limestones were found
near Chewa, Saderkoot Bala, Ajas, and Nadihal. Good exposures of Panjal Traps
are found near Ajas, Saderkoot Payeen, Garoora. Quartzites are mainly found near
Churteng area of S. K. Payeen. The area near Safapora is under Karewa formations. In addition, the Karewa formations were also found near Nusoo and Wudar
Bandipora (Figure 6.3). The highway is crossed by the Himalayan rivers near

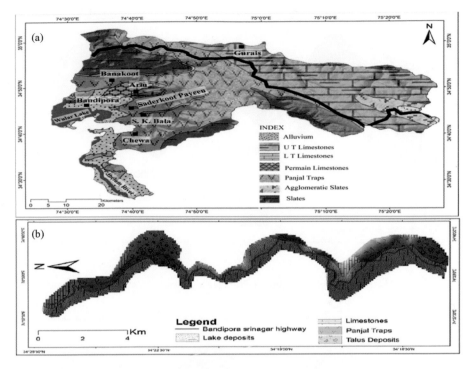

FIGURE 6.3 (a) Geological formations of Bandipora district. (b) Geological formations along Bandipora-Srinagar highway after Thakur and Rawat (1992), Bali et al. (2016), Wani and Bali (2017), Wani et al. (2019).

Bazipora, Garoora, Nadihal and Papacha by streams like Ajas nar, Aragam nar, Rampore nar, and Erin nar, respectively.

6.3 RESULTS AND DISCUSSION

6.3.1 SLOPE

The slope of the area has been categorized into four major classes: low, moderate, high, and very high. Majority of the area falls under high – very high class. High to very high class mostly covers the right side of the road, suggesting the area toward the right side of the road is under high risk, which in turn infers that the area is more vulnerable to landslide hazard. Some of the areas are under moderate to high class from both the right and left side of the highway, inferring the area has moderate chances of landslides in future. This class is under high vegetative cover, which also decreases the chances of landslides. The areas S.K. Bala, Saderkoot Payeen, Garoora, and Ajas come under the category where there is a greater chance of landslides. S.K. Bala and the area between Ajas and Bazipora, between Ajas and Saderkoot Payeen and Nadihal are the areas that have greater risk in terms of landslides because these settlements are built along barren and steep slopes with the rock formations dipping toward these settlements with high dip angle, hence enhancing the vulnerability (Figure 6.4).

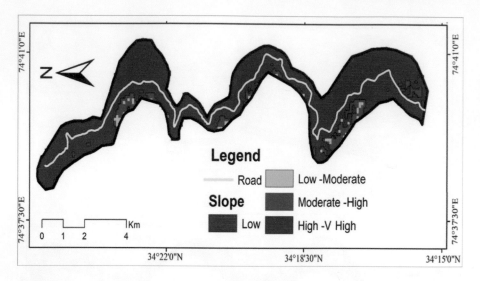

FIGURE 6.4   Slope map along Bandipora-Srinagar highway.

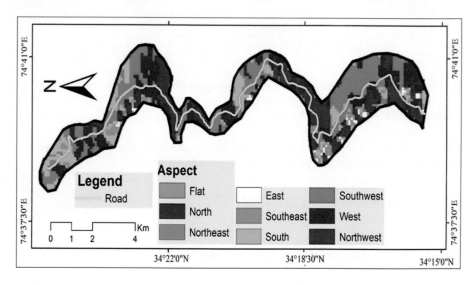

FIGURE 6.5   Aspect map along Bandipora-Srinagar highway

6.3.2   ASPECT

Aspect is the orientation of slope, measured clockwise in degrees from 0 to 360, where 0 is north-facing, 90 is east-facing, 180 is south-facing, and 270 is west-facing. After calculating the aspect along the highway, the majority of the aspects in the study area were facing south-west and northwest i.e. facing toward roadside (Figure 6.5). The overall results of the aspect analysis infer that whenever there is instability in the area, the area will slide toward roadside, which will create great loss of life and property along roadside because the areas falling under this category are densely populated. The areas lying near Saderkoot Payeen, Ajas, and some areas of S. K. Bala are areas that are vulnerable to landslides whenever there is a slight shift in terms of rainfall or other triggering mechanisms.

6.3.3 Geological and Geotechnical Investigations along Bandipora-Srinagar Highway

Geologically, majority of the area is under hard rock terrain. However, some area is under loose sediments near Bazipora and Garoora. But the areas near S. K. Bala and Ajas are under limestones, dipping toward roadside. The high dip degree (slope) and direction (aspect) toward the roadside of bedding planes makes these areas more vulnerable to landslides. In addition to the dip of the bedding planes, the dip direction and angle of the dip of joints are also responsible for the high degree of landslide chances along the Bandipora-Srinagar highway. Moreover, the extensive road widening activities near Sturteng mode without proper engineering consultation have made this area highly vulnerable to landslides and under high landslide risk. Because the roadsides are cut and left at higher angles without proper engineering remedies, this uneven cutting has produced more jointed rocks at high angles making the respective areas more vulnerable to landslides (Figures 6.6a–f). On the basis of geo-technical analysis, three types of failures were observed along the Bandipora-Srinagar highway. Planar, wedge,

FIGURE 6.6 Uneven cutting of slopes.

FIGURE 6.7   Types of failures along Bandipora-Srinagar highway.

and toppling types of failures were observed (Figure 6.7a–d). A high degree of toppling was observed near Bazipora; however, the planar failure was observed near S. K. Bala, and the majority of the wedge failures were observed near S. K. Bala.

On the basis of geological and geotechnical studies, a number of sliding patterns were observed along this highway. Both the planar and wedge failures in addition to the toppling were observed along the Bandipora-Srinagar highway. All the datasets was evaluated from the joint pattern of the study area.

A historical toppling was observed between S. K. Bala and Dodwan where big boulders have been tumbled down toward the roadside. However, the gneiss of this topple has not yet been discussed and documented.

6.3.4   LAND COVER

The information about the land cover was acquired from the Google Earth Pro. and Google Earth Engine. Three classes were identified along the Bandipora-Srinagar highway. Settlements, barren slopes, and vegetative covered slopes are the three main classes found along the Bandipora-Srinagar highway. The settlements are highly built near the barren slopes with dipping beds at higher angle, have made these settlements under high risk. There are certain areas that are built on moderate slopes with thick vegetative cover slopes (Meraj et al. 2021). These abilities indicate that these areas are less prone to the landslides. In addition to the slope angles, the aspect of the area clearly suggests that the slopes are facing toward northwest and south-west i.e. toward settlements which also direct that the areas are more susceptible to landslide hazards (Figure 6.8).

6.3.5   CONTOUR MAP

Using ASTER DEM 30 m resolution, the contour map of the area was built on an interval of 40m. The contours clearly show the steepness of the area. The areas

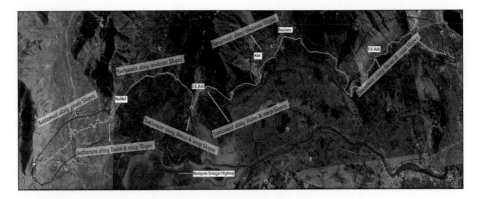

FIGURE 6.8   Land cover map along Bandipora-Srinagar highway.

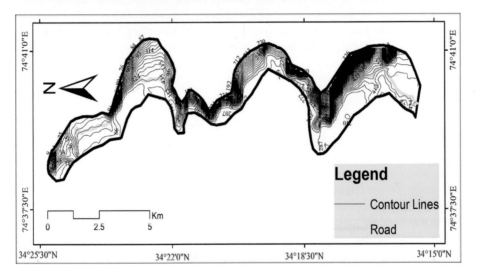

FIGURE 6.9   Contour map along Bandipora-Srinagar highway.

with contours closely spaced infer steep slopes, and the areas with the contour lines sparsely spaced indicate relatively gentle slopes. On the basis of contour lines, the area has been grouped under steep slopes and gentle sloping areas. Black shaded areas indicate very steep slopes, and the areas with the lines sparsely spaced show gentle slopes (Figure 6.9). Thus, the areas S. K. Bala, Ajas, Nadihal, and Bazipora indicate that these areas have very steep slopes and hence are susceptible to landslides, and the rest of the area is gentle to moderate inferring that these areas are moderately to slightly susceptible to landslides. Thus, the analysis of contour map infers that the areas having very steep slopes are relatively more susceptible to landslide hazard risk.

6.3.6   LANDSLIDE HAZARD ASSESSMENT AND INVENTORY MAP

After combining all the layers i.e. slope, aspect, geology, and geotechnical analysis, landcover, and contour map, different areas were categorized in terms of landslide hazard status. The areas with the steep slopes, southwest and northwest aspect, hard rocks with high density of joints and high degree of dip and dipping toward roadside, and slopes without vegetative cover have greater chances of landslides. On the basis of geotechnical analysis of joint sets, three types of failures were observed along the highway i.e. planar, wedge, and toppling. Hence, the areas lying in this class were inferred as highly vulnerable to landslides and having high landslide risk. S. K. Bala, Ajas, S. K. Payeen, Nadihal, and parts of Chewa are the areas that fall under this category. The following properties were identified in moderately landslide-prone areas. These areas have moderate to high slopes covered with vegetative cover, north east facing slopes, less jointed rock formations, moderate dip angles of rock formations, dipping away from the roadside, and sparse density of population. The areas Garoora, Chewa, Bazipora, and Gund Dachina belong to this class. Finally, the least susceptible areas were identified along the highway. These sites have gentle slopes with dense vegetative cover and less jointed rocks, and the road almost crosses along flat areas of the main town of Bandipora, Papchan, Nusoo falls under this category. On the basis of the above results, the final inventory landslide-prone map of the Bandipora-Srinagar highway was prepared by superimposing all the layers (Figure 6.10).

6.3.7 MANAGEMENT PLAN

There are a number of ways to mitigate landslide hazards. As the majority of the area is under hard rock terrain, there should be less chances of landslides. But the high degree of slope and the dip direction of bedding plane and jointing make this area more sensitive and vulnerable. In this case, the first step is to widen the highway under proper engineering considerations. The second step is to not apply extra explosives as the dip direction is toward the roadside. The extra explosions make the joints and bedding planes more active, and there are greater chances of future landslides. As along the highway there are certain places where quarrying is being carried out, we have to convey the certified and concerned persons to make the less use

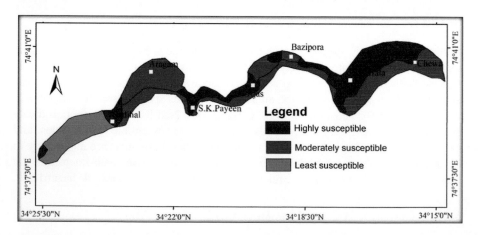

FIGURE 6.10   Landslide inventory map along Bandipora-Srinagar highway.

of explosives used in the quarrying purposes. Retaining walls should be built near Dodwan to protect the chances of landslides because this area is more vulnerable in wet seasons, as the area is highly sheared due to tectonic activity. The tectonic activity has produced highly jointed and sheared rocks with loose and unconsolidated sediment making area more vulnerable to slope instability.

6.4 CONCLUSION

For proper landslide mitigation and management, susceptibility evaluation and hazard zonation are very vital. Landslides are influenced by several inherent causative and external triggering factors that vary significantly from place to place. Proper evaluation of these factors is essential for landslide studies. Each of these factors may influence the landslide process, and in combination, they result in landslide activity. A number of conclusions in the present study have been put forward not to completely mitigate the landslide occurrence. However, many places along the Bandipora-Srinagar highway are highlighted that are vulnerable to landslides. To avoid the landslide occurrence along this particular area, proper consultation from geologists and engineers is required. There are a number of places along this highway where widening of roads is been carried out; as we know, widening of the roads is important, but widening should be done after proper consultation and consideration. However, in this case, the widening has produced road cuts at higher elevations with unstable slope left, which became the hotspots for landslides in future. Due to this activity along this road, all types of failures are possible. The high dip degree of bedding planes, joints, and steep slopes facing toward the highway are the major threat in terms of landslides to the habitat along this highway in addition to the travelers of this highway. As the highway is the commercial and economic unit for the district Bandipora, a management plan is required to mitigate the possibility of landslides along this highway.

REFERENCES

Arshad, A., Zhang, Z., Zhang, W., & Gujree, I. (2019). Long-term perspective changes in crop irrigation requirement caused by climate and agriculture land use changes in Rechna Doab, Pakistan. *Water*, 11(8): 1567.

Bera, A., Taloor, A.K., Meraj, G., Kanga, S., Singh, S.K., Durin, B., & Anand, S. (2021). Climate vulnerability and economic determinants: linkages and risk reduction in Sagar Island, India; A geospatial approach. *Quaternary Science Advances*, 100038. doi:10.1016/j.qsa.2021.100038.

Chandel, R.S., Kanga, S., & Singh, S.K. (2021). Impact of COVID-19 on tourism sector: a case study of Rajasthan, India. *AIMS Geosciences*, 7(2): 224–242.

Crozier, M.J., & Glade, T. (2005). Landslide hazard and risk: issues, concepts, and approach. In: Glade, T., Anderson, M., Crozier, M. (eds) *Landslide Hazard and Risk*. Wiley, Chichester, pp. 1–40.

Dai, F.C., & Lee, C.F. (2002). Landslide characteristics and slope instability modelling using GIS, Lantau Island, Hong Kong. *Geomorphology*, 42: 213–228.

Farooq, M., & Muslim, M. (2014). Dynamics and forecasting of population growth and urban expansion in Srinagar City – a geospatial approach. *The International Archives of Photogrammetry, Remote Sensing and Spatial Information Sciences*, 40(8): 709.

Girma, F., Raghuvanshi, T.K., Ayenew, T., & Hailemariam, T. (2015). Landslide hazard zonation in Ada Berga district, Central Ethiopia – a GIS based statistical approach. *Journal of Geometry*, 9(i): 25–38.

Gujree, I., Arshad, A., Reshi, A., & Bangroo, Z. (2020). Comprehensive Spatial planning of Sonamarg resort in Kashmir Valley, India for sustainable tourism. *Pacific International Journal*, 3(2), 71–99.

Guzzetti, F., Mondini, A.C., Cardinali, M., Fiorucci, F., Santangelo, M., & Chang, K.T. (2012). Landslide inventory maps: new tools for an old problem. *Earth-Science Reviews*, 112(-1): 42–66.

Hamza, T., & Raghuvanshi, T.K. (2017). GIS based landslide hazard evaluation and zonation – a case from Jeldu district, Central Ethiopia. *Journal of King Saud Univesrity Science*, 29(2): 151–165.

Hassanin, M., Kanga, S., Farooq, M., & Singh, S.K. (2020). Mapping of Trees outside Forest (ToF) from Sentinel-2 MSI satellite data using object-based image analysis. *Gujarat Agricultural Universities Research Journal*, 207–213.

Joy, J., Kanga, S., and Singh, S.K. (2019). Kerala flood 2018: flood mapping by participatory GIS approach, Meloor Panchayat. *International Journal of Emerging Technologies*, 10(1): 197–205.

Kanga, S., Singh, S.K., & Sudhanshu. (2017a). Delineation of urban built-up and change detection analysis using multi-temporal satellite images. *International Journal of Recent Research Aspects*, 4(3): 1–9.

Kanungo, D.P., Arora, M.K., Sarkar, S., & Gupta, R.P. (2006). A comparative study of conventional, ANN black box, fuzzy and combined neural and fuzzy weighting procedures for landslide susceptibility zonation in Darjeeling Himalayas. *Engineering Geology*, 85: 347–366.

Kumar S., Kanga S., Singh S.K., Mahendra R.S. (2018). Delineation of Shoreline Change along Chilika Lagoon (Odisha), East Coast of India using Geospatial technique. *International Journal of African and Asian Studies*, 41: 9–22.

Martha, R., & Kerle, N (2008). Segment optimisation for object based landslide detection. *The International Archives of the Photogrammetry, Remote Sensing and Spatial Information Sciences*, XXXVIII-4/C7: 45–59.

Mengistu, F., Suryabhagavan, K.V., Raghuvanshi, T.K., & Lewi, E. (2019). Landslide hazard zonation and slope instability assessment using optical and InSAR data: a case study from Gidole town and its surrounding areas, southern Ethiopia. *Remote Sensing of Land*, 3(1): 1–14.

Meraj, G., Singh, S.K., Kanga, S., & Islam, M.N. (2021). Modeling on comparison of ecosystem services concepts, tools, methods and their ecological-economic implications: a review. *Modeling Earth Systems and Environment*, 1–20. https://doi.org/10.1007/s40808-021-01131-6

Meraj, G., & Kumar, S. (2021). Economics of the Natural Capital and the way forward. Preprints, 2021010083. doi:10.20944/preprints202101.0083.v1.

Muhammad, M., Ilyias, I., Al Sharif, S., Khairul, N., & Mohd, R.T. (2010). GIS based landslide hazard mapping prediction in Ulu Klang, Malaysia. *ITB Journal of Science*, 42 A(2): 163–178.

Nathawat, M.S. et al. (2010). Monitoring & analysis of wastelands and its dynamics using multiresolution and temporal satellite data in part of Indian state of Bihar. *International Journal of Geomatics and Geosciences*, 1(3): 297–307.

Pall, I.A., Meraj, G., & Romshoo, S.A. (2019). Applying integrated remote sensing and field-based approach to map glacial landform features of the Machoi Glacier valley, NW Himalaya. *SN Applied Sciences*, 1(5): 488.

Pan, X., Nakamura, H., Nozaki, T., & Huang, X. (2008). A GIS-based landslide hazard assessment by multivariate analysis landslides. *Journal of the Japan Landslide Society*, 45(3): 187–195.

Parise, M., & Jibson, R.W. (2000). A seismic landslide susceptibility rating of geologic units based on analysis of characteristics of landslides triggered by the 17 January, 1994 Northridge, California earthquake. *Engineering Geology*, 58: 251–270.

Petley, D.N. (2012). Global patterns of loss of life from landslides. *Geology*, 40(10): 927–930.

Raghuvanshi, T. K., Ibrahim, J., Ayalew, D. (2014). Slope stability susceptibility evaluation parameter (SSEP) rating scheme – an approach for landslide hazard zonation. J African Earth Sciences, 99:595–612

Ramakrishnan, S.S., Kumar, V.S., Sadiq, Z., & Venugopal, K. (2002). Landslide zonation for hill area development. *GIS development.net*. Viewed March 29th 2013.

Rather, M.A. et al. (2018). Remote sensing and GIS based forest fire vulnerability assessment in Dachigam National park, North Western Himalaya. *Asian Journal of Applied Sciences*, 11(2): 98–114.

Romshoo, S.A. et al. (2020). Satellite-observed glacier recession in the Kashmir Himalaya, India, from 1980 to 2018. *Environmental Monitoring and Assessment*, 192(9): 1–17.

Thakur, V.C., & Rawat, B.S. (1992). *Geologic Map of Western Himalaya, 1:1,000,000*. Wadia Institute of Himalayan Geology, Dehra Dun.

Tomar, J.S., Kranjčić, N., Đurin, B., Kanga, S., & Singh, S.K. (2021). Forest fire hazards vulnerability and risk assessment in Sirmaur district forest of Himachal Pradesh (India): a geospatial approach. *ISPRS International Journal of Geo-Information*, 10(7), 447.

Varnes, D. (1984). *Landslide Hazard Zonation: A Review of Principles and Practice*. UNESCO, Paris, pp. 1–6.

Wani, A.A., & Bali, B.S. (2017). Geomorphic analysis of relative tectonic activity in the Sindh Basin, Jammu and Kashmir Himalaya, northwest India. *Journal of Himalayan Geology*, 38(2): 171–183.

Wani, A.A., Bali, B.S., & Mohammad, S. (2019). Drainage characteristics of tectonically active area: an example from Mawar Basin, Jammu and Kashmir, India. *Journal of Geological Society of India*, 93(3), 313–320.

7 Causes, Consequences, and Mitigation of Landslides in the Himalayas
A Case Study of District Mandi, Himachal Pradesh

Krishan Chand
Aryabhatta Geo-informatics & Space Application Centre

D.D. Sharma
Himachal Pradesh University

CONTENTS

7.1   Introduction 127
7.2   Study Area 128
7.3   Materials and Methods 131
    7.3.1   Data Preparation and Methodology Framework 132
7.4   Results and Discussion 133
    7.4.1   Field Validation 147
    7.4.2   Policy Imperatives and Mitigations 151
7.5   Conclusions 152
References 153

7.1   INTRODUCTION

A disaster may be the result of a natural or human-induced hazard, resulting in significant physical destruction or damage, loss of life, or a drastic change in the environment at large scale. It can be defined as any awful event resulting from calamities such as earthquakes, floods, catastrophic accidents, fires, or explosions, etc. Hazards are the physical event, phenomenon, or human activity with high damaging potentiality (UNISDR 2009). Disasters are the result of natural hazards or incidents in which a natural phenomenon or a combination of natural phenomenon works. Thus, natural hazards are the unpredicted threats to human and their property. The impacts of

DOI: 10.1201/9781003147107-9

natural hazards are worldwide, and no part of the earth surface is without any effect of natural hazards (Feizizadeh and Blaschke 2011; Kanga et al. 2017b; Hassanin et al. 2020; Kanga et al. 2021).

Various parts of the world with young mountain belts, such as the Himalayas Mountains, are highly vulnerable to landslide, as this type of events is common for the people of mountainous regions (Mukhopadhyay et al. 2012). Landslides are the most frequently occurring natural disasters, most costly, damaging, and landscape building factor in mountainous region. Landslides are one of the major natural hazards that account for hundreds of lives (Onagh et al. 2012); besides that, landslides are the destructive geological processes which cause enormous damage of property, houses, bridges, means of transportation, communication, and agricultural land (Vaani and Sekar 2012) and adversely affect the economy worldwide. Landslides are the complicated geological hazards and complex natural phenomena that are hard to model and pretend and are rarely composed of a single type of movement but are often the result of a combination of many factors (Saha et al. 2005). *"Individual slope failures are generally not so spectacular or as costly as earthquakes, major floods, hurricane or some other may cause more damage to properties than any other geological hazards"* (Varnes 1978). The term "Landslide" is used to describe a wide variety of processes that result in the exterior downward and outward movement of slope—forming materials composed of instable rocks, debris masses, artificial fill, or a combination of these under the action of Gravity (Cruden 1991). The material may move by falling, toppling, sliding, spreading, or flowing. Landslides are a type of mass movement in which the debris moves down the slope as a unit along a glide plane. They occur particularly where the rocks have bedding planes that are roughly parallel to the slope angle. They are activated by the accumulation of large amounts of soil water in the weathered surface material. This added weight increases the stress on the debris and the extra water reduces friction between the particles and lubricates the mass of material. Almost all varieties of mass movements on slopes such as rock fall, topples, and debris flows are termed as "landslide" which may involve little or no true sliding (Varnes 1984; Altaf et al. 2013, 2014; Meraj et al. 2015, 2016).

The growing population and expansion of built-up area over landslide hazardous zones or growing as the built-up environment increases into unstable hill areas has increased the impact of natural disasters (Asthana and Bist 2009; Farooq and Muslim 2014). In general, landslide hazards cannot be completely prevented. However, the intensity and severity of the impacts of the hazards can be reduced if the problem is recognized before the development activity. Hence, there arise needs to identify the unstable slopes, which can be done by the landslide hazard zonation. The landslide hazard zonation of an area aims at identifying the landslide potential zones and ranking them in order of the degree of hazard from landslides (Nithya and Prasanna 2010).

7.2   STUDY AREA

The Mandi district comprises the area of present study. District Mandi is centrally located district of Himachal Pradesh. District Mandi is located between 31°13′50″ to 32°4′30″ North latitudes, and 76°37′20″ to 77°23′15″ East longitudes, lying in the foothills of Shivalik range (Figure 7.1). The total geographical area of the district

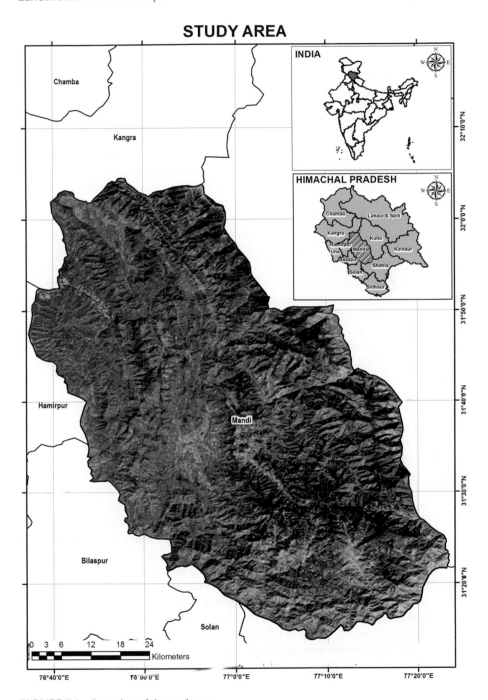

FIGURE 7.1 Location of the study area.

is 3950 square kilometer which makes 7.10% area of the state and ranks seventh in the state Himachal Pradesh. The district is bounded by Kangra districts on the north-west, Hamirpur & Bilaspur districts in the west, Arki tehsil of Solan district in the south, Shimla district in the south-west and Kullu district in the east. The district was formed by merging of two princely states—Mandi and Suket on 15th April 1948, when the State of Himachal Pradesh came into existence.

The climate of the district is subtropical in the valleys and trends to be temperate on the hilltops. The winter starts from the middle of November and continues till the middle of March. Thereafter, the temperature continues rising till the set of Monsoon which starts from the last week of June or early July and continues till the middle of September. During October and November, the nights are pleasant whereas the days are little bit hot. Average minimum and maximum temperatures in the district differ from 3°C to 35°C. The entire Mandi district drains into the Beas River and only the south-east corner situated on the Satluj watershed. Within the district, the main tributaries of the Beas on the north bank are Uhl, Luni, Rana, and Binu and on the south bank are the Hanse, Tirthan, Bakhli, Jiuni, Suketi, Ranodi, Son, and Bakar. The maximum part of the district is mountainous. A main range of mountains is from the north to the south with the system being broken up by numerous transverse spurs. The most noticeable is the Jalori range which is crossed by a high road from Kullu to Shimla by a pass named as Jalori pass. It divides the watersheds of the Satluj and the Beas and on its northern slopes is unusually well wooded with deodar and blue pine forests of great value. The highest peak in the range is Shikari Devi (11,060 feet), its summit being crowned by a shrine to a local goddess. To the north of the Beas is the Nargu range, a continuation of the Bir-Bhangal, separating Mandi from Kullu proper and crossed by the Bhubu pass (9,480 feet). The mountains here run up to 13,000 feet, the slopes often being very precipitous and the valleys deep. The main hill ranges in this district are Dahauladhar, Ghogardhar, Sikandradhar, Dhar Bairkot.

The Dahauladhar has a high elevation range and covers considerable part of the Suket area. It runs with the eastern boundary of the district from north to south. In this range, "Nagru" is the highest peak, with an elevation of 4,400 m above mean sea level. The range joins the Kullu district in the northeast. Ghogardhar has the rock salt mines of Gumma and Drang and is fully covered with forests. It enters the district at Harabagh in Drang block. Sikandra-Dhar range runs from north-west boundary of Suket and Bilaspur and has been divided into subranges i.e. Kamlah and Lindidhar. The range contains some good forests of *chil* but the greater part of it consists of rich grass slopes. Dhar-Bairkot range starts from Rewalsar and extends toward Suket. Some of its branches join Kangra with Sikandra-Dhar. Other hill ranges in the district are Shikari, Kamrunag, Parashar, Bundli, etc. The range of altitudes in the district is high, with the highest point being around 13,000 feet on the Kullu border and the lowest point 1,800 feet near Sandhol where the Beas leaves the district. The only area which is similar to the plains is the Balh valley. Several of the valleys are open and are often irrigated by kuhls, or small water channels, and contain some of the most fertile land in the state. Due to the hilly terrain, some of the hills are so precipitous that cultivation in large scale is impossible. Due to sufficient rainfall, unused land is covered with forests or forms rich grazing land. In the hills the forests are extensive and valuable. Deodhar, blue pine, silver fir, spruce, *chil* and various kinds of oak are plentiful. Below 4,000 feet the forests are not extensive, the only valuable ones consisting of *chil*, but there is considerable scrub jungle and the wide areas of grass covered slopes support the herds of cattle.

Thus, the fragile nature of the topography of the district, along with climatic condition and various other anthropogenic activities has made the district vulnerable to landslide hazard. The problem of landslides is common and frequent in the study area. Almost every year, the district is affected by one or more major landslides affecting society in many ways such as loss of life and damage to houses, roads, means of communication, agriculture, vegetation cover, grasslands, etc.

7.3   MATERIALS AND METHODS

The present study is based on both primary and secondary data. Primary data mainly include the field survey to verify the results on the ground with the help of Assisted-Global Positioning System (A-GPS) mobile phone to get the geo-tagged images and GARMIN Global Positioning System (GPS) for spatial location of existing landslides with in the study area. Secondary data are concerned with the use of remote sensing and GIS to identify potential landslide zones, it involves the generation of thematic maps relating to the causative factors. These thematic maps have been derived from Survey of India (SOI) toposheets at 1:50,000 scale; the Geological Survey of India (GSI) maps, National Bureau of Soil Survey & Land Use Planning (NBSS & LUP) maps and Advanced Spaceborne Thermal Emission and Reflection Radiometer (ASTER) Global Digital Elevation Model (GDEM) (Figure 7.2).

Thematic maps have been processed in GIS environment. Aspect, slope, and elevation maps for the study area have been derived from ASTER DEM raster format through using ArcGIS 3D Analyst Tool. Aspect of slope identifies the downslope direction of the maximum rate of change in value from each cell to its neighbors in the digital elevation model. Thus, aspect is the slope direction and the values of the output raster will be the compass directions of the aspect. Aspect is expressed and measured in positive degrees from 0 to 359.9, clockwise from north. Cells within input raster i.e. an aspect of—1 which is flat with zero slopes—are assigned. Rate of change of elevation for each Digital Elevation Model cell represents slope. Slope tool within ArcGIS calculates the maximum rate of change in value from that cell to its neighbor cells. Burrough and McDonell defined that the maximum change in elevation over the distance between the cells and its eight neighbor cells identifies the steepest downhill descent from the cell. Elevation of the study area has also been derived from ASTER DEM the most common digital data of the shape of the earth's surface which is cell-based digital elevation model and having a DN number in each cell. A Digital Elevation Model is a model of continuous surface, usually referencing the surface of the earth. Similarly, information on hydrological aspects i.e. drainage distribution and drainage density, the SOI toposheets has been used. In order to calculate drainage density for the study area, all drainage lines have been digitized from geo-referenced SOI toposheets, and drainage density is calculated using ArcGIS 3D Analyst tool which is based on the formula i.e. drainage density is equal to the total length of all streams within the study area divided by the total area. Thematic maps on geological conditions and soil properties have been prepared through digitizing geo-referenced maps of Geological Survey of India and National Bureau of Soil Survey & Land Use Planning (NBSS & LUP) respectively. Roads and habitations have been mapped through Survey of India toposheets and have also been verified in Google Earth which is a virtual globe, map, and geographical information program. Metrological data have been derived from Indian Metrological Department official website and then mapped. Land Use/Land Cover and Geomorphological maps have been prepared from Bhuvan official map server of Indian Space Research Organization which allows users to explore a 2D and 3D representation of the earth surface (Nathawat et al. 2010; Kanga et al. 2017a; Gujree et al. 2020).

Therefore, after the preparation of all thematic maps, they have been classified into subthemes and allocated ranks. Spatial overlay analysis technique has been applied in the present study which allows us to apply weights to several inputs and combine them into a single output. Thus, generated vector layers have been processed in GIS environment using Union tool which calculates the geometric union of any number of feature classes and feature layers. Output feature class contains polygon feature layer and represents the geometric union of all the input layers along with the its attribute fields from all the model feature classes. Finally, all resultant values of features classes have been summarized and categorized into three broad categories i.e. low vulnerability, medium vulnerability, and high vulnerability. Output of the model have also been verified by conducting a GPS-based field ground survey (Figure 7.2).

7.3.1 DATA PREPARATION AND METHODOLOGY FRAMEWORK

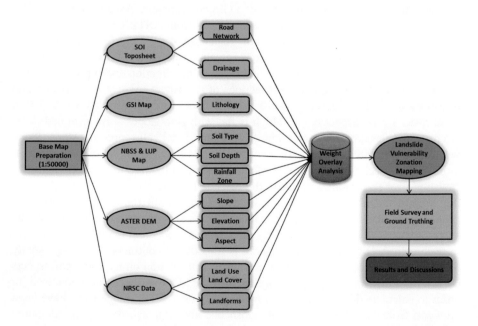

FIGURE 7.2   Flowchart of the methodology used.

7.4 RESULTS AND DISCUSSION

Landslide incidences infrequently involve a single type of movement or root factor but often the result of a combination of several types. Thus, many factors contribute to slides, including geology, gravity, weather, groundwater, wave action, and human actions (Figure 7.3). Landslides usually occur on steep slopes but also in low-relief areas; besides this, in ground failure of river bluffs, cut and fill failures during road and building excavations, slope failures are also associated with collapse of mine waste piles, quarrying and open pit mines (Hassanin et al. 2020; Tomar et al. 2021). Interpretation of upcoming landslide incidence requires an understanding of conditions and processes controlling landslides in a particular area. The whole process of land sliding phenomena is dependent on some of causative factors. The first group consists of preliminary factors related to geology, topography, and environment and second one consists of triggering factors such as extreme rainfall, earthquake, volcanic eruptions, and land use changes.

Geological and topological parameters play a dominant role in the prognosis of landslide. It is an important exercise to model these parameters for accurately delineating the hazard-prone areas. In Himalayas, the lithology has control over the occurrence of landslide. It is seen that the softer rocks like phyllites and shales are more prone to landslide in comparison to the harder rocks like quartzite, granite, etc.

Drainage distribution and density is an important factor in landslide studies (Lee and Choi 2004; Jian and Xiang-Guo 2009; Meraj et al. 2021; Chandel et al. 2021; Bera et al. 2021). Drainage density shows how well or poorly a watershed is drained by stream and also depends upon physical and climatic conditions of the drainage basin. Watershed runoff is affected by the soil infiltration or underlying rock types; impervious ground or exposed bedrock leads to rise in the surface water runoff and therefore to more frequent streams.

The slope considered as the main parameter of the slope stability (Pachauri 2004). It is commonly used in preparing landslide vulnerability analyses (Dai et al. 2002; Coe et al. 2004; Ram et al. 2011) as the slope angle increases the probability of occurrence of landslide increases because the shear stress increases with progressive inclination (Nithya and Prasanna 2010). Singh (2003), defined slope as angular inclinations of terrain between hill tops (crests) and valley bottoms, resulting from the combination of many causative factors like geological structure, climate, vegetation index, relative reliefs, etc. are significant geomorphic attributes in the study of landforms of a (fluvially originated) drainage basin. Upward or downward inclination of earth surface between hills and valleys is slope and form most significant feature of Landscape assemblages. So, slope angle of earth surface is a significant parameter for Landslide Hazard Zonation (LHZ).

Lithology plays an important role in active geomorphological processes such as landslide, as different lithological units have different sensitivity (Bisht et al. 2006; Singh and Devi 2006; Pandey et al. 2008; Naithani and Rawat 2009). Due to the importance of lithology, numerous researchers have used lithology as an input parameter to analyze the landslide susceptibility (Dai et al. 2001; Sarkar and Kanungo 2004; Duman et al. 2005; Lee and Pradhan 2006; Akgun et al. 2008).

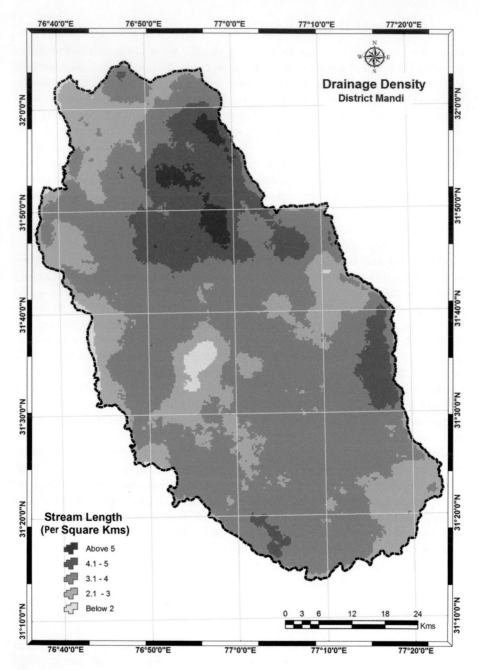

FIGURE 7.3 Parameters for assessing the landslide mapping.

(Continued)

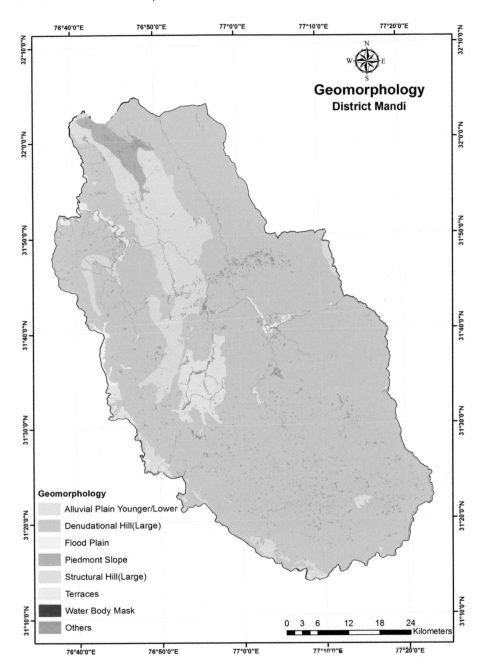

FIGURE 7.3 (*Continued*)   Parameters for assessing the landslide mapping.

(*Continued*)

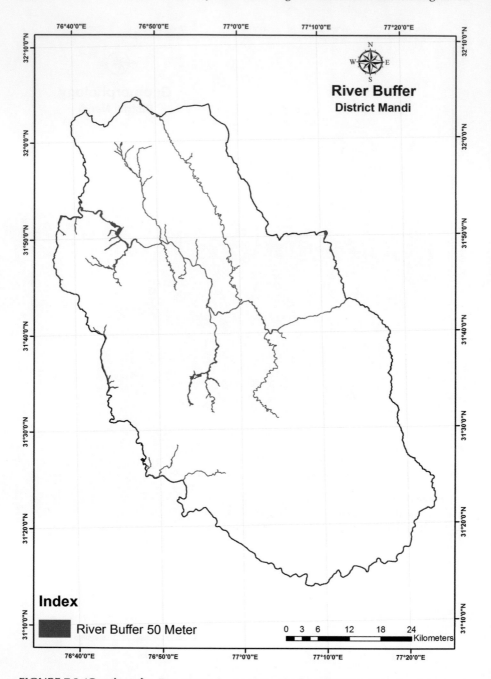

FIGURE 7.3 (*Continued*)   Parameters for assessing the landslide mapping.

(*Continued*)

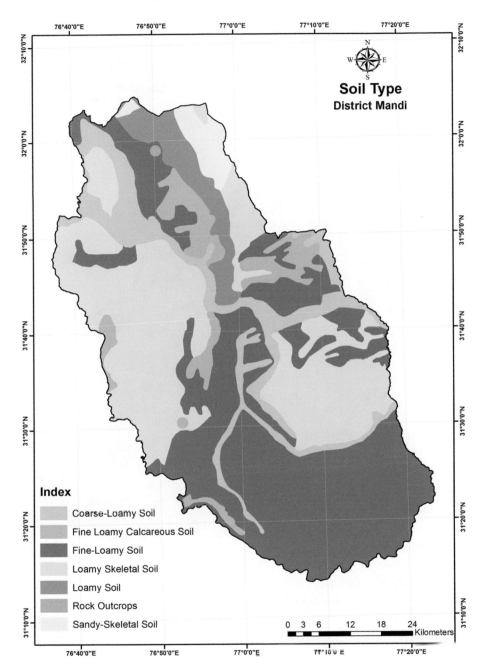

FIGURE 7.3 (*Continued*) Parameters for assessing the landslide mapping.

(*Continued*)

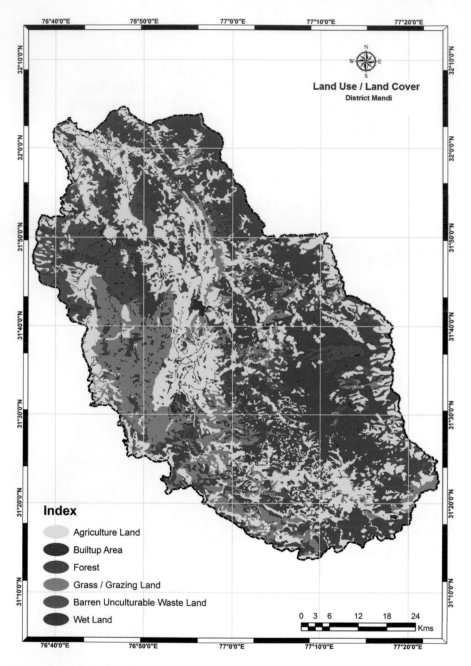

FIGURE 7.3 (*Continued*)   Parameters for assessing the landslide mapping.

(*Continued*)

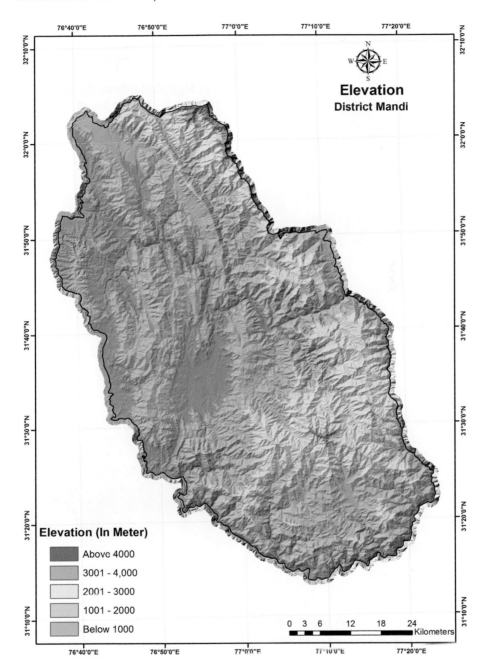

FIGURE 7.3 (*Continued*)   Parameters for assessing the landslide mapping.

(*Continued*)

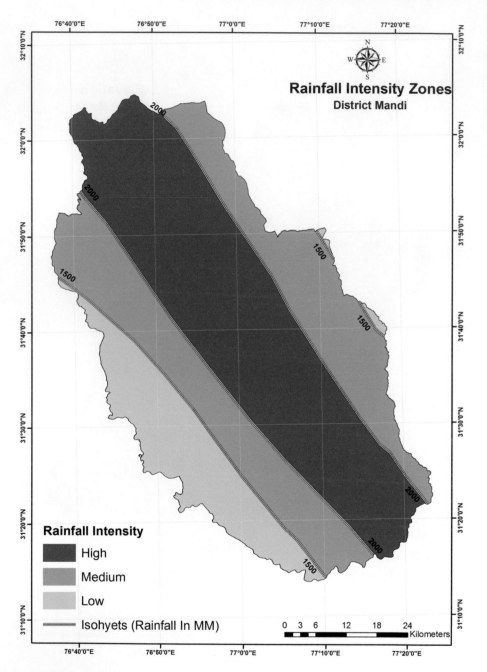

FIGURE 7.3 (*Continued*)   Parameters for assessing the landslide mapping.

(*Continued*)

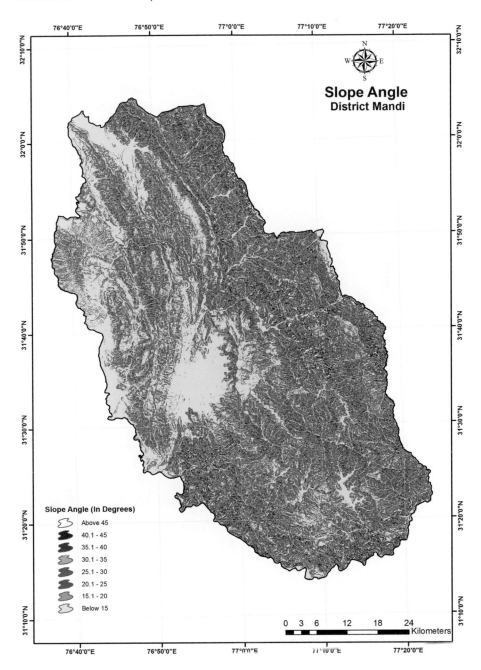

FIGURE 7.3 (*Continued*)   Parameters for assessing the landslide mapping.

(*Continued*)

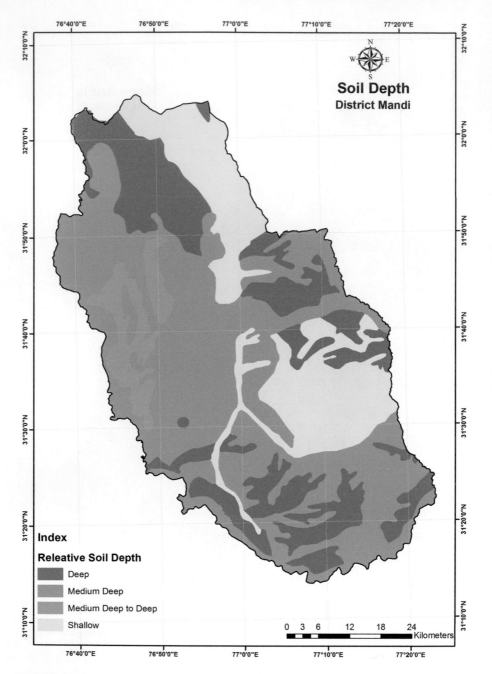

FIGURE 7.3 (*Continued*)   Parameters for assessing the landslide mapping.

(Continued)

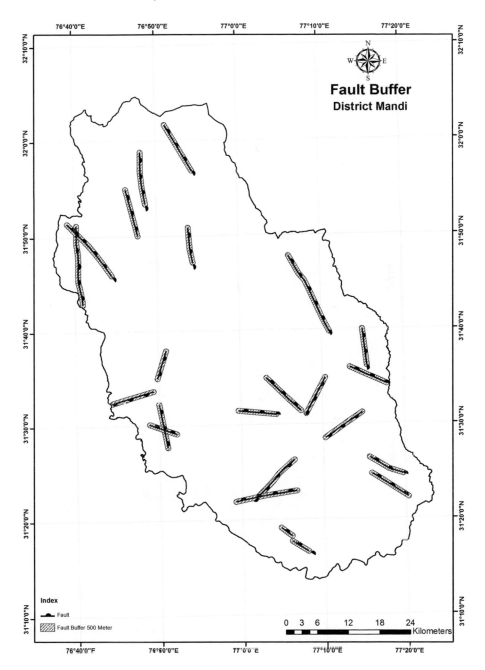

FIGURE 7.3 (*Continued*)   Parameters for assessing the landslide mapping.

(*Continued*)

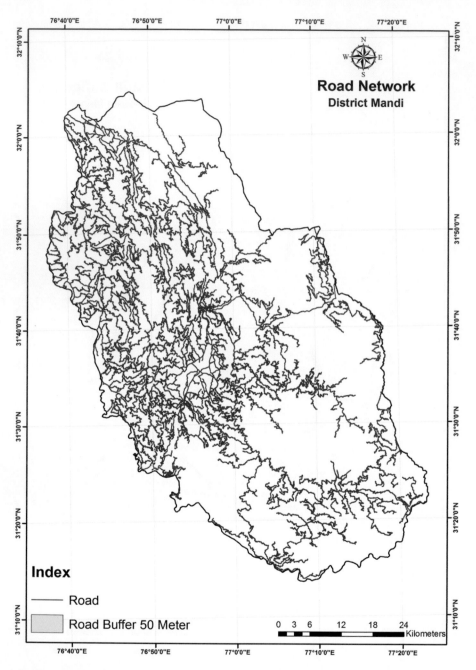

FIGURE 7.3 (*Continued*)   Parameters for assessing the landslide mapping.

A fault can be defined as a displacement along a plan and is a fracture/crack in the crustal rocks (Singh 2004). Along the fault, landslide susceptibility is higher because these are the weaker zones of the earth surface where crustal movements turn out to be operative for longer period of time (Gemitzi et al. 2011). The probability of the landslide hazard increases along the faults which affects the surface material structure as well as makes a contribution to terrain permeability causing slope instability (Kanungo et al. 2005). Geomorphology of the area represents the morphological set-up. This is very important factor because some of the geomorphologic elements give us a clue for the future landslide in that area. The dissection hill, moderately and lowly dissected hills helps in understating the denudation chronology of the area.

Rainfall is a recognized and considered to be the most important landslide triggering parameter causing soil saturation and a rise in pore – water pressure (Sepulveda et al. 2006; Rajbhandari et al. 2002) regions, particularly due to heavy rains.

Elevation also plays an important role in land sliding event. At a very high elevation there are mountain summits that usually consist of weathered rocks, whose shear strength is much higher. After the intermediate elevations of mountainous region, however, slopes tend to be covered by a thin layer of colluvium and therefore are more prone to landslides. On the other hand, at very low elevations where the terrain itself is gentle and covered with thick layer of colluvium/residual soils and required higher perched water to initiate slope failures, therefore, the frequency of landslide is low (Dai and Lee 2002). Landslides may also occur on the road and on the side of the slopes affected by roads (Pachauri et al. 1998; Ayalew and Yamagishi 2005). Construction work along road slopes causes reduction in the load on both the topography and on the heel of the slope. Distance from the river is also an important parameter that controls the stability of the slope which is the saturation degree of the material on the lower part of the slopes. The nearness of the slope to drainage structure is another important factor in terms of stability. Streams may adversely affect the stability by eroding the slopes or by saturating the lower part of the material with the increase in water level (Karsli and Yalcin 1991).

Land use and land cover of any area is one of the key factors which are responsible for the occurrence of landslide hazards. Since, barren slopes are more prone to landslides and vegetative areas tend to reduce the action of climatic agents such as rain, etc., beside this tree roots provides natural anchorage and prevent the erosion and thus, are less prone to landslides (Dahal and Dahal 2013). Different studies have publicized that land use/land cover or vegetation cover exerts a control over landslides, vegetation with strong and large root systems helps to improve stability of slopes. For stability of slopes, vegetation provides hydrological and mechanical effects. Sparsely vegetated slopes are more susceptible to landslide.

Soil is the outcome or end-product of the various climatic, physiographical, organisms, parent materials and time (Mayhew 2004). Soil is defined differently by geologist and engineers. *"Whereas geologists look at soils as the products of the weathering of rock and sediments, engineers consider soil to be a relatively loose agglomerate of mineral and organic materials and sediments found above bedrocks*" (Holtz and Kovacs 1981). Landslide occurrence effected by the topsoil cover of the slopes and to study or describe the landslides soil's definition given by engineers is commonly used. The relative proportion of sand, silt and clay content represents texture of soil. Soils with the availability of high percentage of clay leads to the formation of solid aggregate resistant to detachment. Beside this, sandy or coarse loam like light soils which is having low organic matter content are easy to detach (Das 2011). Hence, the areas having soil with more sand in the high slopes and severe rainfall which constitute most prevailing factors of landslide leads to damage the land cover (Patanakanog 2001).

Directions of the slope faces are known as slope aspect and classify the steepest downslope direction on the surface. This can also be understood as slope direction or the compass direction which hill faces. Like slope, aspect is also an important factor in preparing landslide vulnerability maps and can influence landslide initiation. Aspect related parameters such as exposure to sunlight; winds may dry or wet. Slope aspect or direction may reflect the moisture retention and vegetation, which in turn may also affect the strength and susceptibility of soil, which may lead to landslides. Rainfall distribution and intensity may vary at different slope aspects, if rainfall has a distinct directional component by influence of a prevailing wind (Wieczorek et al. 1987).

The thematic maps of the input data produced for the study are drainage density, slope, distance from fault, geomorphology, rainfall, elevation, distance from road, distance from river, land use/land cover, relative soil depth, soil type and aspect of slope. More than half of the total area has high drainage density which makes study area prone to landslide incidences. In term of slope angle most of the area is having slope above 15° which increases the probability of occurrence of landslide because the shear stress increases with progressive inclination. Geological and morphological set up of the study area is characterized by different landforms features like faults, alluvial plains, denudation hills, flood plains, piedmont hills, structural hills, terraces, and water masks. The study area has claimed highest annual average rainfall of the Himachal Pradesh and rainfall is a recognized to be the most important landslide triggering parameter causing soil saturation and a rise in pore–water pressure. Thus, slope failure is a major concern in hilly regions, particularly due to heavy rains. Around ¼ area of the district is having elevation above 2,000 m from mean seal level and at a very high elevation there are mountain summits that usually consist of weathered rocks, whose shear strength is much higher to accelerate the landslide activity. Road construction along slopes may causes landslide and study area has road length more than 5,100 km. Study area has also a well-drained river system formed by river Beas and Satluj. Streams may also adversely affect the stability by eroding the slopes or by saturating the lower part of the material with the increase in water level. District Mandi area is characterized by different land use/land cover pattern which includes agriculture land, built-up area, forest, grass/grazing land, barren unculturable waste land and wet lands. About ¼ part of the study area is falling under open land or waste land and barren land category which increase the susceptibility of landslide incidences. Further, the area contains soil from shallow to deep relative depth and also represents different soil types which are well drained. Area is having loamy soil conditions which are favorable to landslide occurrences. Further, sunlight, winds may dry or wet, moisture retention vegetation and amount of rainfall on a slope may vary depending on its aspect. So, largest part of the district Mandi is facing south direction slope which is more prone to landslide incidence. At last, it can be concluded that somehow all geo-ecological conditions of the study area are making it susceptible to landslide. Next chapter of the thesis will highlight all the vulnerable areas to landslide by using overlay analysis.

7.4.1 FIELD VALIDATION

Landslide often occurs at a specific location due to the force of gravity with certain geological and topographical conditions. To assess the impact of landslide, the landslide hazard vulnerability map has been prepared under the GIS platform which would be beneficial for the community as well as the administration. The GIS-based analyses were further verified through the field survey by using Global Positioning System (GPS) and geo-tagged photographs in random selected sites of Mandi district. Major landslide-affected areas were examined, and few GPS points and photographs were taken to validate the inventory maps with the help of Global Positioning System (GPS). Thereafter, for ground truthing, the entire district Mandi has been visited randomly to validate the results of overlay analysis model. It was observed that number of landslides of various dimensions occurs in those places where the healthy vegetation was not existing. Majority of landslides are noticed in the form of debris flow and few rockslides besides the major roads. The presence of transverse cracks and scarps is very useful for determining whether the area is suitable for future activity or not.

A massive landslide occurred near the village of Kotrupi in block Darang of district Mandi, Himachal Pradesh on 13th August, 2017 midnight (Figure 7.4). The highway connecting Mandi and Pathankot was swept away, and as per the records of different news agencies, at least 46 people were killed, and many are feared trapped after three vehicles were hit by the landslide. The buses buried under debris for almost 800 meters down into a gorge leaving no trace and one of the bus which were completely buried under the debris.

Pre-event Google Earth satellite data (dated 29 April 2014) of the area show the presence of two existing landslide scars on slope (Figure 7.4). This indicates that the slope was unstable and was already prone to a landslide failure. Post-event satellite data over the landslide-affected area were taken again from Google Earth of dated 19th October 2017. After the comparison of satellite images, it was observed that occurrence of a large landslide in the area where old landslides exists in the pre-event satellite data. The landslide was a "debris flow" type and has a long run out approximately 1,184 m from the Landslide Crown which clearly indicates that the heavy rainfall (monsoonal) was the main cause of its occurrence. The width of the landslide is maximum 234 m on the upper part of NH-154 and 174 m portion of the national highway was washed away. Out of total 13 ha-affected area 1.85 Ha was under agriculture; 4.75 ha area was green cover and having about 500 trees mostly Himalayan Chir Pine; 50% (6.48 ha) land was in the category of open, waste and grassland; only 0.13 ha land was under plantation. Beside this, there were only 2–3 houses and 2–3 sheds which was washed away.

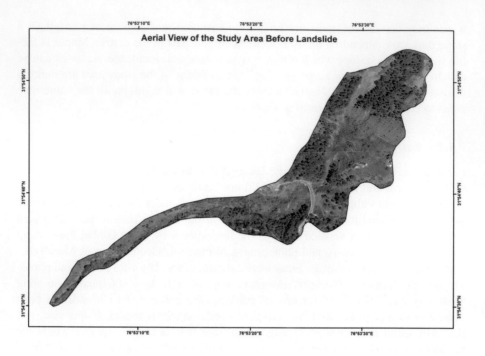

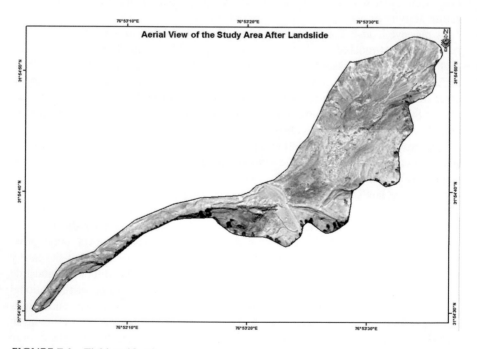

FIGURE 7.4   Field verification.

(*Continued*)

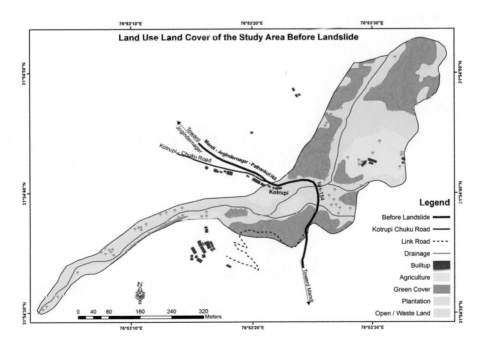

FIGURE 7.4 (*Continued*)   Field verification.

(*Continued*)

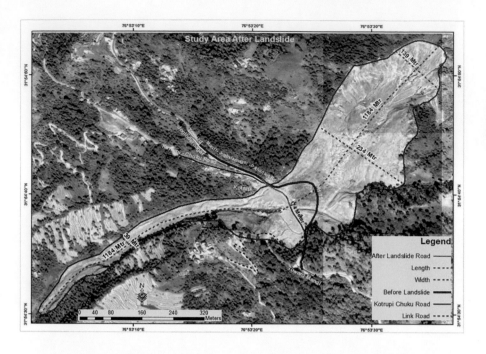

FIGURE 7.4 (*Continued*)   Field verification.

7.4.2 Policy Imperatives and Mitigations

Landslide cannot be prevented fully but its magnitude and impact can be minimized if its probability is recognized on time. In this regard, few suggestions have been given here as under:

- The state of Himachal Pradesh with its young formation of rocks and steep gradient is highly prone to landslides. Therefore, the protection measure for landslide stabilization like retaining wall, covering of slopes, soil reinforcement, soil nailing, rock bolting, surface protection, slope modification, breast walls, wire netting walls, and steel pile should be applied.
- Deforestation activities should be monitored by the stakeholder department with an effective management plan and should be focus on forests conservation.
- Afforestation and reforestation must be done on the deforested slopes and slope base to prevent erosion.
- Vegetation with wide spreading root systems is more effective to improve the shear strength of the material through root networks and protect the soil erosion during rain. Vegetation helps to stabilize slopes using biotechnical methods, commonly referred to as slope bio-engineering.
- In some of the blocks of district Mandi illegal mining and quarrying is leading to landslide incidences which should be strictly banned.
- Expert engineer should be consulted during construction work on slopes so that design can be make which suits the nature of slope.
- Being a hilly terrain, large number of landslides can be seen along the roads of the study area as a result of new construction and widening. In this regards it is suggested that after cutting slope, retention wall with slope modification along road should be there to protect from the slope failure. Plantation along the road side can also minimize the susceptibility to slide and protect from the slope failures especially in National and State Highways.
- The agriculture fields in the study area are in the terrace form so to avoid the penetration of water into surface, it should be drained properly from the field toward natural stream path which will reduce the susceptibility to subsidence.
- The best way to minimize the impact of landslide is the preparation of active or inactive landslide inventory and landslide hazard zonation map of the region which can help in future developmental activities and planning.
- There should be community- and village-level disaster management plans, with specific reference to the management of landslides.
- Modern geo-spatial technology can play a significant role in the identification of landslide vulnerable areas, particularly in unreachable areas of the region where physical monitoring is not possible.

In the present work few sites has been taken to verify the ground realities of the study area in the context of landslides. People's perception can give an important insight in understanding processes, cause and impacts of landslides on the local environs. This can be done by embedding the perception of the local folk with geo-ecological processes of landslides. Present work deals with landslides at regional scale, but it can also be possible on microscale with the high-resolution satellite data. These are some of the research issues and can form the major objectives for the future research.

7.5   CONCLUSIONS

Landslide occurs when the natural surface slope is unable to support its own weight and results in rock fall, earth or debris flows on slope due to gravity. Landslide can be triggered by natural conditions of the area as well as due to human causes such as terrain cutting and filling, unplanned development, etc. The factors affecting landslides can be geo-physical or human induced, they can happen in developed or undeveloped area or any region where the land was transformed for construction of roads, houses, utilities and dam, etc. Some of the conclusions drawn from the present research work are as follows:

- Himachal Pradesh being a hilly state is exposed to various natural hazards and landslides are the common event. Almost every year it affects the society and land surface on various intensity and magnitude which hampers the development and economy of the state. Landslides invariably cause loss of life, damage to property, infrastructure, means of communication, agriculture land and vegetation covers.
- Entire Himachal Pradesh is barring few low-lying valleys prone to landslide. Physico-cultural conditions of the state like unstable steep slopes, weak rock structure and intense rainfall during the monsoon seasons increase the frequency and magnitude of the landslides.
- Landslide is the complex geological phenomena which is the outcome of action and interaction of more than one factor. Study area mostly comprises of unconsolidated sand sediments, clay and conglomerates. Thus, apart from inherent geological weaknesses, ongoing developmental activities such as road construction, deforestation, building construction, debris dumping, etc. has made the area a breeding ground of landslides.
- Spatial overlay analysis reveals that nearly half of the geographical area of the district Mandi falls under moderate to high landslide vulnerability zone, and it varies at block level. At developmental block level Drang, Chauntra, Dharampur, Sadar, and Balh are more susceptible to landslide hazards due to their physical characteristics while Sundarnagar, Gopalpur, Gohar, Seraj, and Karsog blocks have comparatively low vulnerability to landslides.
- It has been observed during the field survey that all ten developmental blocks of the study area are facing different types of landslides as a result of road construction and widening, weathering of rocks, debris slide on the unstable slopes and subsidence of grasslands. Besides, landslide is also affecting the flora of the region at many places.

REFERENCES

Akgun, A., Dag, S., and Bulut, F. (2008). Landslide susceptibility mapping for a landslide—Prone area (Findikli, NE of Turkey) by likelihood—Frequency ratio and weighted linear combination models, *Environment Geology*, Vol. 54, pp. 1127–1143.

Altaf, F., Meraj, G., and Romshoo, S.A. (2013). Morphometric analysis to infer hydrological behaviour of Lidder watershed, Western Himalaya, India. *Geography Journal*, Vol. 2013, pp. 1–14, Article ID 178021, 14 pages. http://dx.doi.org/10.1155/2013/178021

Altaf, F., Meraj, G., and Romshoo, S.A. (2014). Morphometry and land cover based multi-criteria analysis for assessing the soil erosion susceptibility of the western Himalayan watershed. *Environmental Monitoring and Assessment*, Vol. 186(12), pp. 8391–8412.

Asthana, A.K. and Bist, K.S. (2009). Basan slip zone—Causes and mitigation measures, Yamuna Valley, Uttarakhand. *Indian Landslides*, Vol. 2(1), pp. 23–32.

Ayalew, L. and Yamagishi, H. (2005). The application of GIS-based logistic regression for landslide susceptibility mapping in the Kakuda-Yahiko Mountains, Central Japan. *Geomorphology*, Vol. 65, pp. 15–31.

Bera, A., Taloor, A.K., Meraj, G., Kanga, S., Singh, S.K., Durin, B., and Anand, S. 2021. Climate vulnerability and economic determinants—Linkages and risk reduction in Sagar Island, India; A geospatial approach. *Quaternary Science Advances*, 100038. doi:10.1016/j.qsa.2021.100038.

Bisht, M.P.S., Mehta, M., and Nautiyal, S.K. (2006). Geomorphic hazards around Badrinath (Uttaranchal). *Himalayan Geology*, Vol. 27(1), pp. 73–80.

Chandel, R.S., Kanga, S., & Singh, S.K. (2021). Impact of COVID-19 on tourism sector—A case study of Rajasthan, India. *AIMS Geosciences*, Vol. 7(2), pp. 224–242.

Coe, J.A., Godt, J.W., Baum, R.L., Bucknam, R.C., and Michael, J.A. (2004) Landslide susceptibility from topography in Guatemala. *Landslides: Evaluation and Stabilization*, Vol. 1, pp. 69–78.

Cruden, D.M. (1991). A simple definition of landslide. *Bulletin of the International Association of Engineering Geology-Bulletin de l'Association Internationale de Géologie de l'Ingénieur*, Vol. 43(1), pp. 27–29.

Dahal, B.K. and Dahal, R.K. (2013). Probabilistic hazard analysis-induced landslide in the Higher Himalaya, Western Nepal. *International Journal of Landslides and Environment*, Vol. 1(1), pp. 9–10.

Dai, F.C. and Lee, C.F. (2002). Landslide characteristics and slope instability modelling using GIS, Lantau Island, Hong Kong. *Geomorphology*, Vol. 42(3–4), pp. 213–228.

Dai, F.C., Lee, C.F., and Ngai, Y.Y. (2002). Landslide risk assessment and management—An overview. *Engineering Geology*, Vol. 64(1), pp. 65–87.

Dai, F.C., Lee, C.F., Li, J., and Xu, Z.W. (2001). Assessment of landslide susceptibility on the natural terrain of Lantau Island, Hong Kong. *Environmental Geology*, Vol. 40(3), pp. 381–391.

Das, I.C. (2011). Spatial statistical modelling for assessing landslide hazard and vulnerability, ITC Dissertation Number-192, Netherlands.

Duman, T.Y., Can, T., Gokceoglu, C., and Nefeslioglu, H.A. (2005). Landslide susceptibility area (Istanbul, Turkey) by conditional probability. *Hydrology and Earth System Sciences Discussions*, Vol. 2, pp. 155–208.

Farooq, M., and Muslim, M. (2014). Dynamics and forecasting of population growth and urban expansion in Srinagar City—A geospatial approach. *The International Archives of Photogrammetry, Remote Sensing and Spatial Information Sciences*, 40(8), p. 709.

Feizizadeh, B. & Blaschke, T. (2011). Landslide risk assessment based on GIS multi-criteria evaluation—A case study Bostan Abad County, Iran. *Journal of Earth Science and Engineering*, Vol. 1, pp. 66–71.

Gemitzi, A., Falalakis, G., Eskioglou, P., and Petalas, C. (2011). Evaluation landslide susceptibility using environmental factors, fuzzy membership functions and GIS. *Global NEST
Journal*, Vol. 13(1), pp. 28–40.

Gujree, I., Arshad, A., Reshi, A., & Bangroo, Z. (2020). Comprehensive spatial planning of
Sonamarg resort in Kashmir Valley, India for sustainable tourism. *Pacific International
Journal*, Vol. 3(2), pp. 71–99.

Hassanin, M., Kanga, S., Farooq, M., & Singh, S. K. (2020). Mapping of Trees outside Forest
(ToF) From Sentinel-2 MSI satellite data using object-based image analysis. *Gujarat
Agricultural Universities Research Journal*, Vol. 207. 207–213.

Holtz, D.R. and Kovacs, D.W. (1981). *An Introduction to Geotechnical Engineering*. Prentice-
Hall, Inc., Englewood Cliffs, NJ.

Jian, W. and Xiang-Guo, P. (2009). GIS-based landslide hazard zonation model and its applications. *Procedia Earth and Planetary Science*, Vol. 1, pp. 1198–1204.

Kanga, S., Kumar, S., and Singh, S.K. (2017b). Climate induced variation in forest fire using
remote sensing and GIS in Bilaspur district of Himachal Pradesh. *International Journal
of Engineering and Computer Science*, Vol. 6(6), pp. 21695–21702.

Kanga, S., Rather, M.A., Farooq, M., & Singh, S.K. (2021). GIS based forest fire vulnerability
assessment and its validation using field and MODIS data—A case study of Bhaderwah
forest division, Jammu and Kashmir (India). *Indian Forester*, Vol. 147(2), 120–136.

Kanga, S., Singh, S.K., & Sudhanshu. (2017a). Delineation of urban built-up and change detection analysis using multi-temporal satellite images. *International Journal of Recent
Research Aspects*, Vol. 4(3), pp. 1–9.

Kanungo, D.P., Arora, M.K., Gupta, R.P., and Sarkar, S. (2005). GIS based landslide hazard
zonation using neuro-fuzzy weighting, India. *International Conference on Artificial
Intelligence*, Pune, India.

Karsli, F. and Yalcin, M. (1991). Landslide assessment by using digital photogrammetric techniques. *International Association of Engineering Geology*, Vol. 43, pp. 27–29.

Lee, S. and Choi, J. (2004). Landslide susceptibility mapping using GIS and weight-of-
evidence model. *International Journal of Geographical Information Science*, Vol. 18(-
8), pp. 789–814.

Lee, S. and Pradhan, B. (2006). Probabilistic landslide hazards and risk mapping on Penang
Island, Malaysia. *Journal of Earth System Science*, Vol. 115(6), pp. 661–672.

Mayhew, S. (2004). *A Dictionary of Geography*. Oxford University Press, Delhi.

Meraj, G. et al. (2015). Assessing the influence of watershed characteristics on the flood
vulnerability of Jhelum basin in Kashmir Himalaya. *Natural Hazards*, Vol. 77(1),
pp. 153–175.

Meraj, G., Romshoo, S.A., and Altaf, S. (2016). Inferring land surface processes from
watershed characterization. *Geostatistical and Geospatial Approaches for the
Characterization of Natural Resources in the Environment*. Springer, Cham, pp. 741–
744. doi:10.5194/piahs-373-137-2016.

Meraj, G., Singh, S.K., Kanga, S., & Islam, M.N. (2021). Modeling on comparison of ecosystem services concepts, tools, methods and their ecological-economic implications—A review. *Modeling Earth Systems and Environment*, 1–20. https://doi.
org/10.1007/s40808-021-01131-6

Mukhopadhyay, B.P., Roy, S., Chaudhuri, S., and Mitra, S. (2012). Influence of geological parameters on landslide vulnerability zonation of Darjeeling Town, in Eastern
Himalayas. *Asian Journal of Environment and Disaster Management (AJEDM)*, Vol.
4(2), pp. 145–164.

Naithani, A.K. and Rawat, G.S. (2009). Investigations of Bunga Landslide and its mitigation –
A case study from Pindar valley, Garhwal Himalaya, Uttarakhand. *Indian Landslides*,
Vol. 2(1), pp. 9–22.

Nathawat, M.S. et al. (2010). Monitoring & analysis of wastelands and its dynamics using multiresolution and temporal satellite data in part of Indian state of Bihar. *International Journal of Geomatics and Geosciences*, Vol. 1(3), pp. 297–307.

Nithya, S.E. and Prasanna, P.R. (2010). An integrated approach with GIS and remote sensing technique for landslide hazard zonation. *International Journal of Geomatics and Geosciences*, Vol. 1(1), pp. 66–75.

Onagh, M., Kumra, V.K., and Rai, P.K. (2012). Landslide susceptibility mapping in a part of Uttarkashi district (India) by multiple linear regression method. *International Journal of Geology, Earth and Environmental Science*, Vol. 2(2), pp. 102–120.

Pachauri, A.K. (2004). Landslide hazard zonation in Himalayas. *Geophysical Research Abstract*, Vol. 6, pp. 1–2.

Pachauri, A.K., Gupta, P.V., and Chander, R. (1998). Landslide zoning in a part of the Garhwal Himalayas. *Environmental Geology*, Vol. 36(3–4), pp. 325–334.

Pandey, A., Dabral, P.P., Chowdary, V.M., and Yadav, N.K. (2008). Landslide hazard zonation using remote sensing and GIS—A case study of Dikrong river basin, Arunachal Pradesh, India. *Environmental Geology*, Vol. 54, pp. 1517–1529.

Patanakanog, B. (2001). Landslide hazard potential area in 3 dimension by remote sensing and GIS technique, Land Development Department, Thailand. https://www.ecy.wa.gov/programs/sea/landslides/help/drainage.html. Accessed on 21.05.2011.

Rajbhandari, P.C.L., Alam, B.M., and Akther, M.S. (2002). Application of GIS (Geographical Information System) for landslide hazard zonation and mapping disaster prone area—A study of Kulekhani Watershed, Nepal. *Plan Plus*, Vol. 1(1), pp. 117–123.

Ram, M.V., Jeyaseelan, A., Naveen, R.T., Narmatha, T., and Jayaprakash, M. (2011). Landslide susceptibility mapping using frequency ratio method and GIS in south eastern part of Nilgiri district, Tamilnadu, India. *International Journal of Geomatics and Geosciences*, Vol. 1(4), pp. 951–961.

Saha, A.K., Arora, M.K., Gupta, R.P., Virdi, M.L., and Csaplovics, E. (2005). GIS-based route planning in landslide-prone areas. *International Journal of Geographical Information Science*, Vol. 19(10), pp. 1149–1175.

Sarkar, S. and Kanungo, D.P. (2004). An integrated approach for landslide susceptibility mapping using remote sensing and GIS. *Photogrammetric Engineering & Remote Sensing*, Vol. 70 (5), pp. 617–625.

Sepulveda, S.A., Rebolledo, S., Lara, M., and Padilla, C. (2006). Landslide hazard in Santiago, Chile—An overview. *The Geographical Society of London*, IAEG-105, London, pp. 1–8.

Singh, R.K. (2004). National System for Disaster Management, India. *United States Conference on Space Science, Applications and Commerce*, Bangalore, India.

Singh, S. (2003). *Geomorphology*. Prayag Pustak Bhawan, Allahabad.

Singh, T. and Devi, M. (2006). Landslide occurrence and risk assessment in Itanagar Capital Complex, Arunachal Himalaya. *Himalayan Geology*, Vol. 27(2), pp. 145–162.

Tomar, J.S., Kranjčić, N., Đurin, B., Kanga, S., & Singh, S.K. (2021). Forest fire hazards vulnerability and risk assessment in Sirmaur district forest of Himachal Pradesh (India)—A geospatial approach. *ISPRS International Journal of Geo-Information*, Vol. 10(7), p. 447.

UNISDR (2009). *Terminology on Disaster Reduction*. UNISDR, Geneva.

Vaani, N. and Sekar, S.K. (2012). Regional landslide hazard zonation and vulnerability analysis using AHP and GIS—A case study of Nilgiri district, Tamil Nadu, India. *Disaster Advances*, Vol. 5(4), pp. 171–176.

Varnes, D.J. (1978). *Slope Movement and Types and Processes in Landslides. Analysis and Control, Transportation Research Board*. Special Report 176, Chapter 2, Figure 2.1. National Academy of Sciences, Washington, DC.

Varnes, D.J. (1984). *Landslides Hazard Zonation—A Review of Principles and Practice.* UNESCO, France.

Wieczorek, G. et al. (1987). Effect of rainfall intensity and duration on debris flows in central Santa Cruz Mountains, California. *Geological Society of America, Reviews in Engineering Geology*, Vol. VII, pp. 93–104.

8 Landslide Hazard and Exposure Mapping of Risk Elements in Lower Mandakini Valley, Uttarakhand, India

Habib Ali Mirdda and Masood Ahsan Siddiqui
Jamia Millia Islamia

Somnath Bera
Tata Institute of Social Sciences (TISS)

Bhoop Singh
Natural Resource Data Management System

CONTENTS

8.1 Introduction 158
8.2 Study Area 159
8.3 Data Source... 160
    8.3.1 Landslide Inventory 160
8.4 Methodology... 161
    8.4.1 Landslide Spatial Probability Mapping Using Logistic
           Regression Model 161
    8.4.2 Landslide Temporal Probability 163
    8.4.3 Landslide Hazard and Exposed Risk Elements Mapping 163
8.5 Result and Discussion... 164
    8.5.1 Landslide Spatial Probability 164
    8.5.2 Landslide Hazard Analysis.. 166
    8.5.3 Exposed Risk Elements 168
8.6 Conclusions... 168
Conflict of Interest 170
Funding 170
References... 170

DOI: 10.1201/9781003147107-10

8.1   INTRODUCTION

Landslide is a major geological hazard, which poses serious threat to human population and various other infrastructures like highways, rail routes and civil structures like dams, buildings and other structures. Expansion of urban and recreational developments on hill areas result in ever increasing number of residential and commercial properties that are often threatened by landslides (Anbalagan et al. 2015). More than 50,000 people died and 10 million people were affected by landslides around the world in the last century. The increasing trend of landslides and their impact on society and the economy has generated wider interest among scientific community (Bera et al. 2020). There is a growing trend of landslide hazard research all over the world to minimize the adverse impacts of landslides on society.

Landslide hazard can be defined as the occurrence probability of a potentially damaging landslide within a given area and within a specified period of time (Varnes 1984). Thus, landslide hazard includes spatial probability (landslide susceptibility) as well as temporal probability of landslide occurrence. Methods of calculating landslide spatial probability includes numerous methods such as frequency ratio (Jaafari et al. 2014; Anbalagan et al. 2015, Ramesh and Anbazhagan 2015; Hazma and Raghuvansi 2016), information value method (Vijith et al. 2009; Epifânio et al. 2014; Achour et al. 2017), logistic regression (Jaiswal et al. 2010; Mancini et al. 2010; Huang et al. 2015; Sahana and Sajjad 2017), weight of evidence (Neuhäuser et al. 2012; Regmi et al. 2010; Pradhan et al. 2010; Armas 2012; Dahal et al. 2007; Jebur et al. 2015), support vector machine (Yilmaz 2010; Bui et al. 2012; Pradhan 2013), artificial neural networks (Dou et al. 2015; Pradhan and Kim 2017; Chen et al. 2017), random forest (Trigila et al. 2015; Chen et al. 2017), fuzzy logic (Pourghasemi et al. 2012; Kumar and Anbalagan 2014; Bera et al. 2019) spatial multicriteria evaluation (Ahmed 2015; Pradhan and Kim 2016; Sangchini et al. 2016; Mirdda et al. 2020).

The temporal probability of landslides were quantified in this study using a Poisson or Binomial distribution model, rainfall threshold method (Jaiswal et al. 2010a). Poisson distribution model requires detail and sufficient multitemporal landslide inventory data. The rainfall threshold method requires event of landslide distribution data and associated rainfall data.

The risk elements signify their spatial existence in the form of physical, social or economic features under hazards of a particular place. The map of risk elements contributes to understand the potential loss due to disaster. The elements of risk can be exposed as static elements like building, roads, agriculture land, population (Uzielli et al. 2008; Kaynia et al. 2008; Li et al. 2010; Ghosh et al. 2011; Jaiswal et al. 2011) and also as moving elements like vehicles, persons (Fell et al. 2005; Jaiswal et al. 2010) on space. However, the consideration of risk elements varies in the different literature due to the different scales, data availability and purposes of the project (Corominas et al. 2014).

In literature, although landslide susceptibility is increasing, landslide hazard and risk are relatively scarce. Therefore, the main aim of the research is to assess landslide hazard and element of risks. Hence, the broader framework of the study is constituted in two parts first; to carry out modeling of landslide hazard zones and second, determination of risk- elements and its spatial analysis. The present study will provide an effective landslide hazard map and identify the elements exposed to landslide risk in the study area. The research outcome will be beneficial for further study on risk management and spatial planning.

8.2 STUDY AREA

The study area of the present study is Lower Mandakini Valley of Garhwal Himalaya, India (Figure 8.1). The area is located between latitudes 30°17′07″–30°26′11″N and longitudes 78°54′31″–79°03′17″E. It extends over an area of 118 km². The study area is very rugged and highly dissected by river Mandakini and its tributaries. The lowest and highest elevation of the area is 660 and 2,800 m respectively. The slope angle of the area goes up to 77°.

The geology of the study area is very complex. The area is highly folded in nature, with numerous tectonic features like thrust, faults, and shear zones. There are four lithological groups, viz. Damta Group, Janusar Group, Ramgarh Group and Ramgarh Porphyr Group. The region is humid, subtropical in nature and receives an annual rainfall of 150–200 cm. Rainfall mostly occurs during the monsoon period. The average temperature varies from 0°C in January to 35°C in July. The region experiences cold waves in winter due to western disturbances. The relative humidity

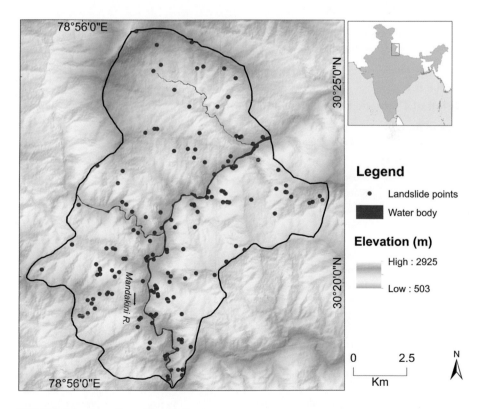

FIGURE 8.1   Location of the study area.

varies between 35% in the premonsoon and 70% during the monsoon period. There are mainly two types of natural vegetation in the area viz. scrub forest and dense evergreen forest.

The region comes under very high seismic risk zone. Almost every year, the area is affected by natural hazards in the form of flood, landslide, forest fire etc. The study area is sensitive to landslides due to complex geological conditions and very steep slopes. At present, various development activities are making the hill slopes unstable and creating favorable conditions for landslides.

8.3   DATA SOURCE

Data for the study are collected through extensive field survey and collection of different government reports, published maps, and satellite images (Table 8.1). The detail of data sources and generation of thematic maps are discussed in this section.

In this study, 11 landslide conditioning factors are taken for susceptibility analysis. These are slope, aspect, curvature, lithology, lineament proximity, drainage proximity, land use land cover, Soil texture, soil thickness, road proximity, and rainfall.

8.3.1   LANDSLIDE INVENTORY

Landslide inventory is past landslide data that provide various information about landslides such as location, type, activity and physical properties etc. Landslide

TABLE 8.1
Detail of Thematic Data Layers and Data Sources

Types of Thematic Data Layers	Sub Classification	Data Types in GIS	Scale/ Resolution	Source
Landslide inventory		Polygon	–	IRS P6 LISS4 (5.8 m)
Topographic maps	Slope gradient	Grid	10×10 m	ALOS DEM (12.5 m)
	Slope aspect	Grid	10×10 m	
	Slope curvature	Grid	10×10 m	
Geological map	Lithology	Polygon	–	Geological map, GSI (1:100,000)
	Lineament Proximity	Grid	10×10 m	IRS P6 LISS4 (5.8 m)
Hydrology	Drainage Proximity	Grid	10×10 m	Toposheet, SOI (1:50,000)
Climatic	Rainfall	Grid	10×10 m	Indian Meteorological Department
Soil	Soil texture	Polygon	–	Soil Map, Govt. of
	Soil thickness	Polygon	–	Uttarakhand
Anthropogenic	Land use and Land cover	Polygon	–	IRS P6 LISS4 (5.8 m)
	Road proximity	Grid	10×10 m	Toposheet, LISS IV

inventory mapping is the very first step in landslide investigation and susceptibility mapping. Landslide inventory data for the study are collected from various sources such as – historical reports and newspaper archives, field survey with GPS, Google Earth and Multitemporal Satellite Images and interaction with local people. Historical records are collected from newspaper agencies, PWD office in Dehradun and its suboffice in Rudraprayag district, Broader Road Organisation, disaster management cell of district magistrate office. Detail landslide information is collected by extensive field survey. Landslide locations are collected with the help of Trimbe JunoSA GPS. About 100 landslide locations are collected in the field and mapped the area by ArcGIS software. Landslide information from secondary sources is also verified during the field visit. Google Earth images of different time periods are compared and analyzed to identify landslide. Similarly, multitemporal satellite images are also used to identify past landslides. In this study IRS LISS-IV and Landsat-7 satellite images of different time periods are used for visual analysis. Important information of many small landslides, their exact date of occurrence, magnitude etc. are collected by interacting with local villagers.

A total of 160 landslides are considered for application in the hazard mapping. The size of the landslide ranges from 28 to 25291.9 m^2. All the landslides are converted to raster format with 10×10 m grid size using conversion tool in Arc GIS 10.3. Total numbers of landslide pixels are 4,392. The whole landslide dataset is divided into training and testing set for further analysis. 70% landslides are considered for training set data on the basis of which models is build. The reaming 30% landslides are used for validation of models.

8.4 METHODOLOGY

Landslide hazard mapping is a two-step process that includes – landslide spatial probability mapping and landslide temporal probability mapping (Figure 8.2).

8.4.1 Landslide Spatial Probability Mapping Using Logistic Regression Model

Logistic regression is a multivariate statistical model used to derive the relationship between a dependent and several independent variables. In this model, prediction of any event is made from a group of predictor variables by finding a best fit function. For this, continuous classes are taken as one single factor while each discrete class is taken as a separate factor. Logistic regression analysis is of two types, viz. binary and multiple logistic regression. In the case of binary logistic regression, there is a dichotomous dependent variable and one or more independent variables that can be either continuous or categorical. In case of multiple logistic regressions, there are more than two categories of dependent variables. The regression coefficient is used for calculating the ratio of each independent variable. The relation between an event and its conditioning factor is expressed as:

$$P = \frac{1}{1 + e^{-z}}$$

(8.1)

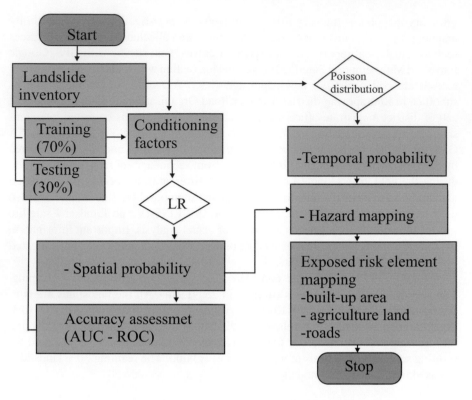

FIGURE 8.2   Methodology for landslide hazard mapping and exposure analysis.

where, P is the probability of an event occurring and z is the linear combination and the value of it is calculated from the following equation:

$$z = b_0 + b_1 x_1 + b_2 x_2 + \ldots + b_n x_n \qquad (8.2)$$

where, b_0 is the intercept of the model, the b_i ($i = 0, 1, 2, \ldots, n$) are slope coefficient of the logistic regression model, and the x_i ($i = 0, 1, 2, \ldots, n$) are the independent variables.

In the present study, binary logistic regression is used and past landslide data is considered as dependent variable whose presence is denoted by "1" and absence is denoted by "0". The "continuous data" for the study are slope, rainfall, lineament proximity, drainage proximity, and road proximity and the discrete data are aspect, lithology, soil texture, soil depth, and land use and land cover.

The probability of landslide occurrence is estimated using the equation:

$$\begin{aligned} Z_{11} = {} & (0.026 \times \text{SLOPE}) + \text{ASPECT}_C + \text{CURVATURE}_C + \\ & (0.001 \times \text{RAINFALL}) + \text{LITHOLOGY}_C + (-0.005 \times \text{LINEAMENT}) + \\ & (0.001 \times \text{DRAINAGE}) + \text{SOIL TEXTURE}_C + \text{SOIL DEPTH}_C + \\ & \text{LULC}_C + (-0.001 \times \text{ROAD}) - 7.205 \end{aligned} \qquad (8.3)$$

In logistic regression, the regression coefficient of landslide conditioning factors is calculated by SPSS software. The coefficient values are shown in Table 8.2. The continuous classes have a common coefficient value while the discrete classes have coefficient values for each class. Independent variables whose significance value is less than 0.05 are added to the model. The Wald chi-square (x^2) value at 5% significant level for the degree of freedom has been used to test the hypothesis.

8.4.2 LANDSLIDE TEMPORAL PROBABILITY

After spatial probability mapping, the temporal probability of landslides was measured for final hazard mapping. We determined the exceedance probability of landslides using a Poisson distribution probability model. In this model, the exceedance probability during the time "t" is described as follow:

$$P[N(t) \geq 1] = 1 - \exp\left(-\frac{t}{u}\right)$$

(8.4)

where, P is the probability, N is the number of landslides and u is the future average recurrence interval between successive events. For the calculation of temporal probability, the study area is divided into 12 units. The exceedance probability of landslides in each unit was analyzed based on the Poisson probability model for a return period of 5 years. The model considered 26 year time period from 1990 to 2015 assuming that the rate of future landslide occurrence will not change.

8.4.3 LANDSLIDE HAZARD AND EXPOSED RISK ELEMENTS MAPPING

Landslide hazard map is developed by multiplication of landslide spatial probability and temporal probability. The landslide hazard map is classified into five hazard zones, viz. very low, low, moderate, high and very high. The hazard map is validated by seed cell area index method (SCAI). In this method, Seed Cell Area Index (SCAI) values are derived based on the relationships between past landslides and the areal extent of each hazard class (Equation 8.5).

$$SCAI_i = \frac{l_i / L}{s_i / S}$$

(8.5)

Where, $SCAI_i$ = seed cell index of ith hazard class, l_i = landslide pixel in an ith hazard class, L = Total number of landslide pixels in all hazard class, s_i = Number of pixels in ith hazard class, S = Total number of pixels in all hazard classes.

The built-up area, agricultural land, and transportation routes have been considered as the elements of risk as they are directly or indirectly linked to the life and socio-economic conditions of people. The map of the built-up area and agricultural land is prepared from LISS IV satellite image, and further, it is verified through field survey. The details of road data are derived from the published road map of Rudraprayag District. The roads are categorized into four types viz., national highway, major district road, other district road and department roads.

The map of exposed risk elements (built-up area, agricultural land, major roads) is generated by overlying in the GIS environment.

8.5   RESULT AND DISCUSSION

8.5.1   Landslide Spatial Probability

Coefficients of logistic regression reveals that slope, curvature, drainage proximity, soil texture are positively correlated while land use, linear proximity shows inverse influence in landslide occurrence (Table 8.2). In the aspect, south-east and south facing group have highest value i.e. has highest influence to slope instability. Among the lithological groups, Ramgarh group has highest influence in causing landslides. In the soil, Loamy-skeletal group contributes higher on landslide occurrence.

The landslide spatial probability map is classified into five zones viz., very low, low, moderate, high and very high (Figure 8.3). The result shows that 74.34% of the

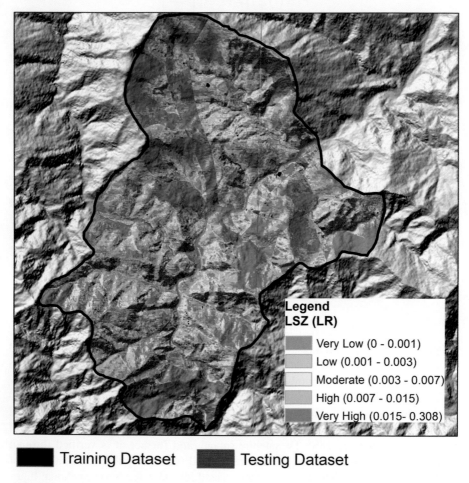

FIGURE 8.3   Landslide spatial probability map of the study area.

TABLE 8.2
Logistic Regression Coefficient Values for Conditioning Factors

Conditioning Factors	Class	Regression Coefficient
Slope		0.026
Aspect	Flat	−15.375
	N	−1.430
	NE	−1.038
	E	0.000
	SE	0.779
	S	0.593
	SW	0.081
	W	−0.296
	NW	−0.048
Curvature	Concave	0.264
	Plain	0.088
	Convex	0.000
Rainfall		0.001
Lithology	Damta Group	0.000
	Jaunsar Group	−15.254
	Ramgarh Group	0.716
	Ramgarh Group Porphyr	−0.658
Lineament proximity		−0.005
Drainage proximity		0.001
Soil texture	Loamy	0.000
	Mixed loamy	2.883
	Coarse-loamy	0.606
	Fine-loamy	1.196
	Loamy-skeletal	18.831
Soil depth	Shallow(<50)	0.000
	Moderate shallow (50–75)	−1.810
	Moderate deep (75–100)	0.748
	Deep(>100)	−1.054
LULC	Water body	−17.050
	Built up area	−2.185
	Agricultural land	−1.267
	Barren land	−0.739
	Scrub forest	0.000
	Evergreen forest	−0.877
Road proximity		−0.001

Constant is −7.205.

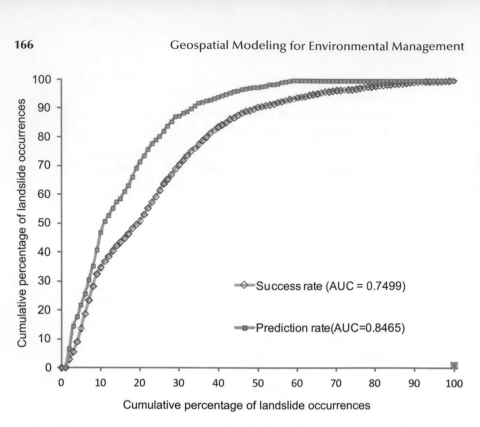

FIGURE 8.4   AUC-ROC curve for accuracy assessment.

observed landslides fall in 27.67% of the predicted high and very high susceptibility zones. The accuracy of the spatial probability model is measured by the area under curve (AUC) method. The AUC value of the success rate curve is 0.7669. The AUC value of prediction rate curve is 0.8563 for the model (Figure 8.4). Both the results reflect the robustness of the model.

8.5.2   LANDSLIDE HAZARD ANALYSIS

The result shows that the temporal probability of landslides varies from 0.26 to 0.61 (Figure 8.5). The value is different in different spatial units. The temporal probability is maximum in zone 5. Whereas temporal probability is low in zones 3, 10, and 11. However, the spatial probability of landslides is not always agreed with the temporal probability. Therefore, the landslide hazard zone also varied across the study area (Figure 8.6).

The result shows the very high hazard zone cover area 10.7%. High, moderate, low and very low covered 26.04%, 19.31%, 17.52% and 26.36% respectively (Table 8.3). The percentage of landslides increases constantly from low hazard zone to very high hazard zone. As a result, seed cell index is maximum in the very high zone (2.62); and its value increases from very low to very high hazard zone. The increasing trend of seed cell index value reflects the reliability of the landslide hazard map.

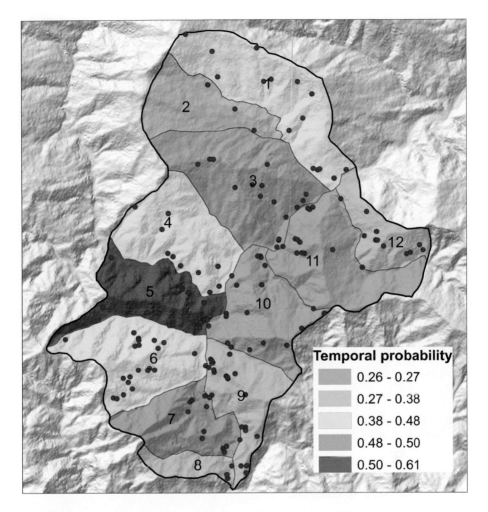

Temporal probability
0.26 - 0.27
0.27 - 0.38
0.38 - 0.48
0.48 - 0.50
0.50 - 0.61

FIGURE 8.5   Spatial units of the study area and temporal probability map.

TABLE 8.3
Seed Cell Index for Hazard Zones

Hazard Zone	No. of Pixels in Domain	% of Pixels in Domain	No. of Landslides in Domain	% of Landslide in Domain	Seed Cell Index
Very low	311,261	26.36	176	4.01	0.15
Low	206,812	17.52	358	8.15	0.46
Moderate	227,963	19.31	747	17.01	0.88
High	307,454	26.04	1,876	42.71	1.64
Very high	126,314	10.70	1,235	28.12	2.62

FIGURE 8.6   Landslide hazard map.

8.5.3   EXPOSED RISK ELEMENTS

The study considers built-up area, agricultural land, and roads as exposed elements. The result of the study shows that 14% of built-up area, 20% of agricultural land, 49% of roads comes under high and very high hazard zone (Table 8.4 and Figure 8.7). Among different road types, 11.26% of National Highway (NH), 46.19% of other district roads and 19% other departmental roads fall under very high hazard zone.

In case of exposure analysis, only the static elements are considered. Moving elements like livestock, vehicles, persons etc. are not considered in the study due to lack of data. Due to similar reasons, the study could not cover the population and their socio-economic attributes in the analysis. When the data will be available in the future, a detailed vulnerability and risk analysis can be carried out based on the study (Table 8.5).

8.6   CONCLUSIONS

The study assesses landslide hazard and exposed risk elements in Lower Mandakini Catchment. The landslide hazard map is prepared by combining two major components

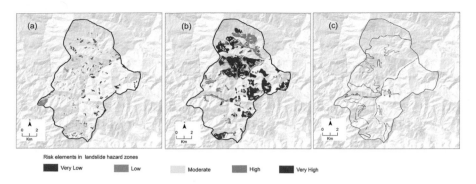

FIGURE 8.7 Exposed risk elements in different landslide hazard zones. (a) Built-up area. (b) Agricultural land. (c) Major roads.

TABLE 8.4
Area of Risk Elements in Landslide Hazard Zones

Hazard Zones	Built Up Area (Sq. m)	% of Built Up Area	Agriculture Land (Sq. m)	% of Agriculture Land	Road Length (m)	% of Road Length
Very low	2620480.85	35.28	15108729.03	46.33	16482.54	17.15
Low	1812718.63	24.40	6749684.84	20.70	14427.88	15.01
Moderate	1963908.88	26.44	4317516.37	13.24	18074.13	18.81
High	946388.052	12.74	5367277.17	16.46	29876.07	31.09
Very High	83790.63	1.12	1063243.91	3.26	17204.41	17.90

TABLE 8.5
Different Roads Exposing in Landslide Hazard Zones

Hazard Zones	National Highway (m)	% of Road Length	Major District Road (m)	% of Road Length	Other District Roads (m)	% of Road Length	Other Department Roads (m)	% of Road Length
Very low	4011.65	25.55	80.44	1.44	4291.99	12.19	13791.52	31.22
Low	1741.98	11.10	0.00	0.00	6298.85	17.89	6866.05	15.54
Moderate	2201.03	14.02	0.00	0.00	6372.91	18.10	9482.53	21.47
High	5976.62	38.07	2925.61	52.37	11243.27	31.94	8303.15	18.80
Very high	1766.99	11.26	2579.92	46.19	6999.44	19.88	5731.71	12.98

of hazard analysis, viz. Spatial probability and temporal probability. The spatial probability map is prepared by the logistic regression model. The temporal probability of landslides is calculated using Poisson distribution model. Different elements that are exposed to landslide hazards are extracted through overlay analysis in GIS. The findings of the study show that 36.74% of the study area comes under high and very high landslide hazard zone. Among risk elements, almost 6% built-up area and 17.09% agricultural land is located in the most probable zone of landslide. Among different road types, 40.61% of National Highway (NH) falls under the high and very high hazard zone. It is recommended from the study that any kind of construction or development activity should be avoided in the high and very high hazard zones as these areas are unstable and vulnerable to slope failure. This study may contribute to planners and decision-makers in risk management and further spatial planning.

CONFLICT OF INTEREST

The authors declare that they have no conflict of interest.

FUNDING

No funding was received from any agency for conducting this study.

REFERENCES

Achour, Y., Boumezbeur, A., Hadji, R., Chouabbi, A., Cavaleiro, V., & Bendaoud, E. A. (2017). Landslide susceptibility mapping using analytic hierarchy process and information value methods along with a highway road section in Constantine, Algeria. *Arabian Journal of Geoscience*, *10*:194. doi: 10.1007/s12517-017-2980-6

Ahmed, B. (2015). Landslide susceptibility mapping using multi-criteria evaluation techniques in Chittagong Metropolitan Area, Bangladesh. *Landslides*, *12*:1077–1095.

Anbalagan, R., Kumar, R., Lakshmanan, K., Parida, S., & Neethu, S. (2015). Landslide hazard zonation mapping using frequency ratio and fuzzy logic approach, a case study of Lachung Valley, Sikkim. *Geoenvironmental Disasters*, *2*:6. doi: 10.1186/s40677-014-0009-y

Armas, I. (2012). Weights of evidence method for landslide susceptibility mapping. Prahova Subcarpathians, Romania. *Natural Hazards*, *60*:937–950.

Bera, S., Guru, B., Chatterjee, R., & Shaw, R. (2020). Geographic variation of resilience to landslide hazard: a household-based comparative studies in Kalimpong hilly region, India. *International Journal of Disaster Risk Reduction*, *46*:101456.

Bera, S., Guru, B., & Ramesh, V. (2019). Evaluation of landslide susceptibility models: a comparative study on the part of Western Ghat Region, India. *Remote Sensing Applications: Society and Environment*, *13*:39–52.

Bui, D.T., Pradhan, B., Lofman, O., Revhaug, I., & Dick, O.B. (2012). Landslide susceptibility assessment in the Hoa Binh province of Vietnam: a comparison of the Levenberg-Marquardt and Bayesian regularized neural networks. *Geomorphology*, *171*:12–29.

Chen, W., Pourghasemi, H. R., & Zhao, Z. (2017). A GIS-based comparative study of Dempster-Shafer, logistic regression and artificial neural network models for landslide susceptibility mapping. *Geocarto International*, *32*:367–385.

Chen, W., Xie, X., Wang, J., Pradhan, B., Hong, H., Bui, D.T., Duan, Z., & Ma, J. (2017). A comparative study of logistic model tree, random forest, and classification and regression tree models for spatial prediction of landslide susceptibility. *Catena*, *151*:147–160.

Corominas, J., van Westen C., Frattini, P., Cascini, L., Malet, J.P., Photopoulou, S., Catani, F., Van Den Eeckhaut, M., Mavrouli, O., Agliardi, F., Pitilakis, K., Winter, M.G., Pastor, M., Ferlisi, S., Tofani, V., Hervas, J., & Smith, J.T. (2014). Recommendations for the quantitative analysis of landslide risk. *Bulletin of Engineering Geology and the Environment*, *73*:209–263. doi: 10.1007/s10064-013-0538-8

Dahal, R.K., Hasegawa, S., Nonomura, A., Yamanaka, M., Masuda, T., & Nishino, K. (2007). GIS-based weights-of-evidence modelling of rainfall-induced landslides in small catchments for landslide susceptibility mapping. *Environmental Geology*, *54*(2):311–324.

Dou, J., Yamagishi, H., Pourghasemi, H.R., Yunus, A.P., Song, X., Xu, Y., & Zhu, Z. (2015). An integrated artificial neural network model for the landslide susceptibility assessment of Osado Island, Japan. *Natural Hazards*, *78*:1749–1776. doi: 10.1007/s11069-015-1799-2

Epifânio, B., Zêzere, J.L., & Neves, M. (2014). Susceptibility assessment to different types of landslides in the coastal cliffs of Lourinhã (Central Portugal). *Journal of Sea Research*, *93*:150–159.

Fell, R., Ho, K.K.S., Lacasse, S., & Leroi, E. (2005) A framework for landslide risk assessment and management. In O. Hungr, R. Fell, R. Couture, & E. Eberhardt (Eds.) *Landslide risk management* (pp. 3–26). London: Taylor and Francis.

Ghosh, S., van Westen, C.J., Carranza, E.J., & Jetten, V.G. (2011). Integrating spatial, temporal, and magnitude probabilities for medium-scale landslide risk analysis in Darjeeling Himalayas, India. *Landslides*, *9*:371–384.

Hazma, T., & Raghuvansi, T.K. (2016). GIS based landslide hazard evaluation and zonation – a case from Jeldu District, Central Ethiopia. *Journal of King Saud University – Science*, *29*:151–165.

Huang, J., Zhou, Q., & Wang, F. (2015). Mapping the landslide susceptibility in Lantau Island, Hong Kong, by frequency ratio and logistic regression model. *Annals of GIS*, *21*(3):191–208.

Jaafari, A., Najafi, A., Pourghasemi, H.R., Rezaeian, J., & Sattarian, A. (2014). GIS-based frequency ratio and index of entropy models for landslide susceptibility assessment in the Caspian forest, northern Iran, *International Journal of Environmental Science and Technology*, *11*:909–926.

Jaiswal, P., van Westen, C.J. & Jetten, V. (2010a). Quantitative assessment of direct and indirect landslide risk along transportation lines in southern India. *Natural Hazards and Earth System Science*, *10*:1253–1267.

Jaiswal, P., van Westen, C.J., Jetten, V. (2010b). Quantitative landslide hazard assessment along a transportation corridor in southern India. *Engineering Geology*, *116*:236–250.

Jaiswal, P., van Westen, C.J. & Jetten, V. (2011). Quantitative estimation of landslide risk from rapid debris slides on natural slopes in the Nilgiri hills, India. *Natural Hazards and Earth System Sciences*, *11*:1723–1743.

Jebur, M.N., Pradhan, B., Shafri, H.Z.M., Yusoff, Z.M., & Tehrany, M.S. (2015). An integrated user-friendly ArcMAP tool for bivariate statistical modelling in geoscience applications. *Geoscientific Model Development*, *8*:881–891.

Kaynia, A.M., Papathoma-Ko¨hle, M., Neuhauser, B., Ratzinger, K., Wenzel, H., & Medina-Cetina, Z. (2008). Probabilistic assessment of vulnerability to landslide: application to the village of Lichten-stein, Baden-Württemberg, Germany. *Engineering Geology*, *101*:33–48.

Kumar, R. & Anbalagan, R. (2014). Landslide susceptibility zonation in part of Tehri reservoir region using frequency ratio, fuzzy logic and GIS. *Journal of Earth System Science*, *124*(2):431–448.

Li, Z., Nadim, F., Huang, H., Uzielli, M., & Lacasse, S. (2010). Quantitative vulnerability estimation for scenario-based landslide hazards. *Landslides*, *7*(2):125–134.

Mancini, F., Ceppi, C., & Rtrovato, G. (2010). GIS and statistical analysis for landslide susceptibility mapping in the Daunia area, Italy. *Natural Hazards and Earth System Sciences*, *10*:1851–1864. doi: 10.5194/nhess-10-1851-2010

Mirdda, H.A., Bera, S., Siddiqui, M.A., & Singh, B. (2020). Analysis of bi-variate statistical and multi-criteria decision-making models in landslide susceptibility mapping in lower Mandakini Valley, India. *GeoJournal*, *85*(3):681–701.

Neuhäuser, B., Damm, B., & Terhorst, B. (2012). GIS-based assessment of landslide suscep-tibility on the base of the weights-of-evidence model. *Landslides*, *9*(4):511–528. doi: 10.1007/s10346-011-0305-5

Pourghasemi, H.R., Pradhan, B., & Gokceoglu, C. (2012). Application of fuzzy logic and analyt-ical hierarchy process (AHP) to landslide susceptibility mapping at Haraz water-shed, Iran. *Natural Hazards*, *63*(2):965–996.

Pradhan, A.M.S. & Kim, Y.T. (2016). Evaluation of a combined spatial multi-criteria eval-uation model and deterministic model for landslide susceptibility mapping. *Catena*, *140*:125–139.

Pradhan, A.M.S. & Kim, Y-T. (2017). Landslide susceptibility mapping of Phewa catchment using multilayer perceptron artificial neural network. *Nepal Journal of Environmental Science*, *4*:1–9.

Pradhan, B. (2013). A comparative study on the predictive ability of the decision tree, sup-port vector machine and neuro-fuzzy models in landslide susceptibility mapping using GIS. *Computer Geoscience*, *51*:350–365.

Pradhan, B., Oh, H.J., & Buchroithner, M. (2010). Weights-of-evidence model applied to land-slide susceptibility mapping in a tropical hilly area. *Geomatics, Natural Hazards and Risk*, *1*(3):199–222. doi: 10.1080/19475705.2010.498151

Ramesh, V. & Anbazhagan, S. (2015). Landslide susceptibility mapping along Kolli hills Ghat road section (India) using frequency ratio, relative effect and fuzzy logic models. *Environmental Earth Sciences*, *73*(12):8009–8021.

Regmi, N.R., Giardino, J.R., & Vitek, J.D. (2010). Modeling susceptibility to landslides using the weight of evidence approach: Western Colorado, USA. *Geomorphology*, *115*(1–2):172–187. doi: 10.1016/j.geomorph.2009.10.002

Sahana, M. & Sajjad, H. (2017). Evaluating effectiveness of frequency ratio, fuzzy logic and logistic regression models in assessing landslide susceptibility: a case from Rudraprayag district, India. *Journal of Mountain Science*, *14*(11):2150–2167.

Sangchini, E.K., Emami, S.N., Tahmasebipour, N., Pourghasemi, H.R., Naghibi, S.A., Arami, S.A., & Pradhan, B. (2016). Assessment and comparison of combined bivariate and AHP models with logistic regression for landslide susceptibility mapping in the Chaharmahal-e-Bakhtiari Province, Iran. *Arabian Journal of Geoscience*, *9*:201. doi: 10.1007/s12517-015-2258-9

Trigila, A., Iadanza, C., Esposito, C., & Scarascia-Mugnozza, G. (2015). Comparison of logis-tic regression and random forests techniques for shallow landslide susceptibility assess-ment in Giampilieri (NE Sicily, Italy). *Geomorphology*, *249*:119–136. doi: 10.1016/j.geomorph.2015.06.00

Uzielli, M., Nadim, F., Lacasse, S., & Kaynia, A.M. (2008). A conceptual framework for quantitative estimation of physical vulnerability to landslides. *Engineering Geology*, *102*:251–256.

Varnes, D.J. (1984). *Landslide Hazard Zonation: A Review of Principles and Practice* (pp. 1–63). Paris: UNESCO.

Vijith, H., Rejith, P.G., & Madhu, G. (2009). Using InfoVal method and GIS techniques for the spatial modelling of landslide susceptibility in the upper catchment of River Meenachil in Kerala. *Journal of the Indian Society of Remote Sensing*, *37*:241–250.

Yilmaz, I. (2010). Comparison of landslide susceptibility mapping methodologies for Koyulhisar, Turkey: conditional probability, logistic regression, artificial neural net-works, and support vector machine. *Environmental Earth Science*, *61* (4):821–836.

Part C

Geospatial Modeling for Climate Change Studies

9 Crop Response to Changing Climate, Integrating Model Approaches
A Review

Arul Prasad S., Maragatham M.,
Vijayashanthi V.A., and Naveen
Tamil Nadu Agricultural University

CONTENTS

9.1   Introduction 175
9.2   Climate Change Projections 176
9.3   Observed Changes 177
9.4   Climate Change Scenarios.. 178
9.5   Climate Change and Agricultural.. 179
      9.5.1   Wheat... 180
      9.5.2   Rice 181
      9.5.3   Pulses 181
      9.5.4   Groundnut 181
9.6   Crop Simulation Model (CSM)... 182
9.7   Global Climate Models.. 183
9.8   Regional Climate Models 183
9.9   Climate Downscaling 184
9.10 Crop Simulation Models and Remote Sensing 184
9.11 Integrated Assessment Through Climate and Crop Models....................... 185
9.12 Conclusion 188
References... 188

9.1   INTRODUCTION

The agricultural sector represents 35% of India's Gross National Product (GNP) and as such plays a crucial role in the country's development. The consequences of climate change on agriculture could lead to food security issues and could risk the lives and health of a large proportion of population. Climate change can have an impact on crop

DOI: 10.1201/9781003147107-12

yield (both positively and negatively) and even on the type of crops grown in certain areas by affecting water for irrigation, amounts of solar radiation that affect plant growth as well as prevalence of pests [1]. Many global studies have shown loss of 10%–40% in crop production in India by the end of the century [2] and for every 1° rise in temperature the yields of mustard, groundnut and soybean can be expected to decrease by 3%–7% [3].

Climate projections are typically presented for a range of conceivable pathways, scenarios or targets that capture the relationships between human decisions, emission concentration and change in temperature. The SRES scenarios were published by IPCC in 2000, the basic policy and economic assumptions for these scenarios were set in 1997. Presently, mainstream researchers have built up a lot of new-emission scenarios named Representative Concentration Pathways (RCPs). In distinction to the SRES scenarios, RCPs represent pathways of radiative forcing, not complete socioeconomic narratives or scenarios. Central to the process is the concept that any single radiative forcing pathway can result from a diverse range of socioeconomic and technological development scenarios. However, RCPs differ from previous sets of standard scenarios they are not scenarios of emissions; they are scenarios of radiative forcing. Each scenario has one esteem: the change in tropopause radiative force by 2100 compared to preindustrial levels. The four RCPs are numbered based on change in radiative forcing by 2100: +2.6, +4.5, +6.0 and +8.5 (W/m²) [4,5].

Crop Simulation model is a scientific approach not only to study the integrated effects, but also to see the consequences of individual factors associated with soil, climate, variety and management. The use of crop models to examine the potential effect of yield under climate change and climate variability gives an immediate connection between models, agrometeorology and concerns of the society. It also deals with yield gap analysis, yield forecast, regional studies, optimization of management practices and genotype studies. The model is capable of growing a crop under simulation process and this simulation process helps to understand the impact on crops [6]. Calibration of models for various popular crop cultivars based on field experiments in varied agro climatic zones is the need of the hour for simulating future climate change impacts and to identify strategies for optimizing resource use, increasing productivity and for reducing adverse impacts.

Impact assessments have recently been analyzed using process-based crop models to predict crop yields under both current and future climate conditions [7]. Simulation modeling is the most reliable tool in assessing the climate change impact on crops. In view of the indispensable role of simulation modeling in global climate change analysis, crop simulation models are considered as helpful tools. To assess the impact of climate change on agriculture, integrated approaches help us to suggest adaptation and mitigation strategies to support and enhance the agricultural production of the region. This book chapter deals with integrated approaches of climate and crop simulation model.

9.2   CLIMATE CHANGE PROJECTIONS

Climate change is an intricate, intuitive framework comprising of the environment, land surface, ice and snow, seas and different water bodies and living things. The IPCC was set up in 1988, by the World Meteorological Organization (WMO) and the United Nations Environment Program (UNEP), to evaluate scientific data on environmental change, just as its ecological and financial effects and to define formulated strategies. IPCC has characterized climate change as any change in climate after some time, regardless of whether because of variability in nature or because of human action. Climate models had experienced many advancements and helped exceptional predictions and projections for multiple scenarios proposed in compliance with IPCC standards. The rise in CO_2 in the atmosphere was projected to increase from 368 µmol/mol in 2000 to between 540 and 970 µmol/mol. By the end of the twenty-first century the temperature increase of around 1.4°C–5.8°C could be expected [8,9].

The global average surface air temperature increases over the twenty-first century; the main reason is increase of anthropogenic greenhouse gas concentrations. It reported an increase of 0.5–1.2°C, 0.88°C–3.16°C and 1.56°C–5.44°C over the near, mid and end centuries over the Indian region. The increase in extreme weather events due to increase temperature leads to an unpredictable state of food production [10,11]. The AR 5 models have projected an increase in the maximum tropical rainfall and also in high altitudes [12]. For the A1B scenario, the multi-model ensembles reported that the Indian monsoon dynamical circulation would likely be weakened by the end of the century [13]. There is no trend in Indian monsoon precipitation which fluctuates randomly over a long period of time [14]. In India, the monsoon precipitation would not experience any change. Several scientists have proved that the impacts of climate change are likely to be severe for the countries like India that have limited arable land but heavy dependence on agriculture. By the end of the twenty-first century, the projections using GCM data represented a slight decline in Indian monsoon with decreased quantities and increased intensity over the East coast.

The recurrence of extreme event is relied upon to change during the following century, with increase in the recurrence of heavy rainfall and heat wave events and decrease in the recurrence of frost days, as an outcome of anthropogenically-constrained climate change [15]. The increase of heavy precipitation and heat wave events is projected in South Asia [16] together with day by day increase of inter-annual variability of rainfall in the Asian summer monsoon. The increase in sea level was projected to be between 70 and 180 mm during 2100 due to ice loss alone but the use of global inventory that gave details of glacial spread and volumes suggested an increase of about 155 ± 41 mm (RCP 4.5) and 216 ± 44 mm (RCP 8.5) with a decrease of about 29%–41% in glacial volume [17]. The annual average water availability and river runoff projected to decline 10%–30% in some part of dry places at mid-latitudes and dry tropics [18]. In Central India Nira basin, studied annual and season rainfall data showed an increasing annual rainfall of 0.28%. Seasonal rainfall in the monsoon and post monsoon shows an increasing trend while in the summer and winter a decreasing trend [19].

9.3   OBSERVED CHANGES

Human activities have led to the increase in atmospheric greenhouse gas (GHG) concentrations by about 40%, 150% and 20% of CO_2, CH_4 and N_2O respectively increase by 2011 from the preindustrial concentrations. The oceans have captured around 30% of the emitted CO_2, leading to ocean acidification. Since the industrial era the ocean pH has decreased by 0.1 [20]. The expected rise in the atmospheric CO_2 concentration was about 700 µmol/mol by the end of the century [21] and could increase at a rate of 1.8 µmol/mol/year [22]. The concentration of CO_2 in the atmosphere increased from 280 to 379 ppm during the preindustrial period to 2005 [23].

Surface temperature of the Indian subcontinent had increased by 0.3°C (1969–2005), i.e., about 0.08°C per decade. The incidence of extreme weather events has been reported to increase [24]. Entire Indian peninsula excluding the extreme southern parts is susceptible to cold waves which prevail during December to March. The occurrence, persistency and the area affected by the cold waves over India is increasing [25]. The increase in the minimum temperature was reported to be higher than that of the maximum temperature and also the winter temperature to be higher than that of summer temperature, with the mean rise in temperature of about 0.74°C during the past 100 years. Whereas, an insignificant increase in the temperature during the monsoon season was also recorded [26–28].

In India, seasonal variation in precipitation showed negative pattern in monsoon season. Increase in heavy rainfall events and decline in low and medium precipitation occasions over India have been watched [28]. Ref. [29] considered temporal variation in month to month, seasonal and yearly precipitation over Kerala, tropical humid climate among the period from 1871 to 2005. The outcomes showed that precipitation during winter and summer seasons showed immaterial increasing pattern. Precipitation amid June and July showed noteworthy decreasing pattern while increasing pattern in January, February and April.

India's land surface ended up dimmer with surface solar radiation falling by about 5% during 1981–2004 [30]. Surface warming quickened in India toward the twentieth century, with least temperature rising 0.025°C/year amid 1981–1990 and 0.056°C/year amid 1991–2000. Other sign of changes in the climate system is a warming in the global oceans. The worldwide sea temperature rises by 0.10°C from the surface to 700 m depth from 1961 to 2003 [31]. Warming makes seawater spread, along these lines adds to ocean level rise. This factor indicated to as warm extension has contributed 1.6 ± 0.5 mm every year to worldwide normal ocean level in the course of the last decade (1993–2003). Different elements adding to ocean level rise in the course of the most recent decade incorporate a decrease in ice tops and glaciers (0.77 ± 0.22 mm every year) [32]. In the coastal regions of Asia, the current rate of sea level increase is reported in the range of 1 and 3 mm/year which is somewhat more than the worldwide mean. The rate of ocean level increase of 3.1 mm/year has been seen over the past decade contrasted with 1.7– 2.4 mm/year over the twentieth century which proposes that the rate of ocean level ascent has quickened with respect to the long-term normal.

9.4   CLIMATE CHANGE SCENARIOS

Scenarios are effective tools for scientific evaluation, to learn about complex behavior of systems and policy making [33]. Climate researchers use emission scenarios and socioeconomic narratives to give conceivable depictions of how the future can develop with respect to a range of factors includes land use, socioeconomic changes, technological change, energy change, greenhouse gas and air pollutant emissions.

The above-mentioned factors are utilized to run the climate model and as a reason for appraisal potential impacts on the climate. For better comparisons between various studies as well as for easier communication of model results it is beneficial to use a common set of scenarios across the scientific world. Before, a few sets of scenarios have performed such a job, including the IS92 situations [34] and more recently, the Special Report on Emission Scenarios (SRES) [35].

IPCC published SRES scenarios in 2000 with A1, B1, A2 and B2 pathways; the underlying economic and policy assumptions for these scenarios were established as early as 1997 [36]. Now the global research community of integrated assessment modeling groups designed the Integrated Assessment Modelling Consortium to develop new scenarios and they named as Representative Concentration Pathways (RCPs) with different emission scenarios namely RCP 2.6 (very low forcing level), RCP 4.5, RCP 6 (medium stabilization) and RCP 8.5 (high emission scenarios). All these scenarios are proposed based on radiative forcing target level for 2100. Here, representative means that each RCP only provides only one of several possible scenarios leading to the relevant pathway of radiative forcing. It is a measure of the earth systems' additional energy consumptions due to increased pollution from climate change.

RCP 8.5 was developed in Austria by the International Institute for Applied System Analysis and is described by increased emission of greenhouse gases leading to high concentration of greenhouse gas over the time, increased pathway of radiative force leading to $8.5 \, W/m^2$ (~1370 ppm CO_2 eq) by 2100 [37]; this scenario is mostly similar with A1F1 SRES scenario. RCP 6 was developed in Japan by the National Institute for Environmental Studies for stabilization after 2100 without overshooting path to $6 \, W/m^2$ (~850 ppm CO_2 eq) [38,39], which is reliable with the application of a range of greenhouse gas emission reduction technologies and strategies. This scenario is mostly similar with B2 SRES scenario. RCP 4.5 was developed in the US by Pacific Northwest National Laboratory for stabilization after 2100 without overshooting path to $4.5 \, W /m^2$ (~650 ppm CO_2 eq) [40–42] compatible with a relatively ambitious future with emission reductions; this scenario is mostly similar to B1 SRES scenario. PBL Netherlands Environmental Assessment Agency developed RCP 2.6, here peak radiative forcing before 2100 at ~3 W/m^2 (~490 ppm CO_2 eq) and then decrease (by 2100 the selected pathway decrease to 2.6 W/m^2) [43]. To achieve such forcing levels, it would require ambitious reductions in greenhouse gas emissions over time.

9.5   CLIMATE CHANGE AND AGRICULTURAL

Production of agriculture also depends on climate and weather. It is expected that potential changes in temperature, precipitation and concentration of CO_2 will have a significant impact on crop growth. With successful adaptation and adequate irrigation, the overall impact of climate change on global food production is considered to be low to moderate [44]. Due to doubling the effect of CO_2 fertilization, global agricultural production could be increased. The increasing climatic variability associated with global warming will result in significant annual/seasonal food production fluctuations [45]. Even today, all agricultural commodities are sensitive to this variability. Floods, heavy rainfall, droughts, heat waves and tropical cyclones are known to have an impact on agricultural production and farmer's livelihood negatively. The projected rise in these events will lead to increased food production instability and threaten farmers' livelihoods. In India, dry areas where farm crops encounter heat stress, even tiny changes in temperature will have disastrous impact on agricultural production. The same temperature rises in cooler places as near the Himalayas, however, could have a positive effect on agricultural production.

The climate-related impacts on the quantity and quality of water resources will also affect crop yield and livelihoods. As temperature rises, the areas in Middle East Asia are projected to become drier; South-East Asia in particular will suffer increased stress on water, expressed as a high-water withdrawal ratio to renewable water resources [46]. The Changes in moisture and temperature rates may result in changes in fertilizer and other mineral absorption rates that determine yield. In short, temperature rise along with rainfall reduction reduces agricultural productivity when both are beyond the limit suitable for crop production [47]. Crop productivity is projected to increase slightly for local mean temperature increase of 1°C–3°C based on the crop at mid- to high latitudes and then decrease beyond that in some regions. At lower latitudes, particularly in seasonal dry and tropical regions, crop performance is projected to decrease even for slight increase in local temperature of 1°C–2°C that could increase risk. It was expected that warmer weather would bring longer growing seasons in northern areas and that plant would benefit from carbon fertilization everywhere [48].

In 12 food-insecure regions around the world, changing climate could have a major impact on food security and agricultural production up to 2030, especially for South Asia and sub-Saharan Africa, due to the change in mean rainfall and temperature as well as increased variability connected with both.

The impact of climate change on water resources will also affect agriculture [49]. Wheat yields are forecasted to fall by 5%–10% with every 1°C increase and overall crop yield in South Asia by the mid–twenty-first century could decrease by up to 30%. India's agricultural productivity could decline by 40% by the 2080s. Rising temperatures will affect growing regions of wheat, putting hundreds of millions at the brink of chronic hunger [45]. Climate change has an impact on crop evapotranspiration. A 14.8% increase in total evapotranspiration was projected in arid regions of Rajasthan with temperature increase [50]. The study also shows that a marginal increase in evapotranspiration due to global warming would have a great influence on the fragile water resources of Rajasthan's arid zone ecosystem [51,52]. Ref. [53] reported that rise in sea level will increase the risk of permanent or seasonal encroachment of saline into rivers and groundwater that will influence water quality and its potential use of agricultural uses.

9.5.1 WHEAT

In wheat crop, decline of 600–650 grains/m^2 in crops with each 1°C increase in mean temperature above 17°C–17.7°C during the anthesis to terminal spikelet initiation has been observed [54]. The Indian Agricultural Research Institute shows the potential for future loss of 4–5 million tons of wheat production with every 1°C increase in temperature over the growing period [55]. An increase in winter temperature of 0.5°C would reduce the duration of wheat crops 7 days with the yield by 0.45

tons/ha and 10% decrease in wheat production in high yielding states of Northern India [56]. It also presumes that, irrigation at today's levels would be available in future. Losses for other crops remain uncertain, but they are expected to be relatively smaller, particularly for kharif crops.

9.5.2 Rice

The integrated impact of increase in temperature and concentration of CO_2 on crop yield can be negative [57]. They estimated that rice yields could be reduced by around 0.75 tons/ha in high yield areas and around 0.06 tons/ha in low yield coastal regions by an increase in mean air temperature of 2°C. Simulated irrigated rice yield under increased temperature and doubled CO_2 for Pantnagar district concluded that the effect on rice production will be positive in the absence of water and nutrient constraints.

Ref. [58] reported that rise of 4°C in temperature will lead to decline 41% of rice yield in Tamil Nadu and 1°C rise in minimum temperature will lead to decline 3% of rice yield in Punjab. In Bihar, due to climate change 31% of rice production could fall in 2080 [59].

9.5.3 Pulses

Projections of climate change have already shown that winter temperatures will increase by 3.2°C and summer temperatures by 2.2°C by 2050. This increase was analyzed in order to reduce the duration of the crop and consequently the economic growth of the pulse [60]. Increase in temperature enhanced the growth phases of the plant [59] and thus reduced the thermal-based vegetative growth phases [61].

In winter, the maximum day temperature goes beyond 40°C during the reproductive phase, which could result in a substantial failure of pod formation, flowering and induce seed hardening in pulse crops [62]; at the same time, pollen germination in other pulse crops is highly sensitive to high temperature [63].

9.5.4 Groundnut

Ref. [64] studied the impact of climate change on groundnut crop using the CROPGRO model for Anantapur, Junagadh and Mahboobnagar districts. They found that by 2050, a significant increase in temperature ($p < 0.05$) decreases the groundnut pod yield in all the districts. The net effect of rainfall, temperature and CO_2 changes, declines 4% yield in Anantapur and 11% increase in Mahboobnagar and Junagadh.

The pod yield of groundnut is likely to decline in different districts of Gujarat (Kesod, Rajkot, Bhuj and Bhavnagar) by 20%–34% over the projected period (2071–2100) of A2 scenario. Further delay in sowing (15 days after the beginning of the monsoon) reduces the yield of the pods by 2%–7% [65]. Groundnut biomass may decline by 20%–36% and the leaf area index may decline even more nearly 42%–47%. The duration of groundnut anthesis date may be reduced by 23%–36%, while the maturity date may be reduced by 7%–16% in different Saurashtra districts within the projected period compared to the baseline period. Decrease in groundnut phenological duration, development and growth under climate projection can cause major yield loss in groundnut because of above optimum soil and air temperature [65].

9.6   CROP SIMULATION MODEL (CSM)

The bio-system consists of a complex relationship between the soil, the plants and the atmosphere in it. Change in one element can produce both a desirable and an unwanted consequence [66]. Crop models and decision support systems can be really useful tools to help extension educators, planners, teachers, policy makers and scientists to assess alternative management practices. Most of the current crop models respond to soil characteristics, local weather, crop management practices and genetic difference. Crop Simulation models reduce the experimental time and cost [67]. Many impact assessment studies in developed countries used simulation models as a tool for assessing the impact on agriculture of global climate.

During the 1970s the utilization of crop simulation models and their application were already started. They are computerized representation of crop development, improvement and yield, simulated through mathematical conditions as elements of soil conditions, management practices and weather [68]. The quality of the CSM is their capacity to extrapolate the temporal pattern of crop yield and growth beyond single experimental plot. CSM can be utilized to new logical information of crop physiological procedures or to assess the effect of agronomic practices on farmers' incomes and environment. They are just an estimation of present reality and do not represent significant factors, for example, weeds, insects, phosphorus and tillage [69].

Efficiency prediction of any yield in a season has a significant economic importance for a country. For yield enhancements of crops, data about appropriate management practices is rapidly increasing. The generation of new information through agronomic research techniques is deficient and time consuming to address these issues. It is significant for a country, where efficiency of yields in any season may fluctuate enormously relying upon the overall climate states of that season [6]. Crop yield prediction by model and their sensitivity help in midcourse modification, so farmers can adopt strategic measures to maintain potential production of crops. The fundamental objective of a crop model is to assess environmental use, crop production and to study environmental effect of soil, weather and management [68]. They are utilized to assess the effects of climate variability [70] and climate change on crop production [71]. Mostly models are utilized to study the response of crops on atmosphere carbon dioxide and temperature.

Crop models are helpful tools to estimate the impacts of climate change on crop production and their application limits the long-term experimentation and cultivation cost. They are also used to assess the impact of climate change on agriculture and many developed countries have employed crop simulation models to study impact assessment. Various kinds of models (Table 9.1) are progressively open for professionals with various levels of expertise and exposure. The most normally utilized models are Decision Support System for Agro-Technology Transfer (DSSAT) model, the Environmental Policy Integrated Climate (EPIC) model and CROPWAT model.

TABLE 9.1

List of Crop Simulation Model

Crop Model	• Website to Download the Model
DSSAT	• http://dssat.net/
• APSIM	• http://www.apsim.info
• Aquacrop	• http://www.fao.org/aquacrop
• EPIC	• https://epicapex.tamu.edu/
• GLAM	• https://www.see.leeds.ac.uk/research/icas/research-themes/climate-change-andimpacts/climate-impacts/glam/
• WOFOST	• http://www.wageningenur.nl/en/Expertise-Services/Research-Institutes/alterra/-Facilities-Products/Software-and-models/WOFOST.htm
• ORYZAv3	• https://sites.google.com/a/irri.org/oryza2000/about-oryza-version-3
• SUCROS	• http://models.pps.wur.nl/node/966
• HERMES	• http://www.zalf.de/de/forschung_lehre/software_downloads/Seiten/default.aspx
• Cropsyst	• http://modeling.bsyse.wsu.edu/CS_Suite_4/CropSyst/index.html
• FarmSim	• http://models.pps.wur.nl/node/961
• SWAP	• http://www.swap.alterra.nl/
• Fasset	• http://www.fasset.dk/

9.7 GLOBAL CLIMATE MODELS

A popular way to acquire future climate projections is to use output from the global climate model. Climatologists use GCMs to understand better climate system behavior and obtain climate projections by implying various RCPs. The GCMs are generally used to depict relevant physical phenomena in the atmosphere, land and ocean. The present GCM generation gives a 100–350 km horizontal resolution. However, since the GCM spatial resolution is often too coarse to reproduce subgrid processes such as convective rainfall, they have to use parameterizations to reproduce the subgrid interactions better. In addition, many policy makers and shareholders consider the spatial resolution of the GCMs to be ineffective for their applications, which means that the GCMs are often reduced to particular points, such as weather stations, or to observer-based data sets of higher resolution than that of the GCMs of choice.

RCM data are acquired from GCMs, which venture future global climate conditions. The data produced by the regional climate model generally have high-resolution data, downscaled such as about 10 km resolution from GCMs output; this could enhance climate data uncertainty [72].

9.8 REGIONAL CLIMATE MODELS

They are considered to be more beneficial in evaluating the effects of climate change scenarios in agriculture than GCMs. Mostly GCMs have spatial resolution, which significantly reduces the accuracy of local climate change projections, especially in the area with sophisticated terrain. Therefore, RCMs will become essential for large countries with diversified agricultural practices and climate such as India. It results in better projections of future climate conditions because they have much finer horizontal resolutions than that of GCMs [73].

9.9 CLIMATE DOWNSCALING

Downscaling is generally a process where GCM outputs of a coarse scale were developed into local information. They are classified into dynamic downscaling and statistical downscaling methods [74]. Both are broadly used in impact-oriented research work and performed on different temporal and spatial scales. There are two steps to approach statistical downscaling [75]. The first step is to establish empirical relationships by interlinking local climate parameters to predictors on a large scale; the second option is to apply these relationships to output of GCM simulation in order to simulate future local climate characteristics. One constraint of this method is that it implies that when introduced to future situations the statistical relationships extracted from the historical period will hold; however these interactions are probable to change [76].

Dynamic downscaling is generally based on the use of RCMs, which produce finer resolution output based on atmospheric physics over a region using GCM fields as boundary condition [77]. The physical coherence between GCMs and RCMs is regulated by the agreement of their large-scale circulations [78]. RCMs provide a better description of topographic occurrences including orographic effects due to the higher spatial resolution output [79]. In addition, RCMs' finer dynamic processes produce more realistic patterns of mesoscale circulation [80]. It is not expected that RCMs will be studied for extreme spatial precipitation on a fine cell scale. Several studies [81] noticed that RCMs' skill enhancement relies not only in the resolution of RCM, but also in the season and region. Even though RCMs can provide feedback on their driving GCMs, several dynamic downscaling techniques are focused on one-way nesting strategies and have no feedback from the RCM to the driving GCM [82]. The output generated by RCM often needs a bias correction.

9.10 CROP SIMULATION MODELS AND REMOTE SENSING

Remote sensing is the process to collect the information with help of sensors from a distance without any physical conduct [83]. In recent years the satellite images are seen to be a valuable method for many agricultural applications like Soil mapping, water resources mapping, crop yield forecast, pest and disease surveillance, soil survey, drought and flood monitoring, crop area estimation, nutrient and weed estimation, land use land cover change analysis. The different sensors play a major role in collecting huge data from various spectral windows for the above-mentioned application. For example, crop yield forecast by processing biophysical quantities from remote sensing data.

The key objective of interfacing GIS and crop simulation model is employed to manage spatiotemporal analysis for region crop growth with crop model and spatial dimension to generate real-time output. Interfacing the GIS and the model helps to visualize the result spatially. Nowadays scientists use a standard method to integrate crop simulation model with remote sensing to estimate the crop yield and crop monitoring over a larger area at a spatial scale resolution [84] which will be fast and time consuming. In the 1970s Wiegand et al. suggested the remote sensing techniques to improve the reliability of crop simulation models. There are various approaches to interlink the crop simulation models with data obtained from remote sensing were initially reported by [85]; later the method was checked by [86,87] and identified five methods for linking GIS & remote sensing with crop simulation models.

1. The driving variable obtained from remote sensing is directly used in the model.
2. Forcing strategy – Minimum one state variable adjusted using data obtained from remote sensing and Leaf Area Index (LAI) is the most frequently updated state variable.
3. Model re-initialization – Initial condition of state variable is adjusted in order to reduce the error from derived parameters and its simulation.
4. Model re-calibration – Simple it is called as re-parameterization, i.e., to reduce the error from derived state parameters and its simulation by adjusting the model parameters.
5. Corrective approach – Statistical relationship is developed between error in the state variable derived from remote sensing measurements and final yield error. In this method, few known sites are chosen where the remote sensing is used to evaluate the state variable and final yield values are known. Therefore, state variable error in simulation is estimated along with final yield error; between these two errors an empirical relationship is developed.

9.11   INTEGRATED ASSESSMENT THROUGH CLIMATE AND CROP MODELS

The Global climate models simulate the climate system at continental scales at coarser resolution but for impact studies the climatic features at local scale would serve the purpose. Hence, data obtained from GCM outputs are forced as lateral boundary conditions to different Regional Climate Model thereby minimizing the effects of uncertainty. These downscaled data are used as an input in crop simulation models for climate change impact assessment. Figure 9.1 is an illustration for Integrating Regional Climate Model to Crop Simulation Model.

Ref. [70] used simulations from a regional climate model (PRECIS) with a crop simulation model (GLAM) to examine crop growth under simulated base period (1961–1990) and future period (2071–2100) climate to study crop adaptation to climate change through genotypical responses to mean and extreme temperatures over India. They observed that the effect of mean and extreme temperature varied regionally, depending on the genotypical properties being simulated. High temperature stress influenced the variability and mean of yield under changing climate. Changes in average temperature in certain locations had a similar effect on mean to high temperature stress and their impacts were more widespread. For a certain simulation, they mean yield decline was as large as 70% showing the significance of genotypical to changes in both mean and extreme temperature under the changing climate. The climate changing scenarios and CERES – Wheat simulation model evaluated the

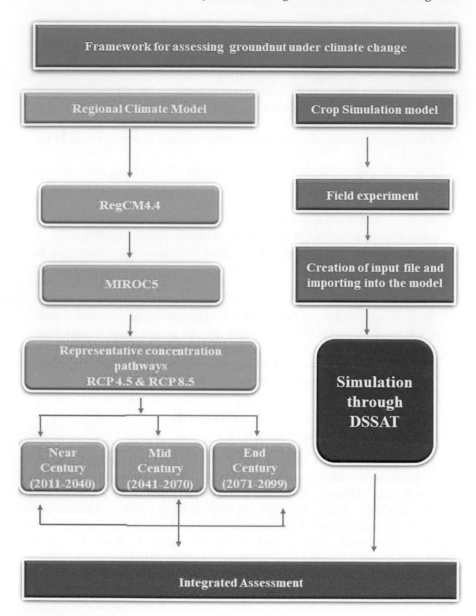

FIGURE 9.1   Illustration for assessment of climate change impact on groundnut.

effects of concurrent CO_2 and temperature changes. They noticed a rise in wheat yield between 29%–37% and 16%–28% under rainfed and irrigated situations, particularly under a modified climate in different genotypes. A temperature increase of 3°C or even more will eliminate the positive impacts of CO_2 [88].

Ref. [89] used PRECIS output to study the impact of climate change on crop yield and quality with CO2 fertilization in China. Results from this study suggested that at the end of the twenty-first century the average annual temperature may rise between 3°C and 4°C. Modeling results revealed that climate change without CO2 fertilization could reduce the rice, maize and wheat yields by up to 37% in the next 20–80 years. Interactions of CO2 with limiting factors, especially water and nitrogen, are capable of modulating observed growth response in crops. By using the CROPGRO-soybean model, Lal et al. [90] projected a 50% increase in Soybean yield for CO_2 doubling in Central India, but increase of 3°C in surface air temperature almost cancels the positive effects of CO_2 and reduced the duration of the crop by triggering early flowering and shortening the filling of grain period. In Central India, soybean is found to become more prone to temperature increases than they are at minimum temperature. A 10% in the daily amount of rainfall limits grain yield to about 32%. They concluded that even under the positive effects of high CO_2, severe water stress due to prolonged dry spells during the monsoon season could be a critical factor for (limiting) soybean productivity.

Ref. [90] reported that the high yielding Maize variety DK 647 in Chile had shown a 15%–28% reduction due to climate change for the A1F1 and B2B scenarios respectively. They attributed the reduction in yield to shortening the maize growth period by as much as 40 and 28 days. As an adaptation criterion under the B2B scenario, early sowing and reduction of fertilizer use were recommended. Climate change impacts assessment on groundnut yield projected using the INFOCROP model at Indore, Anand and Anantapur areas under A1B, B2 and A2 situations for periods of 2021–2050 and 2071–2100 which brought about huge increase in crop season mean air temperature when contrasted with baseline (1961–1990) temperatures. The increase in temperature extended from 1.8°C to 5.1°C in various scenarios and for various time spans. Increase in precipitation adds up to a tune of 10%–25% which was moreover anticipated in various scenarios and time periods. There was 4%–7% increase in rainfed groundnut yields were projected except A1B scenarios for the period 2071–2100 wherein the yield was projected to decrease by 5%. Over the areas the rainfed groundnut yields showed significant positive relationship with season precipitation yet poor or nonsignificant association was seen with crop season mean air temperature.

Ref. [91] assessed the impact of climate change and adaptation strategies to sustain rice production in Cauvery basin of Tamil Nadu. The results of the projected climate change over Cauvery basin of Tamil Nadu for A1B scenario using regional climate models showed an increasing trend for maximum, minimum temperatures and rainfall. The yields of ADT 43 rice simulated by the decision support system for agricultural technology transfer (DSSAT) with CO2 fertilization effect had shown a reduction of 135 Kg/ha/decade for providing regional climates for impact studies (PRECIS) output while there was an increase in yield by 24 Kg/ha/decade for regional climate model system 3 (RegCM3) output. Suggested adaptation strategies include a System of Rice Intensification (SRI), using temperature-tolerant cultivars and using green manures/biofertilizers for economizing water and increasing the rice productivity under warmer climate [92].

Ref. [57] studied climate change on groundnut yield for two varieties cv. Robust 33-1 and GG-2. The yields have been examined for Anand station of Gujarat Agro-climatic regions using PRECIS Regional climate model and PNUTGRO

DSSAT v4.5 crop simulation model for yield simulation. Almost 21% and 31% pod yield decrease were noted in Robut 33-1 and GG-2 when compared with their base yield during the projected period. In terms of yield, GG-2 was found to be more resistant to heat and moisture than Robut 33-1. In early sowing, i.e., 15 days prior to the onset of monsoon, the lowest yield reduction was found in cv GG-2 by supplying one presowing irrigation. During the stage of late sowing of both varieties, the moisture stress during pod formation and seed formation was negatively correlated with yield.

Ref. [92] analyzed Data related to agriculture crop sown area of five major crops were collected from Punjab statistical reports for the period of 1981–2015 and forecasted using linear exponential smoothing based on the historical rate. Results indicated that the cropping patterns in the Rechna Doab, Pakistan will vary with time and proportion of area of which sugarcane, wheat and rice may exhibit an increasing trend, while a decreasing trend with respect to the baseline scenario was found in maize and cotton. Crop sown area is then multiplied with CIR of individual crops derived from CROPWAT to simulate Net-CIR (m3) in three subscenarios S1, S2 and S3. It was observed that under the S2 subscenario (with changing agriculture land-use), total CIR may increase by 12.13 BCM (H3A2) and 12.04 BCM (H3B2) in the 2080s with respect to the baseline (1961–1990) which is greater as compared to S1 (with changing climate).

9.12 CONCLUSION

Changing climate and agriculture are interconnected processes that happen in global scale and their relationship is particularly important as the unbalance among world's population and global food production. Based on few projections, change in temperature, extreme weather and rainfall in several developing regions are intended to reduce crop yield. We need to know the change that could happen in the future to ascertain its impact and workout adaptation and mitigation strategies. The GCMs play an important role in predicting the climate of the future. Since GCMs will not provide adequate finer-scale information as needed for impact assessment by various sectors, scientists finally started using the RCMs to acquire fine-scale local outputs using the GCM output.

REFERENCES

1. Prasada Rao, G. S. L. H. V. (2008). Climate change and agriculture over India. In *International Symposium on Agrometeorology and Food Security (2008: CRIDA). AICRP on Agrometeorology*, Hyderabad.
2. Aggarwal, P. K. (2007). Climate change: implications for Indian agriculture. *Jalvigyan Sameeksha, 22*(1), 37–46.
3. Aggarwal, P. K., Singh, A. K., Samra, J. S., Singh, G., Gogoi, A. K., Rao, G. G. S. N., & Ramakrishna, Y. S. (2009). *Global climate change and Indian agriculture.* Indian Council of Agricultural Research, New Delhi.
4. Van Vuuren, D. P., Edmonds, J., Kainuma, M., Riahi, K., Thomson, A., Hibbard, K.,... & Masui, T. (2011). The representative concentration pathways: an overview. *Climatic Change, 109*(1–2), 5.

5. Thomson, A. M., Calvin, K. V., Smith, S. J., Kyle, G. P., Volke, A., Patel, P.,... & Edmonds, J. A. (2011). RCP 4.5: a pathway for stabilization of radiative forcing by 2100. *Climatic Change*, *109*(1–2), 77.
6. Jones, P. G., & Thornton, P. K. (2003). The potential impacts of climate change on maize production in Africa and Latin America in 2055. *Global Environmental Change*, *13*(1), 51–59.
7. Contreras, D. A., Hiriart, E., Bondeau, A., Kirman, A., Guiot, J., Bernard, L.,... & Van Der Leeuw, S. (2018). Regional paleoclimates and local consequences: integrating GIS analysis of diachronic settlement patterns and process-based agroecosystem modeling of potential agricultural productivity in Provence (France). *PloS One*, *13*(12), e0207622.
8. McCarthy, J. J., Canziani, O. F., Leary, N. A., Dokken, D. J., & White, K. S. (Eds.). (2001). *Climate change 2001: impacts, adaptation, and vulnerability: contribution of Working Group II to the third assessment report of the Intergovernmental Panel on Climate Change* (Vol. 2). Cambridge University Press, Cambridge.
9. Change, I. P. O. C. (2007). Summary for policymakers. In *Climate Change*, IPCC, Geneva, 1–18.
10. Ainsworth, E. A., & Ort, D. R. (2010). How do we improve crop production in a warming world?. *Plant Physiology*, *154*(2), 526–530.
11. Battisti, D. S., & Naylor, R. L. (2009). Historical warnings of future food insecurity with unprecedented seasonal heat. *Science*, *323*(5911), 240–244.
12. Meehl, G. A., Washington, W. M., Collins, W. D., Arblaster, J. M., Hu, A., Buja, L. E.,... & Teng, H. (2005). How much more global warming and sea level rise?. *Science*, *307*(5716), 1769–1772.
13. Ueda, H., Iwai, A., Kuwako, K., & Hori, M. E. (2006). Impact of anthropogenic forcing on the Asian summer monsoon as simulated by eight GCMs. *Geophysical Research Letters*, *33*(6), 1.
14. Guhathakurta, P., & Rajeevan, M. (2008). Trends in the rainfall pattern over India. *International Journal of Climatology: A Journal of the Royal Meteorological Society*, *28*(11), 1453–1469.
15. Seneviratne, S., Nicholls, N., Easterling, D., Goodess, C., Kanae, S., Kossin, J.,... & Reichstein, M. (2012). Changes in climate extremes and their impacts on the natural physical environment. *Managing the Risks of Extreme Events and Disasters to Advance Climate Change Adaptation*, Cambridge University Press. https://www.ipcc.ch/site/assets/uploads/2018/03/SREX-Chap3_FINAL-1.pdf
16. Kamiguchi, K., Kitoh, A., Uchiyama, T., Mizuta, R., & Noda, A. (2006). Changes in precipitation-based extremes indices due to global warming projected by a global 20-km-mesh atmospheric model. *Sola*, *2*, 64–67.
17. Radić, V., Bliss, A., Beedlow, A. C., Hock, R., Miles, E., & Cogley, J. G. (2014). Regional and global projections of twenty-first century glacier mass changes in response to climate scenarios from global climate models. *Climate Dynamics*, *42*, 37–58.
18. Philander, G., & Philander, S. G. (2008). *Encyclopedia of global warming and climate change: AE* (Vol. 1). Sage, London.
19. Murumkar, A. R., & Arya, D. S. (2014). Trend and periodicity analysis in rainfall pattern of Nira basin, Central India. *American Journal of Climate Change*, *3*, 11.
20. Raven, J., Caldeira, K., Elderfield, H., Hoegh-Guldberg, O., Liss, P., Riebesell, U.,... & Watson, A. (2005). *Ocean acidification due to increasing atmospheric carbon dioxide*. The Royal Society, London.
21. Houghton, J. T. (2001). *Climate change 2001: The scientific basis*. Cambridge University Press, Cambridge.
22. Keeling, C. D., & Whorf, T. P. (2005). *Atmospheric carbon dioxide record from Mauna Loa*. Carbon Dioxide Research Group, Scripps Institution of Oceanography, University of California La Jolla, San Diego, 92093–0444.

23. Change, I. P. O. C. (2007). Climate change 2007: the physical science basis. *Agenda*, 6(7), 333.
24. Krishnan, R., Sanjay, J., Gnanaseelan, C., Mujumdar, M., Kulkarni, A., & Chakraborty, S. (2020). Assessment of climate change over the Indian region: a report of the Ministry of Earth Sciences (MoES), Government of India.
25. De, U. S., Dube, R. K., & Rao, G. P. (2005). Extreme weather events over India in the last 100 years. *Journal of Indian Geophysical Union*, 9(3), 173–187.
26. Menon, A., Levermann, A., Schewe, J., Lehmann, J., & Frieler, K. (2013). Consistent increase in Indian monsoon rainfall and its variability across CMIP-5 models. *Earth System Dynamics*, 4, 287–300.
27. Mondal, A., Khare, D., & Kundu, S. (2015). Spatial and temporal analysis of rainfall and temperature trend of India. *Theoretical and Applied Climatology*, 122(1–2), 143–158.
28. Gujree, Ishfaq, et al. (2017). Evaluating the variability and trends in extreme climate events in the Kashmir Valley using PRECIS RCM simulations. *Modeling Earth Systems and Environment*, 3(4), 1647–1662.
29. Goswami, B. N., Venugopal, V., Sengupta, D., Madhusoodanan, M. S., & Xavier, P. K. (2006). Increasing trend of extreme rain events over India in a warming environment. *Science*, 314(5804), 1442–1445.
30. Krishnakumar, K. N., Rao, G. P., & Gopakumar, C. S. (2009). Rainfall trends in twentieth century over Kerala, India. *Atmospheric Environment*, 43(11), 1940–1944.
31. Padmakumari, B., Jaswal, A. K., & Goswami, B. N. (2013). Decrease in evaporation over the Indian monsoon region: implication on regional hydrological cycle. *Climatic Change*, 121(4), 787–799.
32. Bindoff, N. L., Willebrand, J., Artale, V., Cazenave, A., Gregory, J. M., Gulev, S.,... & Shum, C. K. (2007). Observations: oceanic climate change and sea level. *Managing the risks of extreme events and disasters to advance climate change adaptation*. Cambridge University Press, Cambridge. https://www.ipcc.ch/site/assets/uploads/2018/03/SREX-Chap3_FINAL-1.pdf.
33. Hanna, E., Mernild, S. H., Cappelen, J., & Steffen, K. (2012). Recent warming in Greenland in a long-term instrumental (1881–2012) climatic context: I. Evaluation of surface air temperature records. *Environmental Research Letters*, 7(4), 045404.
34. Mitchell, V. W. (1999). Consumer perceived risk: conceptualisations and models. *European Journal of marketing*, 33, 163–195.
35. Leggett, J., Pepper, W. J., Swart, R. J., Edmonds, J., Meira Filho, L. G., Mintzer, I., & Wang, M. X. (1992). Emissions scenarios for the IPCC: an update. *Climate Change*, 1040, 75–95.
36. Nakicenovic, N., Alcamo, J., Grubler, A., Riahi, K., Roehrl, R. A., Rogner, H. H., & Victor, N. (2000). *Special report on emissions scenarios (SRES), a special report of Working Group III of the intergovernmental panel on climate change*. Cambridge University Press, Cambridge.
37. Moss, R. H., Edmonds, J. A., Hibbard, K. A., Manning, M. R., Rose, S. K., Van Vuuren, D. P.,... & Meehl, G. A. (2010). The next generation of scenarios for climate change research and assessment. *Nature*, 463(7282), 747–756.
38. Riahi, K., Grübler, A., & Nakicenovic, N. (2007). Scenarios of long-term socio-economic and environmental development under climate stabilization. *Technological Forecasting and Social Change*, 74(7), 887–935.
39. Fujino, J., Nair, R., Kainuma, M., Masui, T., & Matsuoka, Y. (2006). Multi-gas mitigation analysis on stabilization scenarios using AIM global model. *The Energy Journal*, 27(Special Issue# 3), 343–353.
40. Yasuaki Hijioka, Y. M., & Nishimoto, H. (2008). Global GHG emission scenarios under GIC; concentration stabilization targets. *Journal of Global Environment Engineering*, 13, 97–108.

41. Clarke, L., Edmonds, J., Jacoby, H., Pitcher, H., Reilly, J., & Richels, R. (2007). Scenarios of greenhouse gas emissions and atmospheric concentrations. *Managing the risks of extreme events and disasters to advance climate change adaptation.* Cambridge University Press, Cambridge. https://www.ipcc.ch/site/assets/uploads/2018/03/SREX-Chap3_FINAL-1.pdf.

42. Smith, S. J., & Wigley, T. M. L. (2006). Multi-gas forcing stabilization with Minicam. *The Energy Journal, 27*(Special Issue# 3), 373–391.

43. Wise, M., Calvin, K., Thomson, A., Clarke, L., Bond-Lamberty, B., Sands, R.,... & Edmonds, J. (2009). Implications of limiting CO_2 concentrations for land use and energy. *Science, 324*(5931), 1183–1186.

44. Van Vuuren, D. P., Den Elzen, M. G., Lucas, P. L., Eickhout, B., Strengers, B. J., Van Ruijven, B., & Van Houdt, R. (2007). Stabilizing greenhouse gas concentrations at low levels: an assessment of reduction strategies and costs. *Climatic Change, 81*(2), 119–159.

45. Kumar, R., & Gautam, H. R. (2014). Climate change and its impact on agricultural productivity in India. *Journal of Climatology & Weather Forecasting, 2,* 109.

46. Birthal, P. S., Khan, T., Negi, D. S., & Agarwal, S. (2014). Impact of climate change on yields of major food crops in India: implications for food security. *Agricultural Economics Research Review, 27*(2), 145–155.

47. Bates, B., Kundzewicz, Z., & Wu, S. (2008). *Climate change and water.* Intergovernmental Panel on Climate Change Secretariat, Geneva.

48. Tirado, R., & Cotter, J. (2010). *Ecological farming: drought-resistant agriculture.* Greenpeace Research Laboratories, Exeter.

49. Van Vuuren, D. P., Ochola, W. O., Riha, S., Giampietro, M., Ginzo, H., Henrichs, T.,... & Kuppannan, P. (2009). Outlook on agricultural change and its drivers. In McIntyre, B. D., Herren, H. R., Wakhungu, J., & Watson, R.T. (Eds.), *Agriculture at a crossroads* (pp. 255–305). Island Press, Washington, DC.

50. Solomon, S., Manning, M., Marquis, M., & Qin, D. (2007). *Climate change 2007 – the physical science basis: Working Group I contribution to the fourth assessment report of the IPCC* (Vol. 4). Cambridge University Press, Cambridge.

51. Goyal, R. K. (2004). Sensitivity of evapotranspiration to global warming: a case study of arid zone of Rajasthan (India). *Agricultural Water Management, 69*(1), 1–11.

52. Gopalakrishnan, T., Hasan, M. K., Haque, A. T. M., Jayasinghe, S. L., & Kumar, L. (2019). Sustainability of coastal agriculture under climate change. *Sustainability, 11*(-24), 7200.

53. Saini, A. D., & Nanda, R. (1986). Relationship between incident radiation, leaf area and dry-matter yield in wheat. *Indian Journal of Agricultural Sciences (India), 56,* 638–645.

54. Kalra, N., Chander, S., Pathak, H., Aggarwal, P. K., Gupta, N. C., Sehgal, M., & Chakraborty, D. (2007). Impacts of climate change on agriculture. *Outlook on Agriculture, 36*(2), 109–118.

55. Achanta, A. N. (1993). *An assessment of the potential impact of global warming on Indian rice production. The climate change agenda: an Indian perspective* (pp. 45–60). Tata Energy Research Institute, New Delhi.

56. Swaminathan, M. S., & Kesavan, P. C. (2012). Agricultural research in an era of climate change. *Agricultural Research, 1*(1), 3–11.

57. Geethalakshmi, V., Lakshmanan, A., Rajalakshmi, D., Jagannathan, R., Sridhar, G., Ramaraj, A. P.,... & Anbhazhagan, R. (2011). Climate change impact assessment and adaptation strategies to sustain rice production in Cauvery basin of Tamil Nadu. *Current Science, 101,* 342–347.

58. Haris, A. A., Biswas, S., & Chhabra, V. (2010). Climate change impacts on productivity of rice (Oryza sativa) in Bihar. *Indian Journal of Agronomy, 55*(4), 295–298.

59. Ali, M., Gupta, S., & Basu, P. S. (2009). Higher level of warming in north India will affect crop productivity. *Hindu Survey of Indian Agriculture*, 44–49.

60. Craufurd, P. Q., & Wheeler, T. R. (2009). Climate change and the flowering time of annual crops. *Journal of Experimental Botany*, *60*(9), 2529–2539.

61. Nguyen, C. T., Singh, V., van Oosterom, E. J., Chapman, S. C., Jordan, D. R., & Hammer, G. L. (2013). Genetic variability in high temperature effects on seed-set in sorghum. *Functional Plant Biology*, *40*(5), 439–448.

62. Basu, P. S., Singh, U. M. M. E. D., Kumar, A. N. I. L., Praharaj, C. S., & Shivran, R. K. (2016). Climate change and its mitigation strategies in pulses production. *Indian Journal of Agronomy*, *61*, S71–S82.

63. Kakani, V. G., Reddy, K. R., Koti, S., Wallace, T. P., Prasad, P. V. V., Reddy, V. R., & Zhao, D. (2005). Differences in in vitro pollen germination and pollen tube growth of cotton cultivars in response to high temperature. *Annals of Botany*, *96*(1), 59–67.

64. Singh, P., Singh, N. P., Boote, K. J., Nedumaran, S., Srinivas, K., & Bantilan, M. C. S. (2014). Management options to increase groundnut productivity under climate change at selected sites in India. *Journal of Agrometeorology*, *16*(1), 52–59.

65. Patel, H. R., Lunagaria, M. M., Karande, B. I., Pandey Vyas, Y. S., & Shah, A. V. (2013). Impact of projected climate change on groundnut in Gujarat. *Journal of Agro Meteorology*, *15*(special issue-I), 41–44.

66. Murthy, V. R. K. (2002). *Basic principles of agricultural meteorology* (p. 4). BS Publications, Hyderabad.

67. Loomis, R. S., Rabbinge, R., & Ng, E. (1979). Explanatory models in crop physiology. *Annual Review of Plant Physiology*, *30*(1), 339–367.

68. Hoogenboom, G., White, J. W., & Messina, C. D. (2004). From genome to crop: integration through simulation modeling. *Field Crops Research*, *90*(1), 145–163.

69. Jones, J. W., Hoogenboom, G., Porter, C. H., Boote, K. J., Batchelor, W. D., Hunt, L. A.,.… & Ritchie, J. T. (2003). The DSSAT cropping system model. *European Journal of Agronomy*, *18*(3–4), 235–265.

70. Challinor, A. J., Wheeler, T. R., Craufurd, P. Q., Ferro, C. A. T., & Stephenson, D. B. (2007). Adaptation of crops to climate change through genotypic responses to mean and extreme temperatures. *Agriculture, Ecosystems & Environment*, *119*(1–2), 190–204.

71. Carbone, G. J., Kiechle, W., Locke, C., Mearns, L. O., McDaniel, L., & Downton, M. W. (2003). Response of soybean and sorghum to varying spatial scales of climate change scenarios in the southeastern United States. In Mearns, L. O. (Ed.), *Issues in the Impacts of Climate Variability and Change on Agriculture* (pp. 73–98). Springer, Dordrecht.

72. Mearns, L. O., Hulme, M., Carter, T. R., Leemans, R., Lal, M., Whetton, P.,.… & Wilby, R. (2001). Climate scenario development. In Houghton, T., Ding, Y., Griggs, D., Noguer, M., van der Linden, P. J., Dai, X., Maskell, K., & Johnson, C. A. (Eds.), *Climate change 2001: the science of climate change* (pp. 739–768). Cambridge University Press, Cambridge.

73. Randall, D. A., Wood, R. A., Bony, S., Colman, R., Fichefet, T., Fyfe, J.,.… & Stouffer, R. J. (2007). Climate models and their evaluation. In *Climate change 2007: the physical science basis. Contribution of Working Group I to the Fourth Assessment Report of the IPCC (FAR)* (pp. 589–662). Cambridge University Press, Cambridge.

74. Tsvetsinskaya, E. A., Mearns, L. O., Mavromatis, T., Gao, W., McDaniel, L., & Downton, M. W. (2003). The effect of spatial scale of climatic change scenarios on simulated maize, winter wheat, and rice production in the southeastern United States. In Mearns, L. O. (Ed.), *Issues in the impacts of climate variability and change on agriculture* (pp. 37–71). Springer, Dordrecht.

75. Branzuela, N. E., Faderogao, F. J. F., & Pulhin, J. M. (2015). Downscaled projected climate scenario of Talomo-Lipadas watershed, Davao City, Philippines. *Journal of Earth Science & Climatic Change*, *6*(3), 1.

76. Milly, P. C. D., Betancourt, J., Falkenmark, M., Hirsch, R. M., Kundzewicz, Z. W., Lettenmaier, D. P., & Stouffer, R. J. (2008). Stationarity is dead: whither water management. *Earth*, *4*, 20.

77. Giorgi, F., & Mearns, L. O. (1999). Introduction to special section: regional climate modeling revisited. *Journal of Geophysical Research: Atmospheres*, *104*(D6), 6335–6352.

78. von Storch, H., Langenberg, H., & Feser, F. (2000). A spectral nudging technique for dynamical downscaling purposes. *Monthly Weather Review*, *128*(10), 3664–3673.

79. Christensen, J. H., & Christensen, O. B. (2007). A summary of the PRUDENCE model projections of changes in European climate by the end of this century. *Climatic Change*, *81*(1), 7–30.

80. Buonomo, E., Jones, R., Huntingford, C., & Hannaford, J. (2007). On the robustness of changes in extreme precipitation over Europe from two high resolution climate change simulations. *Quarterly Journal of the Royal Meteorological Society: A Journal of the Atmospheric Sciences, Applied Meteorology and Physical Oceanography*, *133*(622), 65–81.

81. Rauscher, S. A., Coppola, E., Piani, C., & Giorgi, F. (2010). Resolution effects on regional climate model simulations of seasonal precipitation over Europe. *Climate Dynamics*, *35*(4), 685–711.

82. Maraun, D., Wetterhall, F., Ireson, A. M., Chandler, R. E., Kendon, E. J., Widmann, M.,... & Venema, V. K. C. (2010). Precipitation downscaling under climate change: Recent developments to bridge the gap between dynamical models and the end user. *Reviews of Geophysics*, *48*(3), 1-34.

83. Fischer, W. A., Hemphill, W. R., & Kover, A. (1976). Progress in remote sensing (1972–1976). *Photogrammetria*, *32*(2), 33–72.

84. Liang, S., & Qin, J. (2008). Data assimilation methods for land surface variable estimation. In Liang, S. (Ed.), *Advances in land remote sensing* (pp. 313–339). Springer, Dordrecht.

85. Maas, S. J. (1988). Use of remotely-sensed information in agricultural crop growth models. *Ecological Modelling*, *41*(3–4), 247–268.

86. Delécolle, R., Maas, S. J., Guerif, M., & Baret, F. (1992). Remote sensing and crop production models: present trends. *ISPRS Journal of Photogrammetry and Remote Sensing*, *47*(2–3), 145–161.

87. Moulin, S., Bondeau, A., & Delecolle, R. (1998). Combining agricultural crop models and satellite observations: from field to regional scales. *International Journal of Remote Sensing*, *19*(6), 1021–1036.

88. Attri, S. D., & Rathore, L. S. (2003). Simulation of impact of projected climate change on wheat in India. *International Journal of Climatology: A Journal of the Royal Meteorological Society*, *23*(6), 693–705.

89. Erda, L., Wei, X., Hui, J., Yinlong, X., Yue, L., Liping, B., & Liyong, X. (2005). Climate change impacts on crop yield and quality with CO_2 fertilization in China. *Philosophical Transactions of the Royal Society B: Biological Sciences*, *360*(1463), 2149–2154.

90. Lal, M., Singh, K. K., Srinivasan, G., Rathore, L. S., Naidu, D., & Tripathi, C. N. (1999). Growth and yield responses of soybean in Madhya Pradesh, India to climate variability and change. *Agricultural and Forest Meteorology*, *93*(1), 53–70.

91. Meza, F. J., Silva, D., & Vigil, H. (2008). Climate change impacts on irrigated maize in Mediterranean climates: evaluation of double cropping as an emerging adaptation alternative. *Agricultural Systems*, *98*(1), 21–30.

92. Yadav, S. B., Patel, H. R., Patel, G. G., Lunagaria, M. M., Karande, B. I., Shah, A. V., & Pandey, V. (2012). Calibration and validation of PNUTGRO (DSSAT v4. 5) model for yield and yield attributing characters of kharif groundnut cultivars in middle Gujarat region. *Journal of Agrometeorology*, *14*(Special Issue), 24–29.

10 Snow and Glacier Resources in the Western Himalayas
A Review

Asif Marazi and Shakil Ahmad Romshoo
University of Kashmir

CONTENTS

10.1 Introduction 195
10.2 Snow and Glacier Resources 197
10.3 Melt Water from Snow and Glaciers.. 200
      10.3.1 Changing Streamflow Pattern Under Climate Change................ 201
10.4 Recommendations 203
10.5 Conclusions.. 204
Acknowledgment 204
References.. 205

10.1 INTRODUCTION

Hydrological cycle is the most important recycling system that exists on the Earth, because existence of life on this planet depends on water. The hydrological cycle is a continuous process by which water is transported from oceans to atmosphere or from land to atmosphere and back either directly to oceans or via land to oceans through various physical processes e.g. precipitation, evapotranspiration, subsurface flows and river runoff. The water gets temporarily stored in different reservoirs in between while transporting from one component of the earth system to other. However, the residence time of water molecule per reservoir differs widely, e.g., on an average, the water molecule stays 2–5 weeks in a natural river, and the residence time of water molecule in atmosphere is ~10 days (Oki and Kanae, 2006). Fischer et al. (2013) reported that the water molecule in the Antarctic ice sheet resides more than 120, 00 years, though in some parts of ice sheet, the residence time of the water molecule could be millions of years. In contrast to ice sheets, glaciers hold water molecules for shorter time, i.e., 20–100 years; however, velocity and size of the glacier determine the residence time (Marshak, 2015). Despite the fact that snow and glaciers outside the poles store much lesser water than ice sheets, these resources are extremely important because of their proximity to the human society and contribution to river systems.

DOI: 10.1201/9781003147107-13

In mid-latitudes, snow and glacier reservoirs reside at higher elevations where high level of precipitation occurs mostly as snow due to lower temperatures. Mountains often store precipitated water in the form of snow fields and glaciers, and these resources contribute significantly to rivers and provide water to the downstream areas (Viviroli and Weingartner, 2004; Immerzeel et al., 2010; Lutz et al., 2014). The snow and glaciers contribute exactly the time of the year when the water is most needed (e.g., for irrigation purposes and hydropower generation), and when the contribution from other runoff source components is feeble. Thus, the perennial characteristic of the rivers is maintained by the delayed release of water, which is actually stored as snow and glacier fields in the mountains.

The snow and glacier melt plays a crucial role in regulating the hydrology of basins in Hindu Kush-Karakoram-Himalayan (HKH) region. The HKH region hosts the largest snow and glacier resources outside the poles, with ~50% of all the glacial area falling outside the polar region (Mayewski and Jeschke, 1979). According to Bajracharya and Shrestha (2011), the total number of glaciers in the HKH region is 54,252, which covers an area of around $60,000\,km^2$. This vast snow and glacier reserve is the origin of numerous rivers including the 10 largest rivers in the Asia which supply water to more than 1.3 billion people downstream (~18% of the global population), hence referred to as the water tower of Asia or the third pole (Armstrong, 2010; Shrestha et al., 2015). The three major river systems in the South Asia the Indus, Ganges and Brahmaputra are formed due to the convergence of several rivers and streams that flow downwards from the large snow and glacier fields of the Himalayas and the melt water being the primary driver of the annual river flows, though the contribution varies seasonally (Schaner et al., 2012). These rivers nourish the most highly irrigated region in the world, e.g., Indus basin has ~$144,900\,km^2$ of irrigated land with the largest irrigation network in the world, Ganges basin has $156,300\,km^2$, and Brahmaputra has $6,000\,km^2$ of irrigated land (Immerzeel et al., 2010). The South Asian region, which is the combined catchment of Indus, Ganges and Brahmaputra, is heavily dependent on agricultural economy (Devendra, 1997). Moreover, export of rice produced in the Indus basin using the snow and glacier melt waters to other parts of the world, e.g., Africa, regulates the economic and political relationship between South Asia and the world. The Indus basin with an area of ~1.1 million km^2 is nourished by about 18,495 glaciers which cover a total area of $21,193\,km^2$ (~35% of the total glacier area in the HKH) with an estimated ice reserve of $2,696\,km^3$ (Bajracharya and Shrestha, 2011). According to Immerzeel et al. (2009), snow and glacier melt play a dominant role in regulating the hydrology of the Upper Indus Basin (UIB) by contributing ~72% to annual streamflow. Moreover, ~50% of water that flows in the Indus river irrigation scheme is contributed by snow and glaciers (Winiger et al., 2005). This specifies the importance of water provided by the snow and glaciers.

Globally, there has been an increased concern about the depletion of snow and glacier resources due to changing climate (Kulkarni et al., 2011; Bolch et al., 2012; Immerzeel et al., 2010). Signals of climate change world over are clear in terms of increasing temperature and changes in the precipitation pattern (decrease in snowfall) (Maisch, 2000; Singh and Bengtsson, 2005; Solomon 2007). In the HKH, temperature changes have been reported in numerous studies. According to Fowler and Archer

(2006), winter mean and maximum temperature records in the UIB have increased significantly while decreasing trends in mean and minimum summer temperatures have been observed. Furthermore, Khattak et al. (2011) reported a strong increasing trend in winter maximum temperatures at higher elevations with non-significant decreasing trend in precipitation. Pertinently, strong increasing trend in temperature has been reported for the Tibetan Plateau (Liu and Chen, 2000), the central Himalaya (Shrestha et al., 1999), and the Jhelum Basin (one of the five sub-basins of Indus) (Romshoo et al., 2015). Furthermore, Pepin and Lundquist (2008) and Rangwala and Miller (2012) have reported that mountainous regions have warmed more strongly compared to other land surfaces, e.g., the Himalayan region (Bhutiyani et al., 2007). Moreover, recent studies suggest that the solid precipitation has reduced over the Himalayan region, which has resulted in less snow accumulation and negative glacier mass balance (Singh and Bengtsson, 2005; Romshoo et al., 2015). Similarly, in the Lidder basin, the most glaciated watershed in the Jhelum basin, strong warming along with decrease in snow precipitation and glacier cover, and subsequent decrease in streamflows have been reported by several studies (Romshoo et al., 2015; Marazi and Romshoo, 2018; Murtaza et al., 2017). Lidder basin is one of 24 watersheds/sub-basins of the Jhelum Basin, hosting the highest number of glaciers in the Jhelum basin, and is drained by the Lidder River which is one of the largest tributaries of the Jhelum River. However, Lidder basin, a famous tourist destination, faces serious implications of climate change as the economy of people heavily depends on tourism and agricultural activities which in turn depends on its water resources (Dar et al., 2014). Barnett et al. (2005) and Bajracharya et al. (2008) have described the serious consequences of increasing temperatures on the hydrology of the regions where streamflows are dominated by snow and glacier melt. Moreover, enhanced shrinking of snow and glacier resources in response to rising temperatures might affect the water availability and could reduce the volume of water significantly in the rivers in long terms (Thayyen and Gergan, 2010; Bolch et al., 2012; Immerzeel et al., 2012).

10.2 SNOW AND GLACIER RESOURCES

Snowfields and glaciers serve as the vital sources of fresh water (Barnett et al., 2005). Essentially, the hydrological basins with 20% or more of their area covered by glaciers have approximately 50% greater summer runoff and significantly lower annual runoff variability compared to the non-glaciated basins (Fountain and Tangborn, 1985). Pertinently, glacierized basins are hydrologically more stable than the glacier-free basins. In the Himalayan region, huge resources of snow and glaciers (largest outside the polar region) play a primary role in controlling the water flow in the major river systems like the Indus, the Ganges and the Brahmaputra (Schaner et al., 2012). According to Nuimura et al. (2015), glaciers in the Himalayan region cover an area of ~19,460 km^2. Recent estimates of the total glacier area in the Himalayan region are given in Table 10.1. The difference in the estimates could be due to the use of different criteria and mapping approaches.

In general, glaciers worldwide are in continuous shrinking phase (Bajracharya et al., 2008; Ding et al., 2006; Oerlemans, 2005). According to Walther et al. (2002), average global snow and glacier extent has decreased by 10% since the late 1960s. Pertinently,

TABLE 10.1

Recent Estimates of Glacierized Area in the Himalayan Region

Study	Glacier Area (km²)
Cogley (2011)	21,973
Bolch et al. (2012)	22,829
Kääb et al. (2012)	29,000
Nuimura et al. (2015)	19,460

most of the studies are attributing it to the global warming. Global warming is now a top scientific and political agenda, and snow and glaciers are widely considered as the key indicators of the climate change (Houghton et al., 2001; Robinson et al., 1993). The most immediate and important effect of the climate change on the alpine environments is its impact on snow and glacier cover. Furthermore, studies have suggested that the air temperature will gradually increase in the coming decades, and subsequently, the number of snowfall days will decrease (Moen and Fredman, 2007; Romshoo et al., 2015; Schneider, 1990; Ye et al., 2008). Dimri and Dash (2012) reported that the snowfall in the western Himalayas has substantially decreased in response to the winter warming, and the trend is expected to continue in the future as well. Pertinently, Romshoo et al. (2015) observed a strong decrease in the winter time snowfall in the Lidder basin. Subsequently, decrease in snow accumulation during the winter season and the concurrent increase in average temperature shall result in earlier snowmelt peak and shortage of water in later months of the hydrological year (Marazi and Romshoo, 2018; Rashid et al., 2017). Moreover, the less snow accumulation over the glacier surface shall have a negative impact on the health of glaciers. The contemporaneous accelerated glacier mass loss and the increase in temperature in recent past decades (Dyurgerov and Meier, 2005; Ye et al., 2006) incited some studies to suggest the near-complete or complete deglaciation of some glacierized regions of the world in coming decades (Höelzle et al., 2007; Lüthi et al., 2008). Notably, Solomon et al. (2007) reported the fastest glacier receding rate in the Himalayan region and suggested that the deglaciation of the Himalayan region would occur by 2035. Although the statement was highly controversial, this report impelled the scientific community to explore the scarcely studied Himalayan glaciers.

Numerous studies have been carried in the past decade to map the glacier area changes in the Himalayan region. Pertinently, recent studies revealed that the Himalayan glaciers in general are receding and losing mass (Gardelle et al., 2013; Kääb et al., 2012). Contrarily, several studies have reported that the glaciers in some parts of the Himalayan region are in stable state or showing advancements (Hewitt, 2011; Iturrizaga, 2011; Sarikaya et al., 2012); nonetheless, others stated that glacier recession rates in the Himalayan region have slowed down significantly over the past decade (Bolch et al., 2010; Kumar et al., 2008). Contrasting statements from different studies portrayed a contradictory picture of the current status of the Himalayan glaciers. Moreover, it concurrently complicated the picture of understanding the response of Himalayan glaciers to climate change and thus prompted the need of assiduous glacier monitoring efforts in the region.

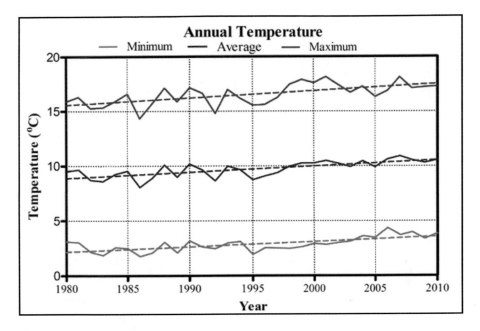

FIGURE 10.1 Temperature changes from 1980 to 2010 in Lidder Basin (Marazi and Romshoo, 2018).

Recent advancements in the remote sensing and GIS techniques expedite and facilitate the efforts of continuous monitoring of snow and glacier resources world-wide. Several studies have used satellite imageries to map the glacier area changes in the Himalayan region (Bhambri et al., 2011; Frey et al., 2014; Kulkarni et al., 2007). Predominantly, imageries from the Landsat series have been widely used to monitor the glaciers due to their free-of-cost availability, better spatial resolution and better coverage (Bajracharya et al., 2015; Bhambri et al., 2011; Racoviteanu et al., 2008). Furthermore, the snow cover data available till date is mostly based on MODIS satellite products and has been used in several studies to monitor the snow cover changes (Hall et al., 2002; Parajka and Blöschl, 2008; Rittger et al., 2013; Salomonson and Appel, 2004; Zhu et al., 2017). However, the data is only available for less than 20 years, and thus, a long-term trend cannot be recognized. Nevertheless, several studies have monitored the snow cover extend over the Himalayan region and reported decreasing/increasing snow cover extent in some basins (Gurung et al., 2011; Immerzeel et al., 2009; Mir et al., 2015; SAC, 2016; Tahir et al., 2016).

The snow and glacier resources in the Kashmir valley particularly in the Lidder basin have been adversely affected by climate change (Marazi and Romshoo, 2018; Murtaza and Romshoo, 2017; Rashid et al., 2017; Romshoo et al., 2015). The temperature in the Kashmir valley has significantly increased during the past 30 years (Figure 10.1). Correspondingly, glaciers in the valley have receded significantly during the past few decades. It is evident from Figure 10.2 that the glaciers in the Lidder basin, most glacier populated sub-basin of Kashmir valley, are small in size and have receded significantly during the past few decades.

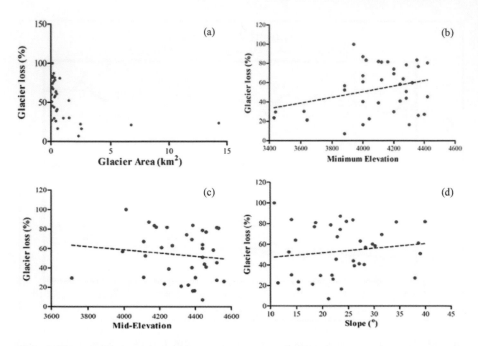

FIGURE 10.2 Relationship between glacier area and topographic parameters in the Jhelum Basin (Marazi and Romshoo, 2018).

10.3 MELT WATER FROM SNOW AND GLACIERS

Water budgeting in the Himalayan region has become a key interest for the scientific and political community, as it is extremely crucial for the sustainable development of the people living in the region. The major river systems in Asia: Indus, Ganges and Brahmaputra originate in the Himalayan mountains and supply water to millions of people living in the South Asian countries for domestic use, drinking, irrigation and power development (Archer et al., 2010; Karim and Veizer, 2002). Pertinently, the runoff in these major river systems is regulated by the combined contribution of snow and glacier melt, rainfall and groundwater. Singh et al. (1997) reported that snow and glaciers contribute significantly to the Himalayan rivers; however, the contribution is dominant at higher elevations (Wulf et al., 2016), e.g., 72% in the UIB (Immerzeel et al., 2009). The flow in the Himalayan rivers varies seasonally and displays a specific seasonal and local pattern depending upon the dominant runoff source component during the period or season. Estimation of the different runoff components in stream-flows is essential for the downstream management of the water for the irrigation and hydropower generation but nonetheless a challenging task because of the remoteness of the Himalayan drainage basins, lack of instrumentation and scarcity of data. However, numerous studies have estimated the streamflow discharge in high Himalayan basins using the modeling approach, which in turn depends upon the accurate estimation of snow and glacier melt in the basins (Hock, 2003; Meraj et al., 2018a, b).

Generally, two types of modeling approaches are used to estimate the snow and glacier melt; the energy balance approach and the temperature index or degree-day approach (Hock, 2003). The energy balance approach requires a large parameter set like the incoming and outgoing shortwave radiation, down-welling long-wave radiation, air and surface temperature, relative humidity, wind speed and wind direction. For estimating the glacier melt, these parameters shall be measured near the glacier surfaces; nevertheless, continuous measurement of these parameters is near impossible (Azam et al., 2014; Fujita and Ageta, 2000; Kayastha et al., 1999; Lejeune et al., 2013; Mölg et al., 2012; Zhang et al., 2013). Conversely, the temperature index or the degree-day approach is comparatively simple and requires less data. Despite the fact that the temperature index models are computationally simple, these models perform reasonably well (Hock, 2003). Furthermore, the air temperature serves as the primary input in these models and is considered an index for snow and ice melt and could be used as an alternative for the components of energy balance models such as long-wave incoming and absorbed shortwave radiation (Ohmura, 2001). Pertinently, several studies have successfully used degree-day or temperature index approach to estimate the melt-runoff in the different Himalayan basins (Ali et al., 2017; Arora et al., 2008; Jain et al., 2010; Pradhananga et al., 2014; Silwal et al., 2016; Singh and Bengtsson, 2004, 2005; Singh et al., 2008). Furthermore, with the development of the distributed temperature index models, it has become possible to regionalize various input parameters from a single location (as in case of the lumped models) to other parts of the catchment (Konz et al., 2007). Furthermore, stable water isotopes in combination with the measurements of electric conductivity of the water have been used by various studies to indicate and estimate the source of water in the Himalayan basins (Jeelani et al., 2017; Maurya et al., 2011). However, these methods have proved costly because the water samples collected from the field need high-class logistic support for their maintenance and analysis.

10.3.1 Changing Streamflow Pattern Under Climate Change

The vast cryosphere in the HKH region provides fresh water to more than 1.3 billion people living in the downstream areas for domestic, agricultural and energy development (Gertler et al., 2016). However, unprecedented changes in the water resources due to changing climatic conditions and the concurrent increasing demand of the water for growing population and burgeoning economy have led to serious impacts on these frozen water resources (Barnett et al., 2005; Vörösmarty et al., 2000). According to Anderson and Mackintosh (2006) and Oerlemans (1994), the sensitivity of snow and glacier resources to changing temperature and precipitation makes them highly vulnerable to climate change. Moreover, several studies have reported that the snow and glacier resources in the HKH region are in continuous state of recession since mid-nineteenth century in response to the long-term temperature changes over the region (Bahuguna et al., 2007; Bhambri et al., 2011; Bhutiyani et al., 2007; Marazi and Romshoo, 2018; Racoviteanu et al., 2008; Salerno et al., 2016). The accelerated loss of snow and glacier resources in the Himalayan region and the anticipated impact on the streamflows have attracted much attention over the past few decades and has become a cause of concern for the scientific and political community (Bolch et al., 2012; Jacob et al., 2012; Marazi and Romshoo, 2018; Murtaza and Romshoo, 2017; Rashid et al., 2017) Regionally, streamflows in the

UIB (where the Lidder basin is located) are dominated by the snow and glacier melt with almost >70% contribution to annual streamflows (Immerzeel et al., 2009). Pertinently, snow and glacier melt play a dominant role in regulating the river flows in the Lidder basin by contributing more than 60% to the annual flows (Chapter 4). Furthermore, studies have suggested that increasing temperature is expected to have serious consequences on the hydrological setup of the regions where streamflows are dominated by the snow and glacier melt (Bajracharya et al., 2008; Barnett et al., 2005; Altaf et al., 2013, 2014; Meraj et al., 2015, 2016). Furthermore, researchers linked the recent unprecedented changes in the flow of the Himalayan rivers to the accelerated shrinking of the snow and glacier resources in response to increasing temperature (Immerzeel et al., 2009).

Climate change impact on the streamflows is clearly evident in the HKH region as has been reported by Ali et al., 2017; Barnett et al., 2005; Immerzeel et al., 2012; Lutz et al., 2016; Ragettli et al., 2016; Rees and Collins, 2006; Romshoo et al., 2015; Thayyen and Gergan, 2010. Pertinently, several studies have successfully forced the hydrological models with the historical and the projected hydrometeorological data to quantify the streamflow changes under the changing climate (Ali et al., 2017; Lutz et al., 2016). In order to predict the hydrological changes at the regional scale, the global circulation models (GCMs) have been downscaled to better resolutions to generate the future climate change projections at the regional level (Lutz et al., 2014). Correspondingly, Lutz et al. (2016) suggested that the snow and glacier melt contribution to streamflows in the HKH will decrease significantly by the end of twenty-first century in response to increasing temperature. Nevertheless, the studies carried out in the Himalayan region have predicted that with increasing temperature, the streamflows shall increase during the first half of twenty-first century and shall decrease thereafter (Ali et al., 2015; Immerzeel et al., 2009; Soncini et al., 2015). Furthermore, Barnett et al. (2005) suggested that in the melt-dominated basins, at constant precipitation, increase in temperature will lead to earlier runoff peaks in the spring due to early melting of snow cover; however, the summers will have reduced flows.

In the Kashmir Himalayas, the indicators of global warming are clear in terms of increase in temperature, enhanced receding of snow and glacier resources, decrease in snow precipitation, and depleting streamflows (Murtaza and Romshoo, 2017; Rashid et al., 2017; Romshoo et al., 2015; Ali et al., 2017; Marazi and Romshoo, 2018). Romshoo et al. (2015) investigated the historical trends in the annual temperature and precipitation in the Lidder basin and reported that the average annual temperature has increased significantly in contrast to the precipitation which showed statistically nonsignificant decreasing trend. In other words, precipitation in the Lidder basin has not changed over the past decades, though the form of precipitation has changed, which means the area receives more liquid precipitation (Romshoo et al., 2015). Similar nonsignificant precipitation trends have been reported for the other basins in the HKH region (Immerzeel et al., 2009; Romshoo et al., 2017). Furthermore, the decreasing snow precipitation over the Lidder basin as a result of increasing temperature could seriously affect the snow and glacier resources and correspondingly the streamflow pattern (Romshoo et al., 2015). It is pertinent to mention here that the Lidder basin hosts the largest number of glaciers in the Jhelum basin and forms the largest tributary of the Jhelum river. Further, the dominance of snow and glacier melt in the streamflows of the Lidder basin and the increasing downstream water demand for various activities makes this basin a focus of tremendous societal importance. Correspondingly, the depletion of frozen water in the Kashmir Himalayas will lead to major socioeconomic and environmental problems like decline of hydropower potential, decrease in agricultural productivity, fall in tourism activities and drinking water scarcity (Dar et al., 2014; Romshoo et al., 2015). Pertinently, snow and glacier resources in the Kashmir Himalayas are extremely vulnerable to the climate change, and the impacts on cryosphere and water resources are potentially serious. Nevertheless, the region has remained highly under-studied in terms of characterizing and understanding various glaciological, hydrological and climatological processes. Therefore, it is important to study the current state and future response of glaciers and streamflow to changing climatic parameters.

10.4 RECOMMENDATIONS

Since water resources are extremely important for our economy and livelihood, the continuous and detailed water resource monitoring programs shall be the priority of the government and concerned departments and institutes. Technically, the monitoring programs should include high-resolution (both temporal and spatial) remote sensing imageries and continuous well-equipped field-based programs to understand the details and dynamics of the snow and glacier resources under the changing climatic conditions. Furthermore, the network of automatic weather stations should be installed on or near the glaciers so that various physical factors that affect the melting process of the glaciers can be monitored. Moreover, measurements of the hydro-meteorological parameters, e.g., temperature, humidity, evapotranspiration, incoming and outgoing solar radiations, snow depth, wind speed, etc. should be taken continuously and professionally, which have key importance in simulating the hydrological processes. Further, the streamflow discharge must be measured at daily or sub-daily frequency. These improvements can only be achieved by appointing the serious and highly professional people and expanding the network of climate and discharge gauging stations in the region especially at higher elevations. Besides this, accessibility to the existing observed data must be improved. Pertinently, these developments will help in understanding the nature of changing climatic parameters and the corresponding impact on the water resources in the region, and eventually, the accuracy of the decisions regarding the water resource management will improve.

Keeping in view the present and projected impacts of climate change on the water resources, government should frame effective plans for the development of water storage infrastructure across the Kashmir valley to tackle the water shortage during the lean period. Furthermore, a better irrigation canal system should be constructed under the supervision of scientists, engineers and planners that should be designed not only to transfer water from the water storage units to agricultural fields but should hold excess water during the floods.

Similar climate change impact studies need to be carried out in other parts of the Jhelum basin, so that the response of water resources to changing climate in different climatic and topographic setting could be quantified. Pertinently, such studies will help in improving the knowledge and understanding of the status of climate change in the Jhelum basin. Eventually, these studies will improve the adaptation strategies and capability of planning and decision-making. Moreover, similar type of study can be extended to adjacent basins and sub-basin in the region to get an insight into the broader climatic and geopolitical issues and thus can be helpful in designing the solution strategies to avoid any geopolitical catastrophe.

10.5 CONCLUSIONS

Hydrological cycle is the most important cyclic system that exists on the surface of Earth because of the fact that life cannot survive without water. During the cyclic process, water gets stored in different temporary reservoirs, e.g., oceans, lakes, rivers, snow fields, glaciers, ground water, atmospheric water, etc. In the mountainous regions, a major portion of the precipitation occurs in the form snow, and the climatic conditions are favorable enough for the development of glaciers and vast snow fields. Pertinently, the vast snow and glacier resources in the HKH region support the livelihood of huge downstream population by contributing dominantly to annual river flows in the region, and thus, the HKH region is often referred to as the Water Tower of Asia (Immerzeel et al., 2009). However, the climate change has seriously affected these frozen water resources in the region, and the subsequent response of streamflows is a cause of concern for scientific and political community (Bolch et al., 2012). Furthermore, the increasing demand of water by growing population for agricultural and other socioeconomic activities has straining the already stressed water resources in the region. The future changes in the hydrology of the region under the projected changes in climatic parameters may have serious implications on the environmental and social setup (Lutz et al., 2016). Nevertheless, the consequences of climate change are serious, and the impacts of climate change on water resources in the region are poorly studied. Therefore, the quantification of the impacts of climate change in the region is important to understand the link between the cryosphere, streamflows and changing climatic parameters to establish a long-term strategy for the conservation and management of the water resources.

ACKNOWLEDGMENT

This study is part of the PhD work carried out in the department of Earth Sciences, University of Kashmir.

We acknowledge the University of Kashmir for providing logistic and other supports during the course of this study.

We acknowledge the financial support provided by the Department of Ministry of Environment, Forests and Climate Change (MoEFandCC), Government of India, New Delhi under the National Mission on Himalayan Studies (NMHS) research project titled "Integrated system dynamical model to design and testing alternative intervention strategies for effective remediation and sustainable water management for two selected river basins of Indian Himalaya".

REFERENCES

Ali, I., Shukla, A., and Romshoo, S. A. (2017). Assessing linkages between spatial facies changes and dimensional variations of glaciers in the upper Indus Basin, western Himalaya. *Geomorphology, 284*, 115–129. https://doi.org/10.1016/j.geomorph.2017.01.005

Ali, S. H., Bano, I., Kayastha, R. B., and Shrestha, A. (2017). Comparative assessment of runoff and its components in two catchments of Upper Indus Basin by using a semi distributed glacio-hydrological model. *International Archives of the Photogrammetry, Remote Sensing and Spatial Information Sciences, XLII*(September), 18–22.

Ali, S., Li, D., Congbin, F., and Khan, F. (2015). Twenty first century climatic and hydrological changes over Upper Indus Basin of Himalayan region of Pakistan. *Environmental Research Letters, 10*(1), 014007. https://doi.org/10.1088/1748-9326/10/1/014007

Altaf, F., Meraj, G., and Romshoo, S. A. (2013). Morphometric analysis to infer hydrological behaviour of Lidder watershed, Western Himalaya, India. *Geography Journal, 2013*, 1–14.

Altaf, F., Meraj, G., and Romshoo, S. A. (2014). Morphometry and land cover based multi-criteria analysis for assessing the soil erosion susceptibility of the western Himalayan watershed. *Environmental Monitoring and Assessment, 186*(12), 8391–8412.

Anderson, B., and Mackintosh, A. (2006). Temperature change is the major driver of late-glacial and Holocene glacier fluctuations in New Zealand. *Geology, 34*(2), 121–124. https://doi.org/10.1130/G22151.1

Archer, D. R., Forsythe, N., Fowler, H. J., and Shah, S. M. (2010). Sustainability of water resources management in the Indus Basin under changing climatic and socio economic conditions. *Hydrology and Earth System Sciences, 14*(8), 1669.

Armstrong, R. L. (2010). *The glaciers of the Hindu Kush-Himalayan region: a summary of the science regarding glacier melt/retreat in the Himalayan, Hindu Kush, Karakoram, Pamir, and Tien Shan mountain ranges.* International Centre for Integrated Mountain Development (ICIMOD), Kathmandu.

Arora, M., Singh, P., Goel, N. K., and Singh, R. D. (2008). Climate variability influences on hydrological responses of a large Himalayan basin. *Water Resources Management, 22*(-10), 1461–1475.

Azam, M. F., Wagnon, P., Vincent, C., Ramanathan, A. L., Favier, V., Mandal, A., and Pottakkal, J. G. (2014). Processes governing the mass balance of Chhota Shigri glacier (western Himalaya, India) assessed by point-scale surface energy balance measurements. *Cryosphere, 8*(6), 2195–2217. https://doi.org/10.5194/tc-8-2195-2014

Bahuguna, I. M., Kulkarni, A. V., Nayak, S., Rathore, B. P., Negi, H. S., and Mathur, P. (2007). Himalayan glacier retreat using IRS 1C PAN stereo data. *International Journal of Remote Sensing, 28*(2), 437–442.

Bajracharya, S. R., and Shrestha, B. R. (2011). *The status of glaciers in the Hindu Kush-Himalayan region.* International Centre for Integrated Mountain Development (ICIMOD), Kathmandu.

Bajracharya, S. R., Maharjan, S. B., Shrestha, F., Guo, W., Liu, S., Immerzeel, W., and Shrestha, B. (2015). The glaciers of the Hindu Kush Himalayas: current status and observed changes from the 1980s to 2010. *International Journal of Water Resources Development, 31*(2), 161–173.

Bajracharya, S. R., Mool, P. K., and Shrestha, B. R. (2008). Global climate change and melting of Himalayan glaciers. In Ranade, P. S. (ed) *Melting glaciers and rising sea levels: impacts and implications,* 28–46. Icfai University Press, Hyderabad.

Barnett, T. P., Adam, J. C., and Lettenmaier, D. P. (2005). Potential impacts of a warming climate on water availability in snow-dominated regions. *Nature, 438*(7066), 303–309. https://doi.org/10.1038/nature04141

Bhambri, R., Bolch, T., Chaujar, R. K., and Kulshreshtha, S. C. (2011). Glacier changes in the Garhwal Himalaya, India, from 1968 to 2006 based on remote sensing. *Journal of Glaciology, 57*(203), 543–556.

Bhutiyani, M. R., Kale, V. S., and Pawar, N. J. (2007). Long-term trends in maximum, minimum and mean annual air temperatures across the Northwestern Himalaya during the twentieth century. *Climatic Change, 85*(1), 159–177.

Bolch, T., Kulkarni, A., Kääb, A., Huggel, C., Paul, F., Cogley, J.G., Frey, H., Kargel, J.S., Fujita, K., Scheel, M. and Bajracharya, S. (2012). The state and fate of Himalayan glaciers. *Science, 336*(6079), 310–314.

Bolch, T., Yao, T., Kang, S., Buchroithner, M.F., Scherer, D., Maussion, F., Huintjes, E. and Schneider, C. (2010). A glacier inventory for the western Nyainqentanglha Range and the Nam Co Basin, Tibet, and glacier changes 1976–2009. The *Cryosphere, 4*(3), 419–433.

Cogley, J. G. (2011). Present and future states of Himalaya and Karakoram glaciers. *Annals of Glaciology, 52*(59), 69–73.

Dar, R. A., Rashid, I., Romshoo, S. A., and Marazi, A. (2014). Sustainability of winter tourism in a changing climate over Kashmir Himalaya. *Environmental Monitoring and Assessment, 186*(4), 2549–2562. https://doi.org/10.1007/s10661-013-3559-7

Devendra, C. (1997). *Improvement of livestock production in crop-animal systems in rainfed agro-ecological zones of South-East Asia.* ILRI (aka ILCA and ILRAD), Nairobi.

Dimri, A. P., and Dash, S. K. (2012). Wintertime climatic trends in the western Himalayas. *Climatic Change, 111*(3), 775–800. https://doi.org/10.1007/s10584-011-0201-y

Ding, Y., Liu, S., Li, J., and Shangguan, D. (2006). The retreat of glaciers in response to recent climate warming in western China. *Annals of Glaciology, 43*, 97–105. https://doi.org/10.3189/172756406781812005

Dyurgerov, M. B., and Meier, M. F. (2005). *Glaciers and the changing Earth system: a 2004 snapshot* (Vol. 58). Institute of Arctic and Alpine Research, University of Colorado Boulder, Boulder.

Fischer, M., Huss, M., and Hoelzle, M. (2013, April). Recent changes of very small glaciers in the Swiss Alps. In *EGU General Assembly Conference Abstracts* (Vol. 15), Vienna.

Fountain, A. G., and Tangborn, W. V. (1985). The effect of glaciers on streamflow variations. *Water Resources Research, 21*(4), 579–586.

Fowler, H.J. and Archer, D.R. (2006). Conflicting signals of climatic change in the Upper Indus Basin. *Journal of climate, 19*(17), 4276–4293.

Frey, H., Machguth, H., Huss, M., Huggel, C., Bajracharya, S., Bolch, T., Kulkarni, A., Linsbauer, A., Salzmann, N. and Stoffel, M. (2014). Estimating the volume of glaciers in the Himalayan–Karakoram region using different methods. *The Cryosphere, 8*(6), 2313–2333.

Fujita, K., and Ageta, Y. (2000). Effect of summer accumulation on glacier mass balance on the Tibetan Plateau revealed by mass-balance model. *Journal of Glaciology, 46*(153), 244–252.

Gardelle, J., Berthier, E., Arnaud, Y., and Kääb, A. (2013). Region-wide glacier mass balances over the Pamir-Karakoram-Himalaya during 1999–2011. *The Cryosphere, 7*(4), 1263–1286.

Gertler, C. G., Puppala, S. P., Panday, A., Stumm, D., and Shea, J. (2016). Black carbon and the Himalayan cryosphere: *A review. Atmospheric environment, 125*, 404–417.

Gurung, D. R., Kulkarni, A. V, Giriraj, A., Aung, K. S., Shrestha, B., and Srinivasan, J. (2011). Changes in seasonal snow cover in Hindu Kush-Himalayan region. *The Cryosphere Discussions, 5*(2), 755–777. https://doi.org/10.5194/tcd-5-755-2011

Hall, D. K., Riggs, G. A., Salomonson, V. V., Digirolamo, N. E., and Bayr, K. J. (2002). MODIS snow-cover products. *Remote Sensing of Environment, 83*, 181–194. https://doi.org/10.1016/S0034-4257(02)00095-0

Hewitt, K. (2011). Glacier change, concentration, and elevation effects in the Karakoram Himalaya, Upper Indus Basin. *Mountain Research and Development*, *31*(3), 188–200. https://doi.org/10.1659/MRD-JOURNAL-D-11-00020

Hock, R. (2003). Temperature index melt modelling in mountain areas. *Journal of Hydrology*, *282*(1), 104–115.

Höelzle, M., Chinn, T., Stumm, D., Paul, F., Zemp, M., and Haeberli, W. (2007). The application of glacier inventory data for estimating past climate change effects on mountain glaciers: a comparison between the European Alps and the Southern Alps of New Zealand. *Global and Planetary Change*, *56*(1–2), 69–82.

Houghton, J.T., Albritton, D.L., Meira Filho, L.G., Cubasch, U., Dai, X., Ding, Y., Griggs, D.J., Hewitson, B., Isaksen, I., Karl, T. and McFarland, M., (2001). *Technical summary of working group 1*. Retrieved from http://pubman.mpdl.mpg.de/pubman/item/escidoc:995493/component/escidoc:995492/WG1_TAR-FRONT.pdf

Immerzeel, W. W., Droogers, P., De Jong, S. M., and Bierkens, M. F. P. (2009). Large-scale monitoring of snow cover and runoff simulation in Himalayan river basins using remote sensing. *Remote sensing of Environment*, *113*(1), 40–49.

Immerzeel, W. W., Van Beek, L. P. H., and Bierkens, M. F. (2010). Climate change will affect the Asian water towers. *Science*, *328*(5984), 1382–1385

Immerzeel, W. W., Van Beek, L. P. H., Konz, M., Shrestha, A. B., and Bierkens, M. F. P. (2012). Hydrological response to climate change in a glacierized catchment in the Himalayas. *Climatic Change*, *110*(3–4), 721–736. https://doi.org/10.1007/s10584-011-0143-4

Iturrizaga, L. (2011). Glacier lake outburst floods. In Singh, V. P., Singh, P., and Haritashya, U. K. (eds) *Encyclopedia of snow, ice and glaciers* (pp. 381–399). Springer, Dordrecht.

Jacob, T., Wahr, J., Pfeffer, W. T., and Swenson, S. (2012). Recent contributions of glaciers and ice caps to sea level rise. *Nature*, *482*(7386), 514–518.

Jain, S. K., Goswami, A., and Saraf, A. K. (2010). Snowmelt runoff modelling in a Himalayan basin with the aid of satellite data. *International Journal of Remote Sensing*, *31*(24), 6603–6618.

Jeelani, G., Shah, R. A., Jacob, N., and Deshpande, R. D. (2017). Estimation of snow and glacier melt contribution to Liddar stream in a mountainous catchment, western Himalaya: an isotopic approach. *Isotopes in Environmental and Health Studies*, *53*(1), 18–35. https://doi.org/10.1080/10256016.2016.1186671

Kääb, A., Berthier, E., Nuth, C., Gardelle, J., and Arnaud, Y. (2012). Contrasting patterns of early twenty-first-century glacier mass change in the Himalayas. *Nature*, *488*(7412), 495–498. https://doi.org/10.1038/nature11324

Karim, A., and Veizer, J. (2002). Water balance of the Indus River Basin and moisture source in the Karakoram and western Himalayas: Implications from hydrogen and oxygen isotopes in river water. *Journal of Geophysical Research: Atmospheres*, *107*(D18).

Kayastha, R. B., Ohata, T., and Ageta, Y. (1999). Application of a mass-balance model to a Himalayan glacier. *Journal of Glaciology*, *45*(151), 559–567.

Khattak, M. S., Babel, M. S., and Sharif, M. (2011). Hydro-meteorological trends in the upper Indus River basin in Pakistan. *Climate research*, *46*(2), 103–119.

Konz, M., Uhlenbrook, S., Braun, L., Shrestha, A., and Demuth, S. (2007). Implementation of a process-based catchment model in a poorly gauged, highly glacierized Himalayan headwater. *Hydrology and Earth System Sciences Discussions*, *11*(4), 1323 1339. Retrieved from https://hal.archives-ouvertes.fr/hal-00305075/

Kulkarni, A. V, Bahuguna, I. M., Rathore, B. P., Singh, S. K., Randhawa, S. S., Sood, R. K., and Dhar, S. (2007), Glacial retreat in Himalaya using Indian Remote Sensing satellite data. *Current Science*, *92*(1), 69–74. https://doi.org/10.1117/12.694004

Kulkarni, A. V., Rathore, B. P., Singh, S. K., and Bahuguna, I. M. (2011). Understanding changes in the Himalayan cryosphere using remote sensing techniques. *International Journal of Remote Sensing*, *32*(3), 601–615.

Kumar, K., Dumka, R. K., Miral, M. S., Satyal, G. S., and Pant, M. (2008). Estimation of retreat rate of Gangotri glacier using rapid static and kinematic GPS survey. *Current Science*, 94(2), 258–262.

Lejeune, Y., Bertrand, J.-M., Wagnon, P., and Morin, S. (2013). A physically based model of the year-round surface energy and mass balance of debris-covered glaciers. *Journal of Glaciology*, 59(214), 327–344.

Liu, X., and Chen, B. (2000). Climatic warming in the Tibetan Plateau during recent decades. *International Journal of Climatology*, 20(14), 1729–1742.

Lüthi, D., Le Floch, M., Bereiter, B., Blunier, T., Barnola, J.M., Siegenthaler, U., Raynaud, D., Jouzel, J., Fischer, H., Kawamura, K. and Stocker, T.F., (2008). High-resolution carbon dioxide concentration record 650,000–800,000 years before present. *Nature*, 453(7193), 379–382.

Lutz, A. F., Immerzeel, W. W., Kraaijenbrink, P. D. A., Shrestha, A. B., and Bierkens, M. F. P. (2016). Climate change impacts on the upper indus hydrology: sources, shifts and extremes. *PLoS One*, 11(11), 1–33. https://doi.org/10.1371/journal.pone.0165630

Lutz, A. F., Immerzeel, W. W., Shrestha, A. B., and Bierkens, M. F. P. (2014). Consistent increase in High Asia's runoff due to increasing glacier melt and precipitation. *Nature Climate Change*, 4(7), 587.

Maisch, M. (2000). The long-term signal of climate change in the Swiss Alps: glacier retreat since the end of the little ice age and future ice decay scenarios. *Geografia Fisica e Dinamica Quaternaria*, 23, 139–151.

Marazi, A., and Romshoo, S. A. (2018). Streamflow response to shrinking glaciers under changing climate in the Lidder Valley, Kashmir Himalayas. *Journal of Mountain Science*, 15(6), 1241–1253. https://doi.org/10.1007/s11629-017-4474-0

Marshak, S. (2015). *Earth: portrait of a planet: 5th international student edition*. WW Norton and Company, New York.

Maurya, A. S., Shah, M., Deshpande, R. D., Bhardwaj, R. M., Prasad, A., and Gupta, S. K. (2011). Hydrograph separation and precipitation source identification using stable water isotopes and conductivity: River Ganga at Himalayan foothills. *Hydrological Processes*, 25(10), 1521–1530. https://doi.org/10.1002/hyp.7912.

Mayewski, P. A., and Jeschke, P. A. (1979). Himalayan and Trans-Himalayan glacier fluctuations since AD 1812. *Arctic and alpine research*, 11(3), 267–287. https://doi.org/10.1080/00040851.1979.12004137

Meraj, G., et al. (2015). Assessing the influence of watershed characteristics on the flood vulnerability of Jhelum basin in Kashmir Himalaya. *Natural Hazards*, 77(1), 153–175.

Meraj, G., et al. (2018a). Geoinformatics based approach for estimating the sediment yield of the mountainous watersheds in Kashmir Himalaya, India. *Geocarto International*, 33(10), 1114–1138.

Meraj, G., et al. (2018b). An integrated geoinformatics and hydrological modelling-based approach for effective flood management in the Jhelum Basin, NW Himalaya. *Multidisciplinary Digital Publishing Institute Proceedings*, 7(1), 8.

Meraj, G., Romshoo, S. A., and Altaf, S. (2016). Inferring land surface processes from watershed characterization. In Raju, N. J. (ed) *Geostatistical and geospatial approaches for the characterization of natural resources in the environment* (pp. 741–744). Springer, Cham.

Mir, R. A., Jain, S. K., Saraf, A. K., and Goswami, A. (2014). Glacier changes using satellite data and effect of climate in Tirungkhad basin located in western Himalaya. *Geocarto International*, 29(3), 293–313. https://doi.org/10.1080/10106049.2012.760655

Moen, J., and Fredman, P. (2007). Effects of climate change on alpine skiing in Sweden. *Journal of Sustainable Tourism*, 15(4), 418–437.

Mölg, T., Maussion, F., Yang, W., and Scherer, D. (2012). The footprint of Asian monsoon dynamics in the mass and energy balance of a Tibetan glacier. *The Cryosphere*, 6(6), 1445.

Murtaza, K. O., and Romshoo, S. A. (2017). Recent glacier changes in the Kashmir Alpine Himalayas, India. *Geocarto International, 32*(2), 188–205. https://doi.org/10.1080/10106049.2015.1132482

Nuimura, T., Sakai, A., Taniguchi, K., Nagai, H., Lamsal, D., Tsutaki, S., Kozawa, A., Hoshina, Y., Takenaka, S., Omiya, S. and Tsunematsu, K., (2015). The GAMDAM glacier inventory: a quality-controlled. *Citeseer,* 849–864. https://doi.org/10.5194/tc-9-849-2015

Oerlemans, J. (1994). Quantifying global warming from the retreat of glaciers. *Science, 264*(5156), 243–245. https://doi.org/10.1126/science.264.5156.243

Oerlemans, J. (2005). Extracting a climate signal from 169 glacier records. *Science, 308*(5722), 675–677. https://doi.org/10.1126/science.1107046

Ohmura, A. (2001). Physical basis for the temperature-based melt-index method. *Journal of Applied Meteorology, 40*, 753–761. https://doi.org/10.1175/1520-0450(2001)040<0753:PBFTTB>2.0.CO;2

Oki, T., and Kanae, S. (2006). Global hydrological cycles and world water resources. *Science, 313*(5790), 1068–1072.

Parajka, J., and Blöschl, G. (2008). Spatio-temporal combination of MODIS images–potential for snow cover mapping. *Water Resources Research, 44*(3), 1-13.

Pepin, N. C., and Lundquist, J. D. (2008). Temperature trends at high elevations: patterns across the globe. *Geophysical Research Letters, 35*(14), 1-6.

Pradhananga, N. S., Kayastha, R. B., Bhattarai, B. C., Adhikari, T. R., Pradhan, S. C., Devkota, L. P., … Mool, P. K. (2014). Estimation of discharge from Langtang River basin, Rasuwa, Nepal, using a glacio-hydrological model. *Annals of Glaciology, 55*(66), 223–230.

Raconviteanu, A. E., Williams, M. W., and Barry, R. G. (2008). Optical remote sensing of glacier characteristics: a review with focus on the Himalaya. *Sensors, 8*(5), 3355–3383.

Ragettli, S., Immerzeel, W. W., and Pellicciotti, F. (2016). Contrasting climate change impact on river flows from high-altitude catchments in the Himalayan and Andes Mountains. *Proceedings of the National Academy of Sciences, 113*(33), 9222–9227. https://doi.org/10.1073/pnas.1606526113.

Rangwala, I., and Miller, J. R. (2012). Climate change in mountains: a review of elevation-dependent warming and its possible causes. *Climatic change, 114*(3), 527–547.

Rashid, I., Romshoo, S. A., and Abdullah, T. (2017). The recent deglaciation of Kolahoi valley in Kashmir Himalaya, India in response to the changing climate. *Journal of Asian Earth Sciences, 138*, 38–50. https://doi.org/10.1016/j.jseaes.2017.02.002

Rees, H. G., and Collins, D. N. (2006). Regional differences in response of flow in glacier-fed Himalayan rivers to climatic warming. *Hydrological Processes, 20*(10), 2157–2169.

Rittger, K., Painter, T. H., and Dozier, J. (2013). Assessment of methods for mapping snow cover from MODIS. *Advances in Water Resources, 51*, 367–380. https://doi.org/10.1016/j.advwatres.2012.03.002.

Robinson, D. A., Dewey, K. F., and Heim, R. R. (1993). Global snow cover monitoring: an update. *Bulletin of the American Meteorological Society, 74*(9), 1689–1696. https://doi.org/10.1175/1520-0477(1993)074<1689:GSCMAU>2.0.CO;2

Romshoo, S. A., Dar, R. A., Rashid, I., Marazi, A., Ali, N., and Zaz, S. N. (2015). Implications of shrinking cryosphere under changing climate on the streamflows in the lidder catchment in the Upper Indus Basin, India. *Arctic, Antarctic, and Alpine Research, 47*(4). https://doi.org/10.1657/AAAR0014-088

SAC. (2016). *Monitoring snow and glaciers of Himalayan Region.* Space Applications Centre, ISRO, Ahmedabad.

Salerno, F., Thakuri, S., Guyennon, N., Viviano, G., and Tartari, G. (2016). Glacier melting and precipitation trends detected by surface area changes in Himalayan ponds. *The Cryosphere, 10*(4), 1433–1448.

Salomonson, V. V, and Appel, I. (2004). Estimating fractional snow cover from MODIS using the normalized difference snow index. *Remote Sensing of Environment, 89*(3), 351–360.

Sarikaya, M. A., Bishop, M. P., Shroder, J. F., and Olsenholler, J. A. (2012). Space-based observations of Eastern Hindu Kush glaciers between 1976 and 2007, Afghanistan and Pakistan. *Remote Sensing Letters, 3*(1), 77–84.

Schaner, N., Voisin, N., Nijssen, B., and Lettenmaier, D. P. (2012). The contribution of glacier melt to streamflow. *Environmental Research Letters, 7*(3), 034029. https://doi.org/10.1088/1748-9326/7/3/034029

Schneider, S. H. (1990). The global warming debate heats up: an analysis and perspective. *Bulletin of the American Meteorological Society, 71*(9), 1292–1304.

Shrestha, A. B., Wake, C. P., Mayewski, P. A., and Dibb, J. E. (1999). Maximum temperature trends in the Himalaya and its vicinity: an analysis based on temperature records from Nepal for the period 1971–94. *Journal of Climate, 12*(9), 2775–2786.

Shrestha, M., Koike, T., Hirabayashi, Y., Xue, Y., Wang, L., Rasul, G., and Ahmad, B. (2015). Integrated simulation of snow and glacier melt in water and energy balance-based, distributed hydrological modeling framework at Hunza River Basin of Pakistan Karakoram region. *Journal of Geophysical Research: Atmospheres, 120*(10), 4889–4919.

Silwal, G., Kayastha, R. B., and Mool, P. K. (2016). Application of temperature index model for estimating daily discharge of Sangda River Basin, Mustang, Nepal. *Journal of Climate Change, 2*(1), 15–26.

Singh, P., and Bengtsson, L. (2005). Impact of warmer climate on melt and evaporation for the rainfed, snowfed and glacierfed basins in the Himalayan region. *Journal of Hydrology, 300*(1-4), 140-154.

Singh, P., Haritashya, U. K., and Kumar, N. (2008). Modelling and estimation of different components of streamflow for Gangotri Glacier basin, Himalayas/Modélisation et estimation des différentes composantes de l'écoulement fluviatile du bassin du Glacier Gangotri, Himalaya. *Hydrological Sciences Journal, 53*(2), 309–322.

Singh, P., Jain, S. K., and Kumar, N. (1997). Estimation of snow and glacier-melt contribution to the Chenab River, Western Himalaya. *Mountain Research and Development, 17*(1), 49–56. http://www.jstor.org/stable/3673913

Solomon, S., Qin, D., Manning, M., Averyt, K., and Marquis, M. (Eds.). (2007). *Climate change 2007 – the physical science basis: working group I contribution to the fourth assessment report of the IPCC* (Vol. 4). Cambridge University Press, Cambridge.

Soncini, A., Bocchiola, D., Confortola, G., Bianchi, A., Rosso, R., Mayer, C., Lambrecht, A., Palazzi, E., Smiraglia, C. and Diolaiuti, G. (2015). Future hydrological regimes in the upper indus basin: A case study from a high-altitude glacierized catchment. *Journal of Hydrometeorology, 16*(1), 306–326.

Tahir, A. A., Adamowski, J. F., Chevallier, P., Haq, A. U., and Terzago, S. (2016). Comparative assessment of spatiotemporal snow cover changes and hydrological behavior of the Gilgit, Astore and Hunza River basins (Hindukush–Karakoram–Himalaya region, Pakistan). *Meteorology and Atmospheric Physics, 128*(6), 793–811. https://doi.org/10.1007/s00703-016-0440-6

Thayyen, R. J., and Gergan, J. T. (2010). Role of glaciers in watershed hydrology: a preliminary study of a "Himalayan catchment". *The Cryosphere, 4*(1), 115–128.

Viviroli, D., and Weingartner, R. (2004). The hydrological significance of mountains: from regional to global scale. *Hydrology and Earth System Sciences, 8*(6), 1017–1030. https://doi.org/10.5194/hess-8-1017-2004

Vörösmarty, C. J., Green, P., Salisbury, J., and Lammers, R. B. (2000). Global water resources: vulnerability from climate change and population growth. *Science, 289*(5477), 284–288. https://doi.org/10.1126/science.289.5477.284

Walther, G.R., Post, E., Convey, P., Menzel, A., Parmesan, C., Beebee, T.J., Fromentin, J.M., Hoegh-Guldberg, O. and Bairlein, F., (2002). Ecological responses to recent climate change. *Nature, 416*(6879), 389–395. https://doi.org/10.1038/416389a

Winiger, M. G. H. Y., Gumpert, M., and Yamout, H. (2005). Karakorum–Hindukush–western Himalaya: assessing high-altitude water resources. *Hydrological Processes: An International Journal*, *19*(12), 2329–2338.

Wulf, H., Bookhagen, B., and Scherler, D. (2016). Differentiating between rain, snow, and glacier contributions to river discharge in the western Himalaya using remote-sensing data and distributed hydrological modeling. *Advances in Water Resources*, *88*, 152–169. https://doi.org/10.1016/j.advwatres.2015.12.004

Ye, H., Yang, D., and Robinson, D. (2008). Winter rain on snow and its association with air temperature in northern Eurasia. *Hydrological Processes*, *22*(15), 2728–2736.

Ye, Q., Yao, T., Kang, S., Chen, F., and Wang, J. (2006). Glacier variations in the Naimona'nyi region, western Himalaya, in the last three decades. *Annals of Glaciology*, *43*(1), 385–389.

Zhang, L., Su, F., Yang, D., Hao, Z., and Tong, K. (2013). Discharge regime and simulation for the upstream of major rivers over Tibetan Plateau. *Journal of Geophysical Research: Atmospheres*, *118*(15), 8500–8518.

Zhu, L., Radeloff, V. C., and Ives, A. R. (2017). Characterizing global patterns of frozen ground with and without snow cover using microwave and MODIS satellite data products. *Remote Sensing of Environment*, *191*, 168–178.

11 Detecting Vegetation and Timberline Dynamics in Pinder Watershed Central Himalaya Using Geospatial Techniques

N. C. Pant and Meenaxi
S.S.J. University

Anand Kumar and Upasana Choudhury
Suresh Gyan Vihar University

CONTENTS

11.1 Introduction 213
11.2 Study Area 214
11.3 Materials and Methods 214
11.4 Results and Discussion 215
    11.4.1 Status of Vegetation Line and Vegetation Cover Area 216
    11.4.2 Status of Timberline and Timber Cover Area 218
11.5 Conclusions 220
References 221

11.1 INTRODUCTION

Vegetation line is the upper limit of vegetation in the alpine zone of mountains. Vegetation mapping through remotely sensed images involves various considerations, processes and techniques (Xie et al., 2008; Singh et al., 2017). Natural upper timberlines in high-mountain areas are remarkable indicators of climate conditions. They are the most noticeable altitudinal borderlines of vegetation, which are highly important regarding microclimate, soil formation and erosion. Moreover, timberline ecotones are natural habitats of high biodiversity. Potential shifts of timberlines and other climate-driven natural borderlines are discussed in the recent climate change debate. Remote sensing technology offers a practical and cost-effective means to study vegetation cover changes, especially over large areas (Langley et al., 2001; Kanga et al., 2017a, b; Rather et al., 2018; Hassanin et al., 2020; Kanga et al., 2021).

DOI: 10.1201/9781003147107-14

With the help of high-resolution remote sensing data like Landsat TM, it is generally possible to classify forest and vegetative areas and to detect timberlines and vegetation. A good case in vegetation and timber mapping by using remote sensing technology is the spectral radiances in the red and near-infrared regions, in addition to others. The radiances in these regions could be incorporated into the spectral vegetation indices that is directly related to the intercepted fraction of photo-synthetically active radiation (Galio et al., 1986).

Temporal remotely sensed data are extremely valuable for detecting changes in vegetation and timberline, land use/cover classes, snow, water bodies and other terrestrial features (Bharti et al., 2011; Gujree et al., 2020). In their study, Bharti et al. (2011) have suggested certain precautions while undertaking change detection studies in the Himalayan region using remote sensing techniques. Panigrahy et al. (2010) conducted a timberline change detection study in an area of Nanda Devi Biosphere Reserve (NDBR) in Central Himalaya using satellite imagery. They have reported aloft change of timberline vegetation by 300 m in NDBR, Garhwal Himalaya since 1960–2004. Based on the analysis of multi-temporal remote sensing data, these authors have also concluded that there is 61% decrease in snow cover and 23% increase in vegetation cover between 1986 and 2004 in NDBR. The fundamental objective of the present paper is to study the dynamics of vegetation and timberlines in one of the Central Himalaya watershed viz., the Pinder watershed.

11.2   STUDY AREA

The study area, viz., the Pinder Watershed (Figure 11.1) in the Central Himalaya extends between 29°59′41″N to 30°18′57″N latitudes and 79°05′43″E to 80°04′38″E longitudes and encompasses an area of 1889.93 km². The altitude of the watershed varies between 757 and 6,746 m. The upper part of the watershed contains large alpine pastures called bugyals by the local communities. These alpine pastures extend till the lofty Himalayan peaks that fall under NDBR.

11.3   MATERIALS AND METHODS

To work out the temporal shifting of vegetation and timberline, remotely sensed data are extremely valuable (Ahmadi and Nusrath, 2010; Kanga et al., 2017a, b; Rather et al., 2018; Hassanin et al., 2020; Kanga et al., 2021). To examine the shifting of vegetation and timber in the Pinder watershed, Landsat satellite imageries of three different years were acquired by the Global Land Cover Facility site and Glovis. The first imagery used in the present study was Landsat TM of 15th November 1990 at 30 m resolution. The second and third imageries were of Landsat TM data of 15th October 1999 and 2011 at a resolution of 30m. From Cartosat 1 data, a digital elevation model (DEM) was used to derive the average elevation of vegetation and timberlines. These imageries helped in understanding the process of shifting of vegetation and timberline in the watershed over the last 21 years (i.e., 1990–2011). ERDAS Imagine software was used for processing the satellite imageries. The area of interest was calculated, and finally, the required image was extracted by subsetting the image. The subset image was then reprojected. The upper limit of vegetation and timberline was delineated using

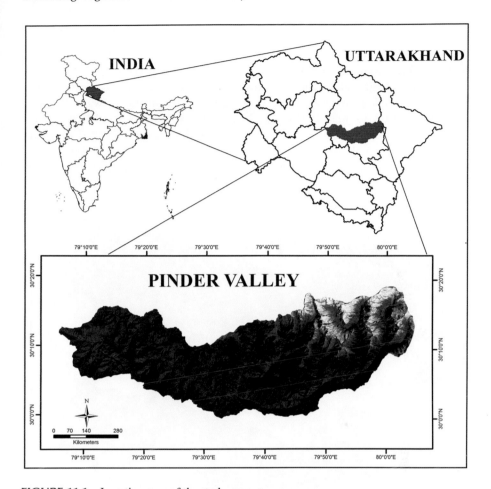

FIGURE 11.1   Location map of the study area.

normalized difference vegetation index (NDVI=NIR-RED/NIR+RED) (Weier and Herring, 2011). The threshold value of NDVI ranging from 0.2 to 0.4 and >0.4 are used to map out vegetation and timberline, respectively (www.earthobservatory.nasa. gov). After displaying the NDVI imagery on the screen of Arc map, the upper limit of vegetation and timberlines in the watershed were digitized for different time periods. The contours generated through Cartosat 1 based on DEM were overlaid on vegetation and timberlines to calculate its average altitudinal ingression.

11.4   RESULTS AND DISCUSSION

The results obtained through the analysis of NDVI imagery are diagrammatically illustrated in Figure 11.2, and data are registered in Tables 11.1–11.3. Figure 11.2 depicts NDVI of different years; Figure 11.3 depicts vegetation and timberlines for different periods; Figure 11.4 and 11.5 depicts vegetation and timber cover for different periods. A brief account of these results is discussed in the following paragraphs.

11.4.1 STATUS OF VEGETATION LINE AND VEGETATION COVER AREA

Figure 11.3 depicts the geographical location of vegetation line of the Pinder watershed in different years, i.e., 1990, 1999 and 2011, which is based on the NDVI values, i.e., >0.2–0.4 (Figure 11.2). The DEM overlay on this vegetation line suggests that the average height of vegetation line in the alpine zone in Pinder watershed was 3,280 m in 1990, 3,307 m in 1999 and 3,331 m in 2011 (Figure 11.3 and Table 11.1).

Table 11.2 reveals that the vegetation line was shifted upward about 27 m during 1990–1999 and 24 m during 1999–2011 in the Pinder watershed. The average shift of vegetation line during 21 years was about 51 m at the rate of 2.42 m/year (Table 11.2). Due to global warming, the snow cover area in the Pinder watershed has been shifted

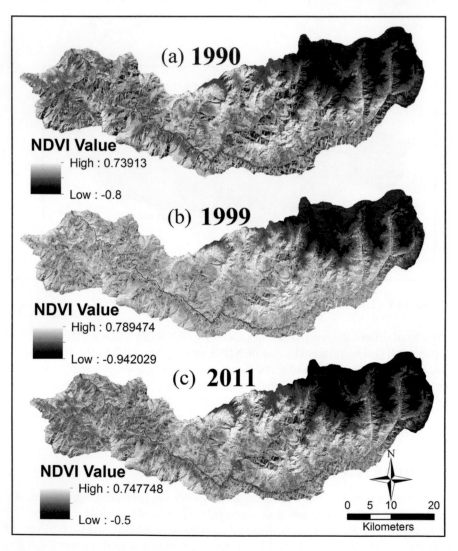

FIGURE 11.2 Geographical distribution of NDVI value in different years in the Pinder watershed.

toward higher elevation and has been depleted considerably during the last two decades (Pant et al., 2014). Study reveals that during 1990–1999 about 13.63 km² area and during 1999–2011 about 12.16 km² area of the Pinder watershed were converted from nonvegetation to vegetation area at an average rate of 1.51 and 1.01 km²/year, respectively. During the last 21 years (i.e., 1990–2011), due to global warming, about 25.79 km² area of the Pinder watershed has been converted from nonvegetation to vegetation area (Table 11.3).

TABLE 11.1
Average Height of Vegetation and Timberlines in Different Years in Pinder Watershed, Central Himalaya, India

Year	Vegetation Line Average Height (m)	Timberline Average Height (m)
1990	3,280	2,795
1999	3,307	2,816
2011	3,331	2,835

TABLE 11.2
Amount and Rate of Vegetation and Timberline Shift in Different Periods in the Pinder Watershed, Central Himalaya, India

Years	Period (years)	Shift of Vegetation Line Amount	Shift of Vegetation Line Rate of Shift (m/year)	Shift of Timberline Amount	Shift of Timberline Rate of Shift (m/year)
1990–1999	9	27	3	21	2.33
1999–2011	12	24	2	19	1.58
1990–2011	**21**	**51**	**2.42**	**40**	**1.90**

Bold face indicates the total change throughout the time series.

TABLE 11.3
Vegetation and Timber Area and Rate of Change in Different Periods in the Pinder Watershed, Central Himalaya, India

Years	Period (years)	Upward Shift of Vegetation Cover Area (km²)	Upward Shift of Vegetation Cover Rate of Upward Shift (km²/year)	Upward Shift of Timber Cover Area (km²)	Upward Shift of Timber Cover Rate of Shift (km²/year)
1990–1999	9	13.63	1.51	11.28	1.25
1999–2011	12	12.16	1.01	10.16	0.85
1990–2011	**21**	**25.79**	**1.22**	**21.44**	**1.02**

Bold face indicates the total change throughout the time series.

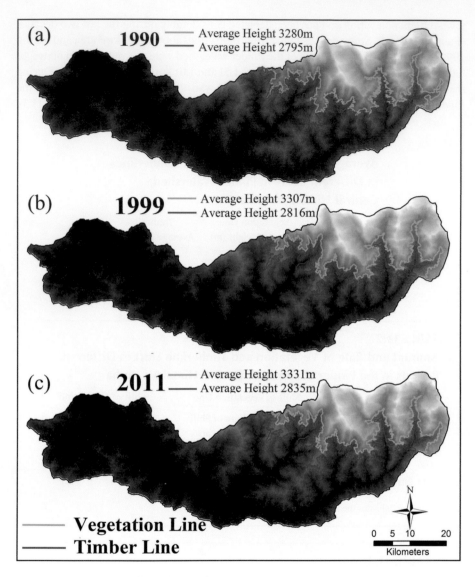

FIGURE 11.3 Geographical distribution of vegetation and timberline based on NDVI values in different years in the Pinder watershed.

11.4.2 STATUS OF TIMBERLINE AND TIMBER COVER AREA

Figure 11.3 depicts the geographical location of timberline of the Pinder watershed of 1990, 1999 and 2011, which is based on the NDVI values, i.e., >0.4 (Figure 11.2). The DEM overlay on this timberline suggests that the average height of timberline in the Alpine zone in Pinder watershed was 2,795 m in 1990, 2,816 m in 1999 and 2,835 m in 2011 (Table 11.1). Due to the shift of timberline upward about 11.28 km² during 1990–1999 and about 10.16 km² area during 1999–2011 were converted from nontimber to timber area (Table 11.3). These data revel that on an average, about

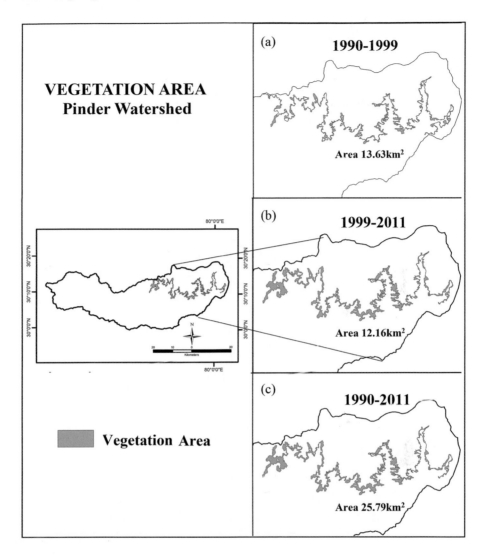

FIGURE 11.4 Geographical distribution of vegetation area in different periods in the Pinder watershed.

21.44 km² area of the Pinder watershed was converted from nontimber to timber area due to global warming (Table 11.3).

Table 11.2 revels that the timberline in the Pinder watershed has been shifted at the rate of 2.33 m/year during 1990–1999 and at the rate of 1.58 m/year during 1999–2011. These data suggest that the timberline in the Pinder watershed shifts upward at an average rate of 1.90 m/year. These data also suggest that due to climate change, the nontimber area was converted into timber area at the rate of 1.25 km²/year during 1990–1999 and 0.85 km²/year during 1999–2011. Based on these data, it can be stated that on an average, about 1.02 km²/year is converted from nontimber to timber area in the Pinder watershed (Table 11.3).

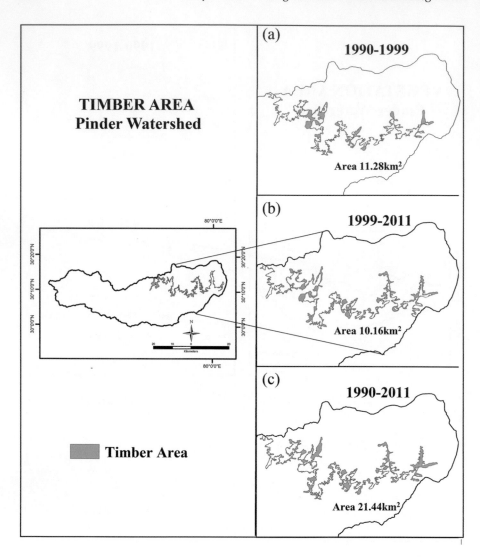

FIGURE 11.5   Geographical distribution of timber area in different periods in the Pinder watershed.

11.5   CONCLUSIONS

With the help of the present study carried out in a Central Himalayan watershed, viz., the Pinder, it can be stated that the vegetation and timberlines in the Central Himalayan region are shifting toward higher elevation due to global warming. The results reveal that during 1990–2011, about 25.79 km² nonvegetative area of the watershed was converted into vegetative area at the rate of 1.22 m/year, whereas during the last 21 years, 1990–2011, about 21.44 km² nontimber area of the watershed was converted into timber area at the average rate of 1.02 m/year. The average shift of timberline during 21 years was about 40 m at the rate of 1.90 m/year. The shifting of both vegetation and timberlines toward higher elevation is sharp evidence that the snow cover area is depleting steadily in the Central Himalaya due to global warming. If the trend of depletion continues, the water resources of the region would be in danger, which may result in severe environmental degradation, social disruption and ecological damages in the region.

REFERENCES

Ahmadi, H. and Nusrath, A., Vegetation change detection of Neka River in Iran by using remote sensing and GIS. *Journal of Geography and Geology*, 2010, 2(1), 58–67.

Bharti, R. R., Rai, I. D., Adhikari, B. S. and Rawat, G. S., Timberline change detection using topographic map and satellite imagery: a critique, International Society for Tropical Ecology. *Tropical Ecology*, 2011, 52(1), 133–137.

Galio, K. P., Daughtry, C. S. T. and Bauer, M. E., Spectral estimation of absorbed photosynthetically active radiation in corn canopies. *Agronomy Journal*, 1986, 78(4), 752–756.

Gujree, I., Arshad, A., Reshi, A. and Bangroo, Z., Comprehensive spatial planning of Sonamarg resort in Kashmir Valley, India for sustainable tourism. *Pacific International Journal*, 2020, 3(2), 71–99.

Hassanin, M., Kanga, S., Farooq, M. and Singh, S. K., Mapping of Trees outside Forest (ToF) From Sentinel-2 MSI satellite data using object-based image analysis. *Gujarat Agricultural Universities Research Journal*, 2020, 207–213.

Kanga, S., Rather, M. A., Farooq, M. and Singh, S. K., GIS based forest fire vulnerability assessment and its validation using field and MODIS data: a case study of Bhaderwah Forest Division, Jammu and Kashmir (India). *Indian Forester*, 2021, 147(2), 120–136.

Kanga, S., Singh, S. K. and Sudhanshu, Delineation of urban built-up and change detection analysis using multi-temporal satellite images. *International Journal of Recent Research Aspects*, 2017a, 4(3), 1–9.

Kanga, S., Kumar, S. and Singh, S. K., Climate induced variation in forest fire using remote sensing and GIS in Bilaspur district of Himachal Pradesh. *International Journal of Engineering and Computer Science*, 2017b, 6(6), 21695–21702.

Langley, S. K., Cheshire, H. M. and Humes, K. S., A comparison of single date and multitemporal satellite image classifications in a semi-arid grassland, *Journal of Arid Environ*, 2001, 49 (2), 401–411.

Panigrahy, S., Anitha, D., Kimothi, M. M. and Singh, S. P., Timberline change detection using topographic map and satellite imagery. *Tropical Ecology*, 2010, 51(1), 87–91.

Pant, N. C., Kumar, M., Rawat, J. S. and Rani, N, Study of snow cover dynamics of Pinder watershed in Central Himalaya using remote sensing and GIS techniques. *International Journal of Advanced Earth Science and Engineering*, 2014, 3(1), 122–128.

Rather, M. A., Farooq, M., Meraj, G., Dada, M. A., Sheikh, B. A. and Wani, I. A., Remote sensing and GIS based forest fire vulnerability assessment in Dachigam National park, North Western Himalaya. *Asian Journal of Applied Sciences*, 2018, 11(2), 98–114.

Singh, S. K., Mishra, S. K. and Kanga, S., Delineation of groundwater potential zone using geospatial techniques for Shimla city, Himachal Pradesh (India). *International Journal for Scientific Research and Development*, 2017, 5(4), 225–234.

Weier, J. and Herring, D., Measuring NDVI, 2011. http://earthobservatory.nasa.gov/Features/MeasuringVegetation (accessed 9th April 2011).

Xie, Y., Sha, Z. and Yu, M., Remote sensing imagery in vegetation mapping: a review. *Journals of Plant Ecology*, 2008, 1(1), 9–23.

12 Climate Change Studies, Permanent Forest Observational Plots and Geospatial Modeling

Swayam Vid
ICFRE

Shanti Kumari
ISRO

CONTENTS

12.1 Introduction 223
12.2 Network of "Preservation Permanent Plot" Established in Indian
Forests Since Its Inception..226
    12.2.1 Why Is There the Need of Long-Term Ecological Research........228
12.3 Establishment of Permanent Long Term Ecological Research Station......228
    12.3.1 The Following Can be Answered Through LTERS 228
12.4 Methods to Establish Permanent Plots to Study Forest Dynamics
(Taken from "The Manual of Instructions for Field Inventory, 2002,
Forest Survey of India, Dehradun.")..228
    12.4.1 Equipment Required for Installation and Layout of POPs 231
12.5 Impacts of Climate Change on Forests of Jharkhand, Bihar and West
Bengal 232
12.6 Climate Change Predictions: Geospatial Species Distribution Model.......235
12.7 Conclusion 237
References...237

12.1 INTRODUCTION

Climate Change and studies on their mechanism and impacts are a crux to the ecologists. In India as well as globally, climate change has been studied in various contexts so far but what is climate change in actual? Climate refers to the average weather condition, prevalent in a particular area. It is assigned to a place by observing and calculating data collected for several years, which is generally taken to be 30 years and more. Climate of an area is determined by temperature, precipitation etc. of that area which are called as the elements of climate. It is generally seen that the climatic elements hence climate show fluctuations about their mean value. This fluctuation is referred to as **climate variability**. But from last few years, a shift in the mean value of these elements is observed, i.e., permanent change in mean values and this change is referred to as **climate change**.

A number of intricate associated physical, biological and chemical processes occurring in the atmosphere, land and water determine the Earth's climate. Biophysical state of the Earth's surface and the atmospheric abundance of a variety of trace constituents affect the radiative properties of the atmosphere, a factor which controls the Earth's climate. These constituents include long-lived greenhouse gases (LLGHGs) such as methane (CH_4), nitrous oxide(N_2O) and carbon dioxide (CO_2), other radiatively active constituents like various aerosol particles and ozone. Anthropogenic and natural emissions of aerosols and gases determine the atmospheric composition, which transfers at a variety of scales, microphysical and chemical alterations, wet scavenging and surface uptake by the terrestrial and land ecosystems, and by the ocean ecosystems. Climate change affects these processes and the biogeochemical cycling rates, and involves interactions between and within the different components of the Earth system. The interactions are generally nonlinear and may result into positive or negative feedback to the climate system (Solomon et al., 2007).

The Fifth Assessment Report (2014) of Intergovernmental Panel on Climate Change (IPCC) has concluded the rise in the global average temperature and warming of the earth's climate system is unequivocal. The global carbon dioxide concentration has increased from 280 (pre-industrial revolution) to 379 ppm (in 2005). These changes could alter the weather events and impact freshwater availability, agriculture, forests and human health. India's development agenda stresses on the development with reduced climate-related vulnerability. Keeping in view the magnitude of uncertainties, the need is to identify and prioritize strategies that help to achieve developmental goals in the face of climate change.

As the frequency of precipitation will change over semiarid and arid regions, a change in these ecosystems will be observed with alteration in precipitation regimes; it is expected that there will be an increase in woody vegetation with rising CO_2 in these semiarid regions. It is expected that due to climate change, the spatial distribution of vegetation will be modified but these changes will also be influenced by some constraints like competition with established plants, limited seed dispersal, habitat fragmentation and soil development rates (Corlett and Westcott, 2013; Arshad et al., 2019; Kanga et al., 2017a, b). It is expected that forests and shrublands will replace alpine vegetation in Tibetan Plateau (Liang et al., 2012; Wang, 2013) and deciduous forests found in North India will be replaced by evergreen forests due to increase in precipitation (Choi et al., 2011; Wang, 2013). In India also one-third area of the deciduous forests will be converted into evergreen due to extreme rainfall events by 2100 (Chaturvedi et al., 2011).

If the Himalayan ecosystem is taken into consideration, i.e., mountain ecosystem, over the last century, rainfall in India is decreased by 68% according to IITM, Pune whereas in Jammu and Kashmir, there is an increase in rainfall (Negi et al., 2012). Again a study shows that the annual mean temperature of the Alaknanda Valley has increased by 0.15°C from 1960 to 2006 (Kumar et al. 2008a; Gujree et al., 2017).

The impact of climate change on socioeconomic and health is also noticed in Himalaya (Chaturvedi et al., 2011). People who are more dependent on subsistence farming and forest products are more affected by climate change as the alteration in the productivity of the agriculture and forest will affect their livelihood. Human health is affected by extreme events occurring due to climate change such as death due to flood events, malnutrition due to insufficient yield and failure of crops, etc. We have seen a shift in tree line in the Himalayan region. Even the composition of species of an area is changing due to this change in climatic variables like temperature, rainfall pattern, etc. In the western Himalayan mountains, early flowering of several members of Rosaceae (e.g., Pyrus, Prunus spp.) and Rhododendrons has often been linked with global warming.

Increased frequency and intensity of extreme weather events are also evident in the Himalayas (Hassanin et al., 2020; Tomar et al., 2021; Meraj and Kumar, 2021; Chandel et al., 2021; Bera et al., 2021). Delayed but heavy rainfall in Jammu and Kashmir in September 2014, frequent cloud bursts of rainfall in Uttarakhand and Nepal in 2013 are also some indicators of climate change in Himalayas. The impact of climate change in Himalaya has local, regional as well as global implications as it poses the unmatched threat not only to the livelihood of local people, wildlife but also to the huge population living in the downstream and ultimately to the global environment. Change in climate is expected to induce shifts in potential natural vegetation and its associated biogeochemical and hydrological cycle. Rhododendron species found in Dhanolti are a very good indicator of climate change and variability. Through real-time observation from field for year 2009–2011 and herbarium records for 1893–2003, early flowering of 88–97 days is observed (Gaira et al., 2014).

Long-term studies on climate change in India have not been quantified well in numbers. For these, long-term observational data is required to see the impacts. Though, few studies based on "Forest Preservation Plots" have been carried out by various research institutions like Indian Council of Forestry Research and Education (ICFRE), Dehradun, Forest Survey of India, Indian Institute of Remote Sensing under different names.

Long-term monitoring and understanding the dynamics of climate change impacts to our forests is required to make the most informed decisions to protect the earth system. An understanding of forest structure, dynamics, productivity and function can contribute immensely to work out long-term strategies of sustainable forest management in a changing global environment by way predicting future impacts of climate change on forests. In order to make prediction of potential climate change impacts on forests more reliable we need to monitor and generate basic information on structure and function of different forest types in a time series for a long period with the help of laying permanent forest observation plots. Establishment of Permanent Preservation Plots (PPPs) in natural forests plays a significant role in assessing the impact of climate change on forests. The ecological studies would help to observe and record the changes in species diversity, composition and growth pattern due to climate change over a period of time.

This article has reviewed various studies on impacts of climate change in Indian forests with the help of Permanent Observational Plots (POPs) more precisely on Himalayan states and few Eastern states of India. The article shows that very less studies have been carried out in this regard, which has resulted into less amount of data availability for these regions. There is potential to conduct LTEM (Long Term Ecological monitoring) studies (Figure 12.1).

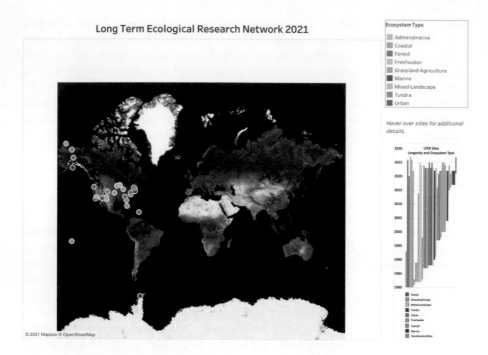

FIGURE 12.1 Figure showing Long Term Ecological Research Network till 2021. (https://lternet.edu/site/.)

12.2 NETWORK OF "PRESERVATION PERMANENT PLOT" ESTABLISHED IN INDIAN FORESTS SINCE ITS INCEPTION

Forest resource information, gathered in Forest Inventory Systems at local, national and global levels, is required for planning and policy decisions. Forest Research Plot Networks provide essential data for studying ecosystem structure and dynamics. Conservation and sustainable use of biological resources based on local knowledge systems and practices is embedded in the Indian culture. India has assigned a special status and protection to biodiversity-rich areas by declaring them as national parks, wildlife sanctuaries, biosphere reserves, ecologically fragile and sensitive areas. For sustainable forest management, Permanent Observational Plots (POPs) are required to lead way to analyze and research. POPs provide service to science and society through reliable information for the identification and solution of ecological problems and sustainable development (Tewari et al., 2014).

Linear tree increment plots, linear increment plots, linear sample plots and permanent preservation plots are actually the names of the same long-term ecological monitoring studies. Actually worldwide growth and yield information coupled with the forest inventory is the base on which forest planning in investment strategy rests. As usual common characteristics of these plots are the fact that tree diameters have been measured at periodic intervals. The *Hopea parviflora* forests in the south and the Sal forest in the north seem to have been the first where observational studies were started during the early years of the twentieth century. The oldest preservation plot was established in 1905 in the Sal forest of Bihar (Tewari et al., 2014). Five plots were laid out in Sal forests of the Daltonganj, Kolhan, Saranda and Porhat forest divisions in then Bihar (Now Jharkhand) during 1936–1968 (annual rainfall between 1,050 and 2,000 mm). These sites are in moist deciduous forests and the area of each plot is 3.2 ha. Measurements were taken at five-year intervals until 1981. Three of the plots are degraded.

During recent years several new observational studies were established by Indian Council of Forestry Research and Education and Forest Survey of India. These plots represent pilot studies demonstrate different methods of field assessment and qualitative analysis in the dense and species rich or open forest in steep or flat terrain using both circular and square plots of different sizes. This contribution presents the history and current state of forest research plots in India, including details of locations and re-measurements (Figure 12.2).

In addition to this, Indian Institute of Remote Sensing (IIRS), ISRO has tried to establish its Long Term Ecological Research Stations (LTERs) in different parts of Western Himalayan Regions primarily in Uttarakhand to study the impacts of climate-driven effects on the Himalayan ecosystem. Some LTERs established by Indian Institute of Remote Sensing, ISRO are:

- Dudhatoli LTER, Uttarakhand
- Pindari LTER, Uttarakhand
- Flux Tower at Barkot, Uttarakhand
- Flux Tower at Haldwani, Uttarakhand.

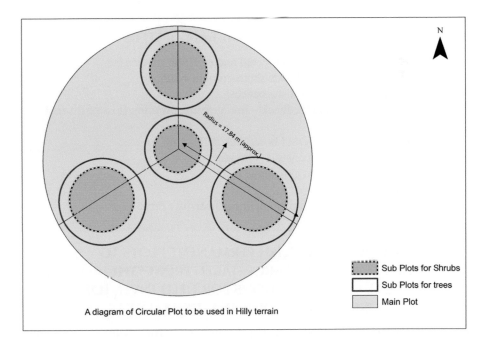

FIGURE 12.2   Circular plot of 0.1 ha for hilly terrain.

12.2.1 Why Is There the Need of Long-Term Ecological Research

- It helps to fill the gaps between theory and modern research
- Understand the impact of climate change on ecosystem structure and function
- Prove the empirical and theories
- Cooperation between empirical and application
- Integrate long-term observations.

12.3 ESTABLISHMENT OF PERMANENT LONG TERM ECOLOGICAL RESEARCH STATION

In view of the significant data requirement of social, biological, cryosphere to understand and forecast various sub-systems of ecosystem, it is necessary to establish a few Long Term Ecological Monitoring Stations (Meraj et al., 2021). These stations could provide the following data to understand the spatial and temporal variations in the ecosystem due to climate change.

- Vegetation data – Phenology, species turn over, productivity
- Disturbance – Grazing, extraction, tourism
- Climate – Rainfall, temperature, Humidity, PAR
- Atmosphere – NO_x and N loading
- Socio-economic – Population density, GDP, Population growth.

12.3.1 The Following Can be Answered Through LTERS

- Species loss – economic, endangered and endemic species etc.
- Succession patterns (grasslands/sholas, alpine systems)
- Patterns and frequency of disturbances
- Ecological damage and feedback – fire, grazing, thinning (loss of diversity, erosion)
- Pest and diseases, weeds (Sal borer/teak defoliator)
- Transient vegetation responses to climate
- Intra/inter annual carbon allocations
- Secondary formations
- Spatial and temporal organization of biodiversity (Table 12.1).

12.4 METHODS TO ESTABLISH PERMANENT PLOTS TO STUDY FOREST DYNAMICS (TAKEN FROM "THE MANUAL OF INSTRUCTIONS FOR FIELD INVENTORY, 2002, FOREST SURVEY OF INDIA, DEHRADUN.")

Generally, for long-term monitoring and research, two types of permanent sample plots are taken:

TABLE 12.1
Parameters Which Can be Studied Through LTERs

Area/Type Structure	Function
Forest cover density	Phytophenology
Vegetation type	Successional patterns
Undergrowth and regeneration	Fragmentation and corridor dynamics
Regional biodiversity inventory	Water balance and erosional processes
Local biodiversity inventory	Productivity, biomass assessment
Species-specific inventories	Decomposition
MFP inventories	Conservation and processes
Impact processes	Macro and micro habitat models
Forest fire proneness and impacts	Conservation zoning
Weed infestation	Agent-based change models
Grazing patterns	Carrying capacity assessment
MFP extractions	Macro and micro economics
Ecosystem services-sustainability	Type transition/species extinction – models
Pest and diseases	Carbon sequestration
Disturbance and climate change sensitivity	

1. Circular Plots (of 0.1 ha with a radius of approx. 17.84 m) preferably in hilly terrain (Himalayan region) (Figure 12.2).
2. Square/Rectangular plots (1 ha, maybe of 5 or 10 ha) in regular terrain (Figures 12.3 and 12.4).

A common protocol has been developed and followed for all the sites to collect data in a standardized manner so that the data could be compared between years and sites.

- Permanent observation plots will be established in Protected Areas of major forest types and grasslands.
- One, five or ten ha size plots will be laid in the site.
- Plots will be fenced to avoid anthropogenic disturbances if required.
- These plots will be gridded into quadrats of 20×20 m. Aluminium/PVC pipes with x and y axis coordinates stamped will be installed as boundary markers.
- For enumeration, each 20×20 m quadrats will be subdivided into 5×5 m subplots with ropes.
- All woody plants above 1 cm DBH will be identified, measured for DBH, numbered with aluminium tags for repeated measurements. For bamboos, the number of culms in each clump will be noted.
- Nested subplots will be laid for herbaceous vegetation.
- Litter traps will be placed for litter collection and decomposition study.
- These plots will be revisited periodically to study the periodic changes in the compositions and structure of the forest.

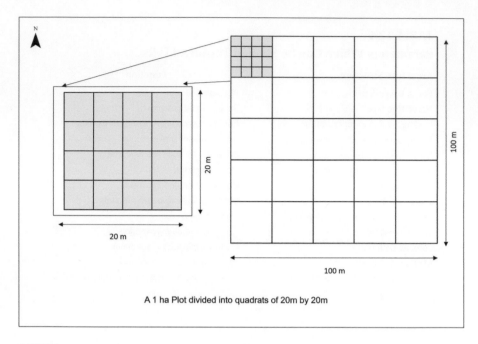

FIGURE 12.3   Figure showing a 1 ha plot subdivided into 20 by 20 m.

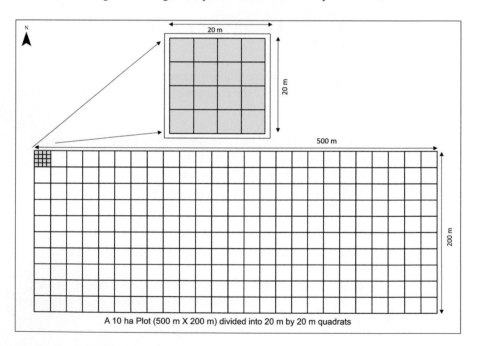

FIGURE 12.4   Figure showing a 10 ha plot (dimension – 500×200 m) subdivided into 20 by 20 m quadrats.

- Natural regeneration and recruitment status of different species will be recorded.
- Microbial diversity, butterfly population and other entomological observations, pest and disease incidence, phenological observations will be recorded from the plots following standard procedures.
- Carbon content in all the components of forest will be measured using CNS analyzer and carbon pool will be estimated accordingly.
- Representative soils from the study area will be analyzed and monitored for physico-chemical characters including bulk density and major soil nutrients (N. P. K. Ca, Mg).
- Precise estimate of biomass carbon (both above ground and below ground) and root-shoot ratio may be worked out for major tree species in all the forest types and allometric equations may be developed if permission for felling and uprooting could be obtained from SFD (otherwise, biomass carbon will be estimated using available equations and expansion factor).
- Dendrochronological study of selected tree species will also be carried out to study past climatic signature.
- Meteorological parameters/weather parameters will be measured through establishing weather station/ meteorological station at the site and will be compared with the observations recorded from nearest meteorological stations for studying microclimatic effect of forests.

12.4.1 Equipment Required for Installation and Layout of POPs

- GPS – for recording latitude/longitude of the origin (preferably SW corner of plot)
- Altimeter – for measuring altitude of the quadrants
- Compass – for measuring azimuth angle of individual tree
- Measuring Tape – for measuring distance between the origin to individual trees and to record the girth of each tree at breast height (Calipers in the case to measure the diameter)
- Range finder/Hypsometer – for measuring height, angle and distance of trees
- Concrete pillars, PVC pipes – for demarcation of boundaries of plot and quadrats
- Metal plates – to assign each quadrant a unique number and to write their altitudes
- Soil auger and soil core – to measure the surface carbon, bulk density of soil, moisture, texture, pH etc.
- Soil testing kit – to measure soil organic carbon, pH, moisture, nutrients in the soil
- Automated Weather Station (AWS) – to record weather parameters like temperature, rainfall, speed and direction of wind, soil radiation etc.

12.5   IMPACTS OF CLIMATE CHANGE ON FORESTS OF JHARKHAND, BIHAR AND WEST BENGAL

Global warming due to climate change usually results in injury on forests because of severe drought and wildfires. Such risks, in an exceedingly dynamical climate, loss of forest cover can have an effect on growth and production of trees and timbers. Further, forest fires, insect outbreaks, wind injury due to cyclonic hazards, and alternative extreme events lead to substantial loss to forest cover (Kanga et al., 2017a, b; Rather et al., 2018; Hassanin et al., 2020; Kanga et al., 2021). Subsequent upon such adverse effects, forest injury results reduction of biodiversity, negative impacts on erosion and geophysical features.

Forests of Jharkhand are mostly Tropical Dry Deciduous. Sal is dominant species of all forests. Only a few studies are done to assess the impact of global climate change on forests of Jharkhand. Sal species exhibits semi-evergreen habit and hence cannot cope up with the drought conditions. The dense forest of Saranda consisting of Sal is observed to be decreasing apace because of warming wherever as the nonforest area is increasing. The dense forest is showing the decreasing trend from 70,084 to 39,421 ha whereas the nonforest area has increased from 39,421 to 53946 ha over 1975 to 2015 (Gujree et al., 2020; Ahmad & Goparaju, 2017). Results from previous work suggest that the natural distribution of Sal has shrunk at a faster rate over the previous couple of decades (Chitale & Behera, 2012). The study by Chitale and Behera (2012) reveals that projected increase in annual precipitation and annual mean temperature might limit the distribution of Sal (the optimum annual mean temperature was 28°C, and annual precipitation ranges from 1,000 to 2,000 mm). Increase in rainfall variability and extreme drought conditions within the central and northern regions of Bangladesh might lead to unsuitable climate conditions for Sal forests.

As per the "Jharkhand – Action Plan on Climate Change", forests of Ranchi, Gumla, Pashchim Singhbhum and Gumla districts are most prone to future climate changes; all the five districts have social groups which are economically backward populations and these populations are vulnerable as they are mostly dependent on forests and traditional agriculture for their sustenance. Districts like Bokaro and Dhanbad are less vulnerable to climate change because of their high reconciling capability and less social vulnerability. The north-western part of Jharkhand, i.e., Garhwa, Palamu, Chatra, Koderma and the northern part of Hazaribag are exposed to moderate to high vulnerability. Except this region, Jharkhand is in less vulnerable states as compared to other states of India. Forests of Jharkhand are in growing condition so they are a good sink of carbon as they sequestered approximately 136.03 and 145.86 million tonnes of carbon in 2009 and 2011 respectively. There are 1,148 FSI grids in the state; it is projected (based on the A1B scenario) that due to climate change about 24.30 grids will get affected in 'long-term' periods, though there is no short-term threat perceived for the forests in the state. There is no negative impact predicted in medium term (by year 2035), whereas by the end of century (projections for year 2085) the forests in the north-western districts will come under severe stress. The A2 scenario predicts similar results projecting that by 2085 the forests in the north-western part of the state will become highly vulnerable due to temperature and rainfall variations. Furthermore, the intensity of fire events has increased in forests of Jharkhand which may be due to increase in the temperature of soil (IPCC AR 4, 2007).

The impact of global warming, through the predicted sea level rise, on Sundarbans estuary, in the Bay of Bengal is an inevitable phenomenon. While the mitigation measures have to be primarily at the global level as well as at local level, the adaptation strategy will also involve participation of global and local communities. As the sea level has increased, salinity in mangrove forests of Sundarbans is increasing which in turn reducing the productivity of these forests. The impact of salinity within the deltaic Sundarbans is important since it controls the distribution of species and productivity of the forest significantly. Due to increase in salinity, *Heritiera fomes* (Sundari) and *Nypa fruticans* (Golpata) are declining (Raha et al., 2013). This caused a significant effect on the biomass of the selected species thriving along these hypersaline river banks; the effects are species–specific. Increased salinity shows the positive influence on *A. alba* and *E. agallocha* whereas reduced growth in *S. apetala*.

In West Bengal, the response of climate change is likely to be multifaceted, particularly in the case of social forestry, where some species more suitable to climate will replace the earlier species which are no longer suitable for the changing climatic condition. Favorable climatic condition with changing temperature and precipitation pattern that produce a direct impact on natural and modified forest will eventually result in abundant growth and forest expansion in the state of West Bengal. Modeling of vegetation in 2004 using the BIOME vegetation model with inputs from a climate change scenario derived from IPCC IS92a, indicates that in the West Bengal region, as major increase in precipitation is expected, a shift in vegetation type toward the wetter, more evergreen type is expected. Since these are rather slow growing, the replacement will take much longer, and increased mortality in the existing vegetation may lead to a decrease in the standing stock except in the Western part of the state near Purulia and Birbhum where the vegetation might become xerophytes. Despite, there is no drastic shift in the biome type, changes in the composition of the assemblages are certainly very likely. Thus few species may show a steep decline in the population and perhaps may become extinct. This in turn will impact other taxa dependent on the different species (i.e., domino effect) because of the interdependent nature of the plant-animal-microbe communities that are known to exist in the forest ecosystems. This could lead to major changes in the forest biodiversity. Increase in CO2 concentration with accompanied temperature alters the Crop-weed competition depending upon their photosynthetic pathway. C3 crop growth would be favored over C4 weeds. Increase may alter the competition depending upon the threshold ambient temperatures. Diseases and insect populations are strongly dependent upon the temperature and humidity. Any increase in them, depending upon their base value, can significantly alter their population, which ultimately results in yield loss. With small changes, the virulence of different pests changes (Ghosh and Ghosal, 2020).

Bihar is an eastern Indian state with great cultural heritage. It is one of the oldest inhabited places in the world. Climate is one of the major factors that made Bihar a major agricultural hub. Due to its geographical location, its climate is greatly affected by the Himalayas and the Ganga plateau. Vegetation of the state is classified under climatic, serial and edaphic types. The climatic forest types are broadly classified into dry deciduous and moist deciduous types. Besides the deciduous forest, the vegetation types also include Sal, gregarious forests, bamboo forests, canebrakes and grasslands, covering relatively less areas. Sal mixed dry deciduous forests (northern dry mixed deciduous forests) occupy major proportion under natural vegetation cover and predominantly found in Kaimur, Rohtas, Aurangabad, Gaya, Nalanda, Nawada, Jamui, Banka and Munger districts. There is no in-depth study of impact of climate change on forests per say in the state. In general, climate change is likely to make the State's meagre forest resources more vulnerable on account of possible shift of the species of forest ecosystems and also on account of increased occurrences of fire, pest/diseases, invasive species, change in species assemblage/forest type, forest die-back and loss of biodiversity. It is also almost certain that forest-dependent livelihoods may get severely affected, increasing the vulnerability of local communities in and around forests in the State (Sagar, 2017).

There is a significant decreasing trend for the annual potential evapotranspiration in Bihar, except for Araria and Kishanganj, which show a statistically significant increasing trend. There is a significant increasing trend for the annual reference crop evapotranspiration for all the districts of Bihar (Roy et al., 2016).

Few studies have been done in India to get an idea of how the forest is changing temporally and spatially. These studies often used geospatial technologies for seeing the change temporally and spatially. Ahmad et al. (2018) observed that the long-term NDVI (1982–2006) evaluation revealed negative change trends in seven northwest districts of Jharkhand state; these were Hazaribag, Ramgarh, Palamu, Lohardaga, Chatra, Garhwa and Latehar. The forests as well as the agriculture of these districts have lost their greenness during this period. The forest fire frequency events were found to be more pronounced in the land use/land cover of "tropical lowland forests, broadleaved, evergreen, <1,000 m" category, and were roughly twice the intensity of the "tropical mixed deciduous and dry deciduous forests" category. The climate change vulnerability was found to be highest in the district of Saraikela followed by Pashchim Singhbhum, whereas agricultural vulnerability was found to be highest in the district of Pashchim Singhbhum followed by Saraikela, Garhwa, Simdega, Latehar, Palamu and Lohardaga. The temperature anomalies prediction for the year 2030 shows an increasing trend in temperature with values of 0.8°C–1°C in the state of Jharkhand. The highest increases were observed in the districts of Pashchim Singhbhum, Simdega and Saraikela.

Chakraborty et al. (2018) also found the same kind of negative change in greenness over years 2001–2014. Dry deciduous forests showed significant decrease in greenness over 1407.73 thousand ha, i.e., 6.5% of the total area the forest type. Jharkhand (51.18 thousand ha) showed a negative trend of its greenness. Kumar et al. (2010) observed significant degradation of forests in Ranchi district from 1996 to 2008. A positive mean of 2008–1996 NDVI differencing is an indication of reduction in above ground biomass within 12 years. This implies a decline in vegetation. Singh and Jeganathan (2016) compared the Enhanced Vegetation Index for 11 years (2001–2011) for the forests of Palamau region. This inter-year comparison about standard anomalies revealed an alarming situation about persistent stress in the vegetation, especially after 2008. However, the degradation rate is slow as per anomaly frequency but steadily increasing after 2008.

Invasive or alien species are known to have a great impact on the native biodiversity of a community. Survival and spread of the alien species are known to be affected both by disturbance and climate change. Exotic Species Invasion causes a little recognized, but very substantial impact to forest ecosystems worldwide. Climatic variability, physiographic range, increasing trade, travel and tourism have accelerated the spread of unwanted nonnative species to conservation areas, making vulnerable to the establishment of ESI. Exotic Invasive Plants (EIPs) are known to displace native plants, alter ecosystems processes, hydrology, primary productivity, nutrient cycling and soil structure and most importantly reduce native biodiversity. There is evidence to suggest that the threats due to ESI may increase with climate change and associated changes in habitats. *Lantana camara* and *Parthenium hysterophorus* are most common invasive species of plants in Jharkhand. They grow more rapidly than the native species and try to invade the area and hence the native species vanishes. Invasive exotic plant species (IEPS) threaten the environment, reduce biodiversity, replace economically important plant species and increase the investment in agriculture and silviculture practices, prevail vegetation dynamics and alter nutrient cycling (Richardson & Higgins, 1998). They can promote hazards like forest fire. Plant invasions dramatically affect the distribution, abundance and reproduction of many native species (Mooney and Ramakrishnan, 1999).

Roy et al. (2016) modeled the spread of Maling Bamboo in Darjeeling, West Bengal. The invasion potential of Maling bamboo in Darjeeling Himalayas has reached alarming potential and is a great concern for the policy makers and forest departments. Checking its potential spread in the region needs information on the dispersal potential as well as available area for its establishment. It has been observed from extensive field surveys as well as literature.

12.6 CLIMATE CHANGE PREDICTIONS: GEOSPATIAL SPECIES DISTRIBUTION MODEL

Anticipating the reaction of biodiversity to environmental change has turned into a tremendously dynamic field of research (Dillon et al., 2010; Beaumont et al., 2011; McMahon et al., 2011; Dawson et al., 2011). These forecasts assume an essential part in alerting researchers and policy makers to prospective future dangers, give a way to reinforce acknowledgment of biological changes to environmental change and can upkeep the advancement of proactive systems to reduce environmental change impacts on biodiversity. Recently, species responses to climate and LULC changes bring up a significant issue for conservationists and environmentalists. Predictive spatial modeling or species distribution models have been used to assess the impacts of overall changes on biodiversity distribution, by anticipating current species distributions and applying fact-based models from current distributions to expand future dispersals under overall global change (Singh et al., 2017b; Kanga et al., 2020a, b). This approach has become very useful for producing biogeographical data that can be connected over an expansive scope of fields, including conservation science, ecology and evolutionary biology.

Species distribution modeling is a useful tool to assess how patterns of species distribution may get modified under climate change. The modeling approach can be categorized into two broad categories: (i) mechanistic models and (ii) correlative models. Mechanistic models are intricate and sophisticated models which simulate complex and detailed relationship and tradeoffs between species and their biotic and abiotic surroundings. However, mechanistic models are very data intensive and need detailed information about various ecological and physiological parameters of a species which is rarely available. On the other hand, correlative models are statistical models which require relatively less data.

Correlative models utilize relationship between ecological factors and identified species location records to distinguish natural conditions inside which populations can be sustained using different algorithms. This correlative relationship can be combined with data on climate change to project movement of species climatic envelops across space in any temporal domain. The models are generally preferred as they considerably require less data and thus can be applied for a variety of species including rare species for expected climate change impacts on species distribution patterns across multiple taxa at large geographical scales (Berry et al., 2002; Thomas et al., 2004; Thuiller et al., 2005). Present-day progresses in species distribution models enable us to possibly forecast anthropogenic consequences on spatial organizations of biodiversity at various spatial scales. The initial attempt on species distribution modeling in the literature is by all accounts the niche-based spatial forecasts of yield species by Henry Nix and associates in Australia. Lately, predictive distribution modeling of species has become a progressively important approach to address different issues in environment, biogeography, and more importantly, in conservation science and climate change research.

Recently, MaxEnt software package (Phillips et al., 2006) has become one of the most widespread and popular models in the studies for species distribution/ environmental niche modeling. The advantage of the program is that the model uses presence-only species records for modeling species distributions. MaxEnt offers many advantages over other models, the most important of which (with regard to experimental design and data accessibility) is that it acknowledges both continuous and categorical predictor variables. The model is also very effective with a comparatively small number of species occurrence records (Pearson et al., 2007).

Deb et al. (2017) modeled climate space suitability for Sal and Garjan species, projected to 2070, using MaxEnt. Annual precipitation was the key bioclimatic variable in all GCMs for explaining the current and future distributions of Sal and Garjan (Sal: 49.97 ± 1.33; Garjan: 37.63 ± 1.19). The models predict that suitable climate space for Sal will decline 24% and 34% (the mean of the three GCMs) by 2070 under RCP 4.5 and RCP 8.5, respectively. In contrast, the consequences of imminent climate change appear less severe for Garjan, with a decline of 17% and 27% under RCP 4.5 and RCP 8.5, respectively. The findings of this study can be used to set conservation guidelines for Sal and Garjan by identifying vulnerable habitats in the region. In addition, the natural habitats of Sal and Garjan can be categorized as low to high risk under changing climates where artificial regeneration should be undertaken for forest restoration. The consequences of climate change may result in the absence of Sal and Garjan either locally or regionally, the disappearance of entire ecosystems, or their replacement by other ecosystem types.

12.7 CONCLUSION

It is clearly seen that there are potentials to study the impact of climate change through long-term monitoring in India. Forests of Jharkhand are moderately vulnerable to climate change but if the rate of climate change accelerates, the forest may not keep pace with the changing climate as the rate of adaptability is quite less. Worldwide chain of LTERs is also not enough for the studies as they are not evenly distributed but confined to particular places only. Local or regional data is extremely required for such studies as global data is easily available but there is a paucity of regional data. Studies based on Permanent Plots needed to be more widespread throughout India which was tried by FSI, but these can be given more boosts by local institutions. Species Distribution Models (SDMs) and Dynamic Vegetation Models (DVMs) have the potential to predict the future scenarios of change in vegetation due to these changing phenomena. Incorporation of these models into permanent plots and regional data can provide better results. Overall there is a potential to explore more in climate change domains.

REFERENCES

Ahmad, F., & Goparaju, L. (2017). Predicting forest cover and density in part of Porhat forest division, Jharkhand, India using geospatial technology and Markov chain. *Biosciences Biotechnology Research Asia, 14*(3), 961–976.

Ahmad, F., Uddin, M. M., & Goparaju, L. (2018). An evaluation of vegetation health and the socioeconomic dimension of the vulnerability of Jharkhand state of India in climate change scenarios and their likely impact: a geospatial approach. *Environmental & Socio-Economic Studies, 6*(4), 39–47.

Arshad, A., Zhang, Z., Zhang, W., & Gujree, I. (2019). Long-term perspective changes in crop irrigation requirement caused by climate and agriculture land use changes in Rechna Doab, Pakistan. *Water,* 11(8), 1567.

Beaumont, L. J., Pitman, A., Perkins, S., Zimmermann, N. E., Yoccoz, N. G., & Thuiller, W. (2011). Impacts of climate change on the world's most exceptional ecoregions. *Proceedings of the National Academy of Sciences,* 108(6), 2306–2311.

Bera, A., Taloor, A. K., Meraj, G., Kanga, S., Singh, S. K., Durin, B., and Anand, S. (2021). Climate vulnerability and economic determinants: linkages and risk reduction in Sagar Island, India; a geospatial approach. *Quaternary Science Advances,* 100038. doi: 10.1016/j.qsa.2021.100038.

Berry, P. M., Dawson, T. P., Harrison, P. A., and Pearson, R. G. (2002). Modeling potential impacts of climate change on the bioclimatic envelope of species in Britain and Ireland. *Global Ecology and Biogeography,* 11(6), 453–462.

Chakraborty, A., Seshasai, M. V. R., Reddy, C. S., & Dadhwal, V. K. (2018). Persistent negative changes in seasonal greenness over different forest types of India using MODIS time series NDVI data (2001–2014). *Ecological Indicators,* 85, 887–903.

Chandel, R. S., Kanga, S., & Singhc, S. K. (2021). Impact of COVID-19 on tourism sector: a case study of Rajasthan, India. *AIMS Geosciences,* 7(?), 224–242.

Chaturvedi, R. K., Gopalakrishnan, R., Jayaraman, M., Bala, G., Joshi, N. V., Sukumar, R., & Ravindranath, N. H. (2011). Impact of climate change on Indian forests: a dynamic vegetation modeling approach. *Mitigation and Adaptation Strategies for Global Change,* 16(2), 119–142. doi: 10.1007/s11027-010-9257-7.

Chitale, V., & Behera, M. (2012). Can the distribution of sal (*Shorea robusta* Gaertn. f.) shift in the northeastern direction in India due to changing climate. *Current Science,* 102, 1126–1135.

Choi, S., Lee, W. K., Kwak, D. A., Lee, S., Son, Y., Lim, J. H., & Saborowski, J. (2011). Predicting forest cover changes in future climate using hydrological and thermal indices in South Korea. *Climate Research*, 49(3), 229–245. doi: 10.3354/cr01026.

Corlett, R. T., & Westcott, D. A. (2013). Will plant movements keep up with climate change? *Trends in Ecology and Evolution*, 28(8), 482–488. doi: 10.1016/j.tree.2013.04.003.

Dawson, T. P., Jackson, S. T., House, J .I., Prentice, I. C., & Mace, G. M. (2011). Beyond predictions: biodiversity conservation in a changing climate. *Science*, 332(6025), 53–58.

Deb, J. C., Phinn, S., Butt, N., & McAlpine, C. A. (2017). The impact of climate change on the distribution of two threatened Dipterocarp trees. Ecology and Evolution, 7(7), 2238–2248.

Dillon, M. E., Wang, G., & Huey, R. B. (2010). Global metabolic impacts of recent climate warming. *Nature*, 467(7316), 704.

Gaira, K. S., Rawal, R., Rawat, B., & Bhatt, I. D. (2014). Impact of climate change on the flowering of Rhododendron arboreum in central Himalaya, India Impact of climate change on the flowering of Rhododendron arboreum in central Himalaya, India Kailash S. Gaira, Ranbeer S. Rawal, June. doi: 10.1093/aobpla/plt015.15.

Ghosh, M. & Ghosal, S. (2020). Climate change vulnerability of rural households in flood-prone areas of Himalayan foothills, West Bengal, India. *Environment, Development and Sustainability*, 1–26.

Gujree, I., Arshad, A., Reshi, A., & Bangroo, Z. (2020). Comprehensive Spatial planning of Sonamarg resort in Kashmir Valley, India for sustainable tourism. *Pacific International Journal*, 3(2), 71–99.

Gujree, I., Wani, I., Muslim, M., Farooq, M., & Meraj, G. (2017). Evaluating the variability and trends in extreme climate events in the Kashmir Valley using PRECIS RCM simulations. *Modeling Earth Systems and Environment*, 3(4), 1647–1662.

Hassanin, M., Kanga, S., Farooq, M., & Singh, S. K. (2020). Mapping of Trees outside Forest (ToF) From Sentinel-2 MSI satellite data using object-based image analysis. *Gujarat Agricultural Universities Research Journal*, 207–213.

IPCC, W. I. (2014). Summary for policymakers. Climate change (2014). Mitigation of climate change. *Contribution of Working Group III to the Fifth Assessment Report of the Intergovernmental Panel on Climate Change*, 1–33. doi: 10.1017/CBO9781107415324.

JAPCC (2014). Jharkhand – Action Plan on Climate Change. http://www.moef.nic.in/sites/default/files/sapcc/Jharkhand.pdf (Accessed on 2nd April 2017).

Kanga, S., Kumar, S., & Singh, S. K. (2017b). Climate induced variation in forest fire using Remote Sensing and GIS in Bilaspur District of Himachal Pradesh. *International Journal of Engineering and Computer Science*, 6(6), 21695–21702.

Kanga, S., Meraj, G., Das, B., Farooq, M., Chaudhuri, S., & Singh, S. K. (2020a). Modeling the spatial pattern of sediment flow in Lower Hugli Estuary, West Bengal, India by quantifying suspended sediment concentration (SSC) and depth conditions using geoinformatics. *Applied Computing and Geosciences*, 8, 100043.

Kanga, S., Rather, M. A., Farooq, M., & Singh, S. K. (2021). GIS based forest fire vulnerability assessment and its validation using field and MODIS data: a case study of Bhaderwah forest division, Jammu and Kashmir (India). *Indian Forester*, 147(2), 120–136.

Kanga, S., Sheikh, A. J., & Godara, U. (2020b). GIscience for groundwater prospect zonation. *Journal of Critical Reviews*, 7(18), 697–709.

Kanga, S., Singh, S. K., & Sudhanshu (2017a). Delineation of urban built-up and change detection analysis using multi-temporal satellite images. *International Journal of Recent Research Aspects*, 4(3), 1–9.

Kumar, K., Joshi, S., & Joshi, V. (2008a). Climate variability, vulnerability, and coping mechanism in Alaknanda catchment, Central Himalaya, India. *AMBIO*, 37, 286–291.

Kumar, P., Rani, M., Pandey, P. C., Majumdar, A., & Nathawat, M. S. (2010). Monitoring of deforestation and forest degradation using remote sensing and GIS: a case study of Ranchi in Jharkhand (India). *Report and Opinion*, 2(4), 14–20.

Liang, T., Feng, Q., Yu, H., Huang, X., Lin, H., An, S., & Ren, J. (2012). Dynamics of natural vegetation on the Tibetan Plateau from past to future using a comprehensive and sequential classification system and remote sensing data. *Grassland Science*, 58(4), 208–220. doi: 10.1111/grs.12000.

McMahon, S. M., Harrison, S. P., Armbruster, W. S., Bartlein, P. J., Beale, C. M., Edwards, M. E., Kattge, J., Midgley, G., Morin, X., & Prentice, I. C. (2011). Improving assessment and modeling of climate change impacts on global terrestrial biodiversity. *Trends in Ecology & Evolution*, 26, 249–259.

Meraj, G., & Kumar, S. (2021). *Economics of the natural capital and the way forward*. Preprints 2021, 2021010083. doi: 10.20944/preprints202101.0083.v1.

Meraj, G., Singh, S. K., Kanga, S., & Islam, M. N. (2021). Modeling on comparison of ecosystem services concepts, tools, methods and their ecological-economic implications: a review. *Modeling Earth Systems and Environment*, 1–20. https://doi.org/10.1007/s40808-021-01131-6

Mooney, H. A. & Ramakrishnan, P. S. (1999). Global change, biodiversity and ecological complexity. In: Walker, B., Steffen, W. L., Canadell, J., & Ingram, J. S. I. (eds) *The terrestrial biosphere and global change: implications for natural and managed ecosystems* (p. 304). Cambridge: Cambridge University Press.

Negi, G. C. S., Samal, P. K., Kuniyal, J. C., Kothyari, B. P., Sharma, R. K., & Dhyani, P. P. (2012). Impact of climate change on the western Himalayan mountain ecosystems: an overview. Tropical Ecology, 53(3), 345–356.

Pearson, R. G., Raxworthy, C. J., Nakamura, M., & Townsend Peterson, A. (2007). Predicting species distributions from small numbers of occurrence records: a test case using cryptic geckos in Madagascar. *Journal of Biogeography*, 34(1), 102–117.

Phillips, S. J., Anderson, R. P., & Schapire, R. E. (2006). Maximum entropy modeling of species geographic distributions. *Ecological Modeling*, 190(3–4), 231–259.

Raha, A. K., Zaman, S., Sengupta, K., Bhattacharyya, S. B., Raha, S., Banerjee, K., & Mitra, A. (2013). Climate change and sustainable livelihood programme: a case study from Indian Sundarbans. *Journal of Ecology*, 107(335348), 64.

Rather, M. A., Farooq, M., Meraj, G., Dada, M. A., Sheikh, B. A., & Wani, I. A. (2018). Remote sensing and GIS based forest fire vulnerability assessment in Dachigam National park, North Western Himalaya. *Asian Journal of Applied Sciences*, 11(2), 98–114.

Richardson, D. M., & Higgins, S. I. (1998). Pines as invasion in the Southern hemisphere. In: Richardson, D. M. (ed) *Ecology and biogeography of Pinus* (pp. 450–473). Cambridge: Cambridge University Press.

Roy, A., Bhattacharya, S., Ramprakash, M., & Kumar, A. S. (2016). Modeling critical patches of connectivity for invasive Maling bamboo (Yushania maling) in Darjeeling Himalayas using graph theoretic approach. *Ecological Modeling*, 329, 77–85.

Roy, L. B., Bhushan, M., & Kumar, R. (2016). Climate change in Bihar, India: a case study. *Journal of Water Resource and Hydraulic Engineering*, 5(3), 140–146.

Sagar, K. (2017). Climate change perception and adaptation strategies among various stakeholders in Nalanda District of Bihar (India), *Disaster Advances*, 10(8), 39 47.

Singh, B., & Jeganathan, C. (2016). Spatio-temporal forest change assessment using time series satellite data in Palamu district of Jharkhand, India. *Journal of the Indian Society of Remote Sensing*, 44(4), 573–581.

Singh, S. K., Mishra, S. K., & Kanga, S. (2017b). Delineation of groundwater potential zone using geospatial techniques for Shimla city, Himachal Pradesh (India). *International Journal for Scientific Research and Development*, 5(4), 225–234.

Solomon S., Qin D., Manning M., Chen Z., Marquis M., Averyt K., Tignor M., & Miller H. (2007). IPCC fourth assessment report (AR4). Climate change, 374.

Tewari, V. P., Sukumar, R., Kumar, R., & Gadow, K. (2014). Forest observational studies in India: past developments and considerations for the future. *Forest Ecology and Management*, 316, 32–46.

The Manual of Instructions for Field Inventory (2002). Forest survey of India, Dehradun.

Thomas, C. D., Cameron, A., Green, R. E., Bakkenes, M., Beaumont, L. J., Collingham, Y. C., Erasmus, B. F., De Siqueira, M. F., Grainger, A., Hannah, L., & Hughes, L. (2004). Extinction risk from climate change. *Nature*, 427(6970), 145.

Thuiller, W., Lavorel, S., Araújo, M. B., Sykes, M. T., & Prentice, I. C. (2005). Climate change threats to plant diversity in Europe. *Proceedings of the National Academy of Sciences*, 102(23), 8245–8250.

Tomar, J. S., Kranjčić, N., Đurin, B., Kanga, S., & Singh, S. K. (2021). Forest fire hazards vulnerability and risk assessment in Sirmaur district forest of Himachal Pradesh (India): a geospatial approach. *ISPRS International Journal of Geo-Information*, 10(7), 447.

Wang, H. (2013). A multi-model assessment of climate change impacts on the distribution and productivity of ecosystems in China. *Regional Environmental Change*, 14(1), 133–144. doi: 10.1007/s10113-013-0469-8.

13 Analyzing the Relationship of LST with MNDWI and NDBI in Urban Heat Islands of Hyderabad City, India

Subhanil Guha and Himanshu Govil
National Institute of Technology

CONTENTS

13.1 Introduction 241
13.2 Study Area and Data.. 242
13.3 Methodology.. 244
    13.3.1 Image Preprocessing... 244
    13.3.2 Extraction of Different LU–LC Types Using MNDWI and NDBI 244
    13.3.3 Retrieving LST from Landsat-8 TIR Band 245
    13.3.4 Mapping UHI... 246
    13.3.5 Delineating the UHS 246
13.4 Results and Discussion 246
    13.4.1 Spatial Distribution of MNDWI and NDBI 246
    13.4.2 Spatial Distribution of LST 247
    13.4.3 Spatial Distribution of UHIs and Non-UHIs.................................... 248
    13.4.4 Identification of UHSs 248
    13.4.5 Relationship of LST with MNDWI and NDBI for Whole City, UHIs, Non-UHIs, and UHSs 248
13.5 Conclusions.. 251
Acknowledgements... 253
Disclosure Statement 253
References.. 253

13.1 INTRODUCTION

The urban heat island (UHI) effect presents the higher air and land surface temperature (LST) in urban areas in comparison to the surroundings, generated by high levels of near-surface energy emission, solar radiation absorption of ground objects, and low rates of evapotranspiration (Buyantuyev and Wu 2010; Kleerekoper, van Esch and Salcedo 2012; Oke 1982, 1997; Rizwan, Dennis and Liu 2008; Farooq and Muslim 2014). The UHI of big cities has increased gradually since the last few decades (Akbari, Pomerantz and Taha 2001; Oke 1976; Stone 2007), with urban concentrations generating modeled and observed changes in regional temperatures (Georgescu et al. 2011; He et al. 2007; Kalnay and Cai 2003; Li et al. 2004). The relationship between landscape pattern and UHI becomes globally considerable (Chen et al. 2014; Coseo and Larsen 2014; Du et al. 2016a, b; Li et al. 2011; Peng et al. 2016). A large number of studies considered that the built-up area and bare land accelerate the effect of UHI, whereas green space and water reduce the UHI intensity (Amiri et al. 2009; Song et al. 2014; Guha et al. 2020a–e; Meraj et al. 2021a, b). Furthermore, LST is controlled by a complex pattern of landscape composition and configuration (Zhou et al. 2014; Asgarian, Amiri and Sakieh 2015; Guha et al. 2018; Guha, Govil and Diwan 2019), and natural and anthropogenic factors simultaneously generate certain effects on LST pattern (Buyantuyev and Wu, 2010; Jenerette et al. 2007; Kuang et al. 2015; Guha and Govil 2020a–c; Kanga et al., 2017a; Kanga et al., 2017b; Rather et al. 2018; Hassanin et al. 2020; Kanga et al. 2021). Landsat thermal infrared (TIR) data have been utilized for local-scale studies of UHI (Weng 2001; Weng and Yang 2004; Chen, Wang and Li 2002; Bendib, Dridi and Kalla 2017; Zhang et al. 2016). A variety of algorithms have been developed to retrieve LST from Landsat data, such as mono-window algorithm (Qin, Karniel and Berliner 2001), single-channel algorithm (Jiménez-Muñoz and Sobrino 2003; Jiménez-Muñoz et al. 2009), etc. Urban hot spots (UHSs), the special urban thermal features, experience extreme heat stress, mainly developed by man-made activities within a UHI zone (Chen et al. 2006; Coutts, White and Tapper 2016; Feyisa et al. 2016; Ren et al. 2016; Lopez, Heider and Scheffran 2017; Pearsall 2017). So, identification of these UHSs for mitigation purposes becomes an important task to maintain the ecological balance within a city. Different scholars in different time periods attempted to draw a correlation between LST in UHIs and some land use/land cover (LULC) indices for different study areas (Chen et al. 2006; Weng and Yang 2004; Ma, Wu and He 2016; Deilami and Kamruzzaman 2017; Tran et al. 2017; Nie et al. 2016; Mushore et al. 2016; Estoque, Murayama and Myint 2017; Mathew et al. 2016; Govil et al. 2019, 2020; Gujree et al. 2020).

The specific objectives of this study are (i) to retrieve LST from Landsat-8 OLI/TIRS data for identifying the UHIs, non-UHIs, and UHSs based on the retrieved LST over the whole Hyderabad city and (ii) to correlate LST with modified normalized difference water index (MNDWI) and normalized difference built-up index (NDBI) for the whole city and for the UHIs, non-UHIs, and UHSs existing within the city.

13.2 STUDY AREA AND DATA

Hyderabad city, capital of Telangana state in Southern India, located along the bank of Musi River, has been selected for the entire research work. Hyderabad is located between 17°12′01″N–17°36′06″N and 78°10′02″E–78°39′02″E (Figure 13.1). It has an average elevation of 542 m. The city is characterized by a number of artificial lakes. Overall, the city is in tropical savannah (Aw) type of climate, but periphery lands are characterized by a hot and semi-arid type of climate (BSh). Summer months (March-June) are characterized by hot and humid climate. The average annual range of temperature is 21°C–33°C, and average annual precipitation is 174 cm. A standard sex ratio (945) and effective literacy rate (>82%) indicate that Hyderabad is under a stable demographic condition. The rate of urbanization is high during the recent years. It is the fourth most populous city and the sixth most populous urban agglomeration of India.

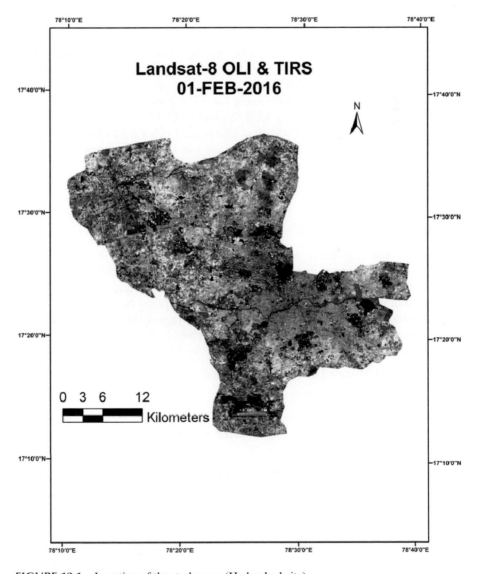

FIGURE 13.1   Location of the study area (Hyderabad city).

TABLE 13.1
Landsat 8 OLI and TIRS Satellite Image

Data	TIR Band Resolution (m)	Path/Row	Time	Date of Acquisition
Landsat 8 OLI/TIRS	100×100	144/048	05:09:43	1 February 2016

Landsat-8 OLI/TIRS data of 1 February 2016 (Figure 13.1) was used to determine the LST and to detect the UHIs and UHSs over whole Hyderabad city due to its cloud-free nature. The dataset was created by the U.S. Geological Survey and was obtained in geographic Tagged Image-File format (GeoTIFF). The data specification of Landsat-8 OLI/TIRS sensor is shown above (Table 13.1). Landsat-8 TIRS dataset has two bands (bands 10 and 11) having thermal characteristics. But, due to the larger calibration uncertainty associated with TIR band 11, it is recommended that users should work with TIRS Band 10 data as a single spectral band and should not attempt a split-window algorithm using both TIR Bands 10 and 11. The optical bands have been used in developing MNDWI and NDBI. A geographic database for Hyderabad city created by the Hyderabad Municipal Corporation was also used for this study. High-resolution Google Earth Image was used as the reference data for LULC identification.

13.3 METHODOLOGY

13.3.1 IMAGE PREPROCESSING

The satellite data acquired from the Landsat-8 sensor was subset to limit the data size. The thermal infrared band for Landsat-8 TIRS image (band 10) has a spatial resolution of 100 m. This thermal band was resampled using the nearest neighbor algorithm with a pixel size of 30 m to match the optical bands. To analyze the changes in temperature and LULC types in the study region, the MNDWI and NDBI were applied to determine the correlation with derived LST.

13.3.2 EXTRACTION OF DIFFERENT LU–LC TYPES USING MNDWI AND NDBI

Normalized Difference Water Index (NDWI) is frequently used for water area extraction. The NDWI proposed by McFeeters (1996) is expressed by the equation as follows:

$$(\text{Green Band} - \text{NIR Band})/(\text{Green Band} + \text{NIR Band})$$

To remove the built-up noise in city area, a modified version of NDWI has been introduced by Xu (2006) by substituting the near-infrared band by middle infrared band. MNDWI has been used in this paper for extracting the water bodies. NDBI (Zha, Gao and Ni 2003) was applied in this study to detect the built-up area. These two indices can be applied to categorize different types of LULC (Chen et al. 2006)

TABLE 13.2

Indices Used for Extraction of Different LULC Types

Indices	Equation	Threshold Values for Different LULC Types			
		Vegetation	Water Body	Built-Up	Bare Land
MNDWI	(Green − MIR)/(Green + MIR)	>0.05	>0	<0	<0
NDBI	(MIR − NIR)/(MIR + NIR)	<0	<0	0.10-0.30	>0.30

by the suitable threshold values (Table 13.2). To get more accurate classification, Boolean operators may be used on the spectral bands of the indices. For example, NDBI < 0 and MNDWI > 0 may be used together to extract water bodies. But these threshold values may differ due to atmospheric conditions. These values may also be integrated for LULC classification.

13.3.3   RETRIEVING LST FROM LANDSAT-8 TIR BAND

LST was retrieved from band 10 of the Landsat-8 OLI/TIRS image of Hyderabad city using the following algorithm (Artis and Carnahan 1982):

$$L_\lambda = 0.0003342 * DN + 0.1 \tag{13.1}$$

where L_λ is the spectral radiance in W/m^2/sr/mm. After that, an atmospheric correction method was applied to obtain the at-surface reflectance (Chavez 1996):

$$T_B = \frac{K_2}{\ln\left(\left(K_1/L_\lambda\right)+1\right)} \tag{13.2}$$

where T_B is the brightness temperature in Kelvin (K), L_λ is the spectral radiance in W/m^2/sr/mm, and K_2 K_1 are calibration constants. For Landsat-8 OLI/TIRS data, K_1 is 774.89, and K_2 is 1321.08.

The surface emissivity ε was estimated using the NDVI threshold method (Sobrino, Munoz and Paolini 2004). The fractional vegetation F_v, of each pixel, was determined from the NDVI using the following equation (Carlson and Ripley 1997):

$$F_v = \left(\frac{\text{NDVI} - \text{NDVI}_{\min}}{\text{NDVI}_{\max} - \text{NDVI}_{\min}}\right)^2 \tag{13.3}$$

where NDVI$_{\min}$ is the minimum NDVI value (0.2) where pixels are considered as bare soil, and NDVI$_{\max}$ is the maximum NDVI value (0.5) where pixels are considered as healthy vegetation.

$d\varepsilon$ is the effect of the geometrical distribution of the natural surfaces and internal reflections. For heterogeneous and undulating surfaces, the value of $d\varepsilon$ may be 2%:

$$d\varepsilon = (1 - \varepsilon_s)(1 - F_v)F\varepsilon_v \tag{13.4}$$

where ε_v is the vegetation emissivity, ε_s is the soil emissivity, F_v is the fractional vegetation, and F is the shape factor whose mean is 0.55 (Sobrino, Munoz and Paolini 2004):

$$\varepsilon = \varepsilon_v F_v + \varepsilon_s (1 - F_v) + d\varepsilon \tag{13.5}$$

where ε is the emissivity. From Equations (13.4) and (13.5), ε may be determined by the following equation:

$$\varepsilon = 0.004 * F_v + 0.986 \tag{13.6}$$

Finally, the LST was derived using the following equation (Weng and Yang 2004):

$$LST = \frac{T_B}{1 + \left(\lambda \sigma T_B / (hc)\right) \ln \varepsilon} \tag{13.7}$$

where λ is the effective wavelength (10.9 μm for band 10 in Landsat8 data), σ is the Boltzmann constant (1.38×10^{-23} J/K), h is Plank's constant (6.626×10^{-34} Js), c is the velocity of light in a vacuum (2.998×10^{-8} m/sec), and ε is the emissivity.

13.3.4 MAPPING UHI

UHI and non-UHI were identified by the range of LST determined by the following equations (Ma, Kuang and Huang 2010; Guha, Govil and Mukherjee 2017):

$$LST > \mu + 0.5 * \sigma \tag{13.8}$$

$$0 < LST \leq \mu + 0.5 * \sigma \tag{13.9}$$

where μ and σ are the mean and standard deviation of LST in the study area, respectively.

13.3.5 DELINEATING THE UHS

In this study, LST maps were used in delineating the UHS over Hyderabad to provide special emphasis for continuous monitoring. These small pockets are too hot and unfavorable for human settlement and mostly developed within the UHI. These UHSs were delineated by the following equation (Guha, Govil and Mukherjee 2017):

$$LST > \mu + 2 * \sigma \tag{13.10}$$

These UHSs may be referred to as the most heated zones in an urban area. These can be generated naturally by a large extension of bare land with exposed rock surface or sand deposits. But these special types of urban features are generally developed by the large industrial activities, the huge flow of transportation, or the concentration of thermal power plants (Meraj and Singh, 2021).

13.4 RESULTS AND DISCUSSION

13.4.1 SPATIAL DISTRIBUTION OF MNDWI AND NDBI

The descriptive statistics (Table 13.3) presents the reliable nature of MNDWI and NDBI (Figure 13.2).

TABLE 13.3

Descriptive Statistics of MNDWI and NDBI for Hyderabad City

	Min	Max	Mean	Std
MNDWI	−0.6554	0.2975	−0.1527	0.0742
NDBI	−0.4174	0.5376	0.0173	0.0557

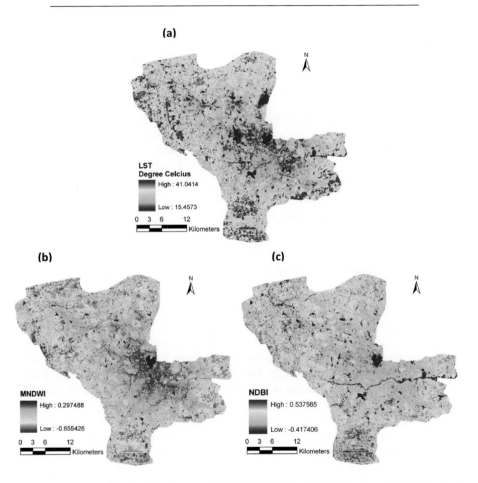

FIGURE 13.2   Spatial distribution map of LST, MNDWI, and NDBI for the whole Hyderabad city. (a) LST 01–02–2016. (b) MNDWI 01–02–2016. (c) NDBI 01–02–2016.

13.4.2   SPATIAL DISTRIBUTION OF LST

The retrieved LST image is shown in (Figure 13.2). LST distribution was classified into appropriate ranges and color-coded to generate a thermal pattern distribution map of LST over the study area. The mean and maximum LST values are 32.0042°C and 41.0414°C. A little heterogeneity in LST was developed due to the LULC

TABLE 13.4

Spatial Distribution of LST (°C) in Hyderabad City

LST (Min)	LST (Max)	LST (Mean)	LST (Std)	Threshold LST for UHI	Threshold LST for UHS
15.45	41.04	32.00	2.12	33.06	36.25

dynamics. The high rate of activities in manufacturing and transport sector along with power generation accelerates the mean LST (Table 13.4).

13.4.3 SPATIAL DISTRIBUTION OF UHIs AND NON-UHIs

The intensity of UHI may be defined as the difference between the average temperature of UHIs and non-UHIs (Table 13.5). In Hyderabad, the UHI zones are consistently concentrated in the north, north-west, south, and south-east peripheral region (Figure 13.3). The threshold value of UHIs is estimated as 33.0665°C. The standard deviation of UHIs is 0.8636, which reflects more concentration of LST. Conversely, the standard deviation value (1.61) of non-UHIs shows more variability. On an average, the calculated mean LST value of UHIs is 3.3746°C more than non-UHIs (Table 13.5).

13.4.4 IDENTIFICATION OF UHSs

In this study, UHSs were delineated to provide special emphasis for continuous monitoring. These small pockets are developed within the UHIs. These places are too hot and unfavorable for human settlement. The highest LST of satellite images is found in those hot spots. UHSs were more abundant over the bare land areas due to lack of both vegetation and shadows despite the higher albedo in the exposed surface. Some scattered hot spots also exist over the built-up areas and are more abundant along the western and southern boundaries of the city (Figure 13.4). The UHSs were identified by a threshold value of 36.2532°C. Bare land, exposed rock, parking area, roadways, power plants, metal roofs, and industrial factories are the most suitable places for the development of UHS. Almost all such hot spots have a little or a negligible amount of vegetation and water bodies.

13.4.5 RELATIONSHIP OF LST WITH MNDWI AND NDBI FOR WHOLE CITY, UHIs, NON-UHIs, AND UHSs

Generally, LST presents a positive relationship with NDBI and an inverse relationship with MNDWI. Table 13.6 shows the correlation coefficient of LST with MNDWI and NDBI for the whole city, for UHIs, for non-UHIs, and for UHSs of Hyderabad city. MNDWI shows strong negative correlation with LST for whole Hyderabad city (−0.73153) and for non-UHIs (−0.67375). This strength becomes weaker for UHIs (−0.17982), and for UHSs, it presents a slightly positive value (0.19062). NDBI

TABLE 13.5

Variation of LST (°C) in UHIs and Non-UHIs

LST (Min)		LST (Max)		LST (Mean)		LST (Std)	
UHIs	Non-UHIs	UHIs	Non-UHIs	UHIs	Non-UHIs	UHIs	Non-UHIs
33.0665	15.4573	41.0414	33.0665	34.2489	30.8743	0.8636	1.6115

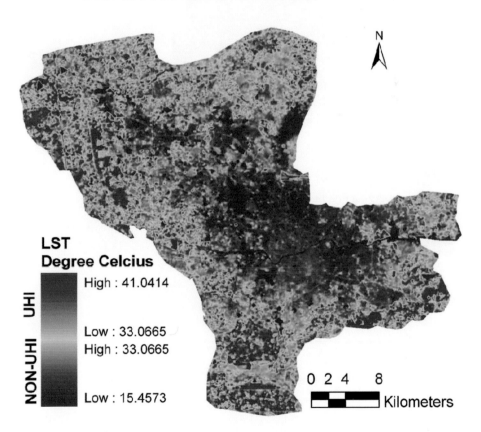

FIGURE 13.3 Spatial extents of UHIs and non-UHIs.

presents highly positive correlation with LST for whole Hyderabad city (0.62778), and for non-UHIs (0.54782). But, in UHIs, the value of correlation coefficient becomes much lower (0.26125), while in UHSs, it gives a slightly negative value (−0.06029). It may be due to the presence of complexity in landscape composition. So, LST generates a strong correlation for a wide area. But, it tends to be weak or even inverse correlation with the increase of LST concentrated in a small area.

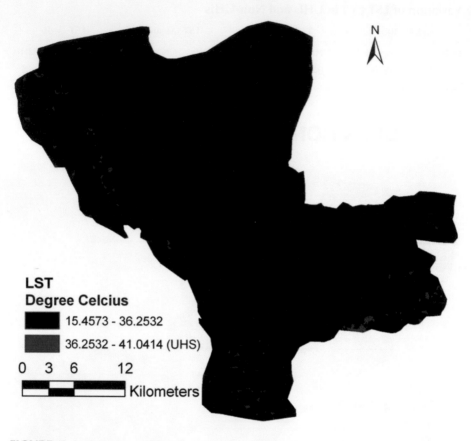

FIGURE 13.4   Location of UHSs.

TABLE 13.6

Correlation Coefficient Values of LST with MNDWI and NDBI

	Whole Hyderabad City	UHIs	Non-UHIs	UHSs
LST with MNDWI	−0.73153	−0.17982	−0.67375	0.19062
LST with NDBI	0.62778	0.26125	0.54782	−0.06029

Figure 13.5 represents the LST map and the corresponding MNDWI and NDBI map for UHIs. It is visually clear that high LST values are characterized by low MNDWI values and high NDBI values. For non-UHIs, the relationship status was almost the same between LST and MNDWI and NDBI (Figure 13.6). Figure 13.7 provides the relationship and distribution of LST with MNDWI and NDBI only

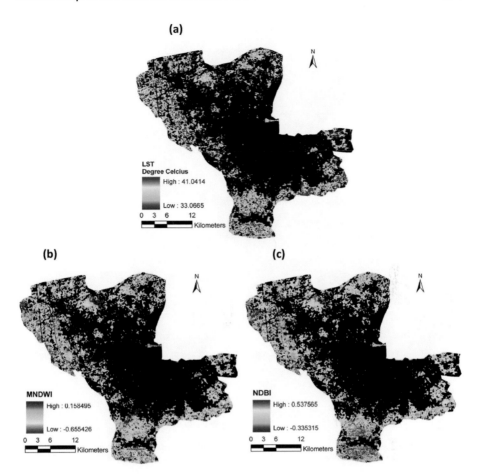

FIGURE 13.5 Relationship between LST with MNDWI and NDBI for UHIs. Black color represents the non-UHIs. (a) UHI 01–02–2016. (b) MNDWI 01–02–2016. (c) NDBI 01–02–2016.

within the UHSs. Due to complexities and uncertainties of composed materials in UHSs, MNDWI and NDBI reveal a slightly inverse relationship with LST.

13.5 CONCLUSIONS

In this chapter, Landsat-8 OLI/TIRS data was used to investigate the UHI intensity effect in Hyderabad city in India and to interpret the dynamic relationship between LST with MNDWI and NDBI. UHIs are identified through LST, which are distributed in the north, north-west, south, and south-east portions of Hyderabad. Bare land and built-up area are mostly responsible for LST generation. The presence of vegetation and water bodies reduces the LST level. Some UHSs were also delineated within the UHIs, which are characterized by high concentrated LST. The total area under UHI remains approximately the same indicating that most urban and industrial activities

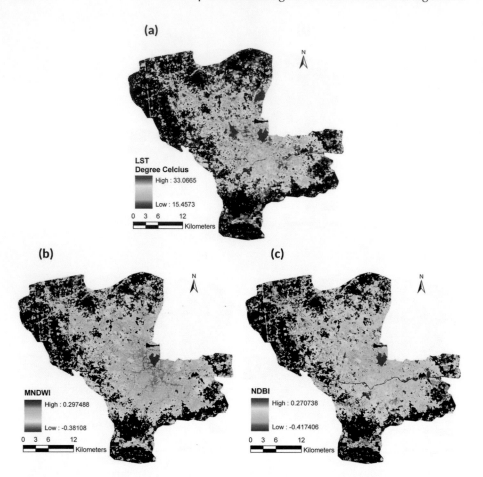

FIGURE 13.6 Relationship between LST with MNDWI and NDBI for non-UHIs. Black color represents the UHIs. (a) Non-UHI 01–02–2016. (b) MNDWI 01–02–2016. (c) NDBI 01–02–2016.

are concentrated in nature. Furthermore, the relationship of LST to MNDWI and NDBI was interpreted quantitatively by linear regression analysis at the pixel level. For whole Hyderabad city, LST shows strong negative correlation with MNDWI and strong positive correlation with NDBI. The relationship becomes weaker or even tends to be inverse for UHIs and UHSs. It may be due to the heterogeneous landscape in the urban area. In future, many additional research works may be included. Firstly, LST may be retrieved using another method or from another satellite image, and then the same methods may be followed to examine the validity of the results. Secondly, the *in situ* LST data may be measured with the same overpass of satellites for the calibration and validation of LST estimation. Thirdly, satellite images with various seasons in a particular year may be analyzed to get more reliability in UHI calculation. Finally, apart from linear regression, several new statistical methods can be applied to determine the correlation between LST and different LULC indices.

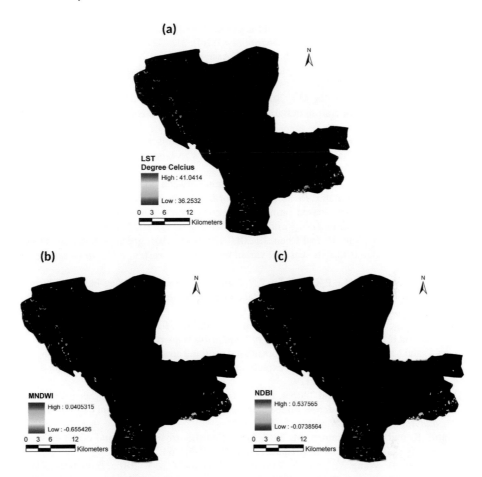

FIGURE 13.7 Relationship between LST with MNDWI and NDBI for UHSs. Black color represents the Non-UHSs. (a) UHS 01–20–2016. (b) MNDWI 01–02–2016. (c) NDBI 01–02–2016.

ACKNOWLEDGEMENTS

The author is thankful to the United States Geological Survey (USGS).

DISCLOSURE STATEMENT

No potential conflict of interest was reported by the author.

REFERENCES

Akbari, H., Pomerantz, M. and Taha, H. 2001. Cool surfaces and shade trees to reduce energy use and improve air quality in urban areas. *Solar Energy* 70: 295–310.
Amiri, R., Weng, Q., Alimohammadi, A. and Alavipanah, S.K. 2009. Spatial-temporal dynamics of land surface temperature in relation to fractional vegetation cover and land use/cover in the Tabriz urban area, Iran. *Remote Sensing of Environment* 113: 2606–2617.

Artis, D.A. and Carnahan, W.H. 1982. Survey of emissivity variability in thermography of urban areas. *Remote Sensing of Environment* 12(4): 313–329.

Asgarian, A., Amiri, B.J. and Sakieh, Y. 2015. Assessing the effect of green cover spatial patterns on urban land surface temperature using landscape metrics approach. *Urban Ecosystem* 18: 209–222.

Bendib, A., Dridi, H. and Kalla, M.I. 2017. Contribution of Landsat-8 data for the estimation of land surface temperature in Batna city, Eastern Algeria. *Geocarto International* 32(-5): 505–513.

Buyantuyev, A. and Wu, J. 2010. Urban heat islands and landscape heterogeneity: Linking spatiotemporal variations in surface temperatures to land-cover and socio-economic patterns. *Landscape Ecology* 25: 17–33.

Carlson, T.N. and Ripley, D.A. 1997. On the relation between NDVI, fractional vegetation cover, and leaf area index. *Remote Sensing of Environment* 62: 241–252.

Chavez, P.S. 1996. Image-based atmospheric correction – Revisited and improved. *Photogrammetric Engineering and Remote Sensing* 62 (9): 1025–1036.

Chen, A., Yao, X.A., Sun, R. and Chen, L. 2014. Effect of urban green patterns on surface urban cool islands and its seasonal variations. *Urban Forestry & Urban Greening* 13: 646–654.

Chen, X.C., Zhao, H.M., Li, P.X. and Yin, Z.Y. 2006. Remote sensing image-based analysis of the relationship between urban heat island and land use/cover changes. *Remote Sensing of Environment* 104 (2): 133–146.

Chen, Y., Wang, J. and Li, X. 2002. A study on urban thermal field in summer based on satellite remote sensing. *Remote Sensing of Land Management and Planning* 4: 55–59.

Coseo, P. and Larsen, L. 2014. How factors of land use/land cover, building configuration, and adjacent heat sources and sinks explain urban heat islands in Chicago. *Landscape and Urban Planning* 125: 117–129.

Coutts, A.M., White, E.C. and Tapper, N.J. 2016. Temperature and human thermal comfort effects of street trees across three contrasting street canyon environments. *Theoretical and Applied Climatology* 124 (1–2): 55–68.

Deilami, K. and Kamruzzaman, M. 2017. Modelling the urban heat island effect of smart growth policy scenarios in Brisbane. *Land Use Policy* 64: 38–55.

Du, H., Wang, D., Wang, Y., Zhao, X., Qin, F., Jiang, H. and Cai, Y. 2016a. Influences of land cover types, meteorological conditions, anthropogenic heat and urban area on surface urban heat island in the Yangtze River Delta Urban Agglomeration. *Science of the Total Environment* 571: 461–470.

Du, S., Xiong, Z., Wang, Y. and Guo, L. 2016b. Quantifying the multilevel effects of landscape composition and configuration on land surface temperature. *Remote Sensing of Environment* 178: 84–92.

Estoque, R.C., Murayama, Y. and Myint, S.W. 2017. Effects of landscape composition and pattern on land surface temperature: An urban heat island study in the megacities of Southeast Asia. *Science of the Total Environment* 577: 349–359.

Farooq, M. and Muslim, M. 2014. Dynamics and forecasting of population growth and urban expansion in Srinagar city – A geospatial approach. *The International Archives of Photogrammetry, Remote Sensing and Spatial Information Sciences* 40(8): 709.

Feyisa, G.L., Meilby, H., Jenerette G.D. and Pauliet, S. 2016. Locally optimized separability enhancement indices for urban land cover mapping: Exploring thermal environmental consequences of rapid urbanization in Addis Ababa, Ethiopia. *Remote Sensing of Environment* 175: 14–31.

Georgescu, M., Moustaoui, M., Mahalov, A. and Dudhia, J. 2011. An alternative explanation of the semiarid urban area "oasis effect". *Journal of Geophysical Research* 116: D24113.

Govil, H., Guha, S., Dey, A. and Gill, N. 2019. Seasonal evaluation of downscaled land surface temperature: A case study in a humid tropical city. *Heliyon* 5(6): e01923. https://doi.org/10.1016/j.heliyon.2019.e01923

Govil, H., Guha, S., Diwan, P., Gill, N. and Dey, A. 2020. Analyzing linear relationships of LST with NDVI and MNDISI using various resolution levels of Landsat 8 OLI/TIRS data. In: Sharma, N., Chakrabarti, A., & Balas, V. (eds) *Data Management, Analytics and Innovation. Advances in Intelligent Systems and Computing*, vol. 1042. Springer, Singapore, pp. 171–184. https://doi.org/10.1007/978-981-32-9949-8_13

Guha, S. and Govil, H. 2020a. An assessment on the relationship between land surface temperature and normalized difference vegetation index. *A Multidisciplinary Approach to the Theory and Practice of Sustainable Development* 23(2): 1944–1963.

Guha, S. and Govil, H. 2020b. Seasonal impact on the relationship between land surface temperature and normalized difference vegetation index in an urban landscape. *Geocarto International*. https://doi.org/10.1080/10106049.2020.1815867.

Guha, S. and Govil, H. 2020c. Land surface temperature and normalized difference vegetation index relationship: A seasonal study on a tropical city. *SN Applied Sciences* 2: 1661.

Guha, S., Govil, H. and Diwan, P. 2019. Analytical study of seasonal variability in land surface temperature with normalized difference vegetation index, normalized difference water index, normalized difference built-up index, and normalized multiband drought index. *Journal of Applied Remote Sensing* 13(2): 024518.

Guha, S., Govil, H. and Diwan, P. 2020a. Monitoring LST-NDVI relationship using premonsoon Landsat datasets. *Advances in Meteorology* 2020: 4539684.

Guha, S., Govil, H. and Mukherjee, S. 2017. Dynamic analysis and ecological evaluation of urban heat islands in Raipur city, India. *Journal of Applied Remote Sensing* 11(3): 036020.

Guha, S., Govil, H. and Besoya, M. 2020e. An investigation on seasonal variability between LST and NDWI in an urban environment using Landsat satellite data. *Geomatics, Natural Hazards and Risk* 11(1): 1319–1345.

Guha, S., Govil, H., Dey, A. and Gill, N. 2018. Analytical study of land surface temperature with NDVI and NDBI using Landsat 8 OLI/TIRS data in Florence and Naples city, Italy. *European Journal of Remote Sensing* 51(1): 667–678.

Guha, S., Govil, H., Dey, A. and Gill, N. 2020b. A case study on the relationship between land surface temperature and land surface indices in Raipur city, India. *Geografisk Tidsskrift-Danish Journal of Geography.* 120(1): 35–50.

Guha, S., Govil, H., Gill, N. and Dey, A. 2020c. Analytical study on the relationship between land surface temperature and land use/land cover indices. *Annals of GIS* 26(2): 201–216.

Guha, S., Govil, H., Gill, N. and Dey, A. 2020d. A long-term seasonal analysis on the relationship between LST and NDBI using Landsat data. *Quaternary International* 575: 249–258.

Gujree, I., Arshad, A., Reshi, A. and Bangroo, Z. 2020. Comprehensive spatial planning of Sonamarg resort in Kashmir Valley, India for sustainable tourism. *Pacific International Journal*, 3(2): 71–99.

Hassanin, M., Kanga, S., Farooq, M. and Singh, S.K. 2020. Mapping of Trees outside Forest (ToF) From Sentinel-2 MSI satellite data using object-based image analysis. *Gujarat Agricultural Universities Research Journal*, 207–213.

He, J.F., Liu, J.Y., Zhuang, D.F., Zhang, W. and Liu, M.L. 2007. Assessing the effect of land use/land cover change on the change of urban heat island intensity. *Theoretical and Applied Climatology* 90: 217–226.

Jenerette, G.D., Harlan, S.L., Brazel, A., Jones, N., Larsen, L. and Stefanov, W.L. 2007. Regional relationships between surface temperature, vegetation, and human settlement in a rapidly urbanizing ecosystem. *Landscape Ecology* 22: 353–365.

Jime´nez-Mun˜oz, J. C., and Sobrino, J. A. 2003. A generalized single channel method for retrieving land surface temperature from remote sensing data. *Journal of Geophysical Research*, 108(D22): 4688. https://doi.org/10.1029/2003JD003480.

Jime´nez-Mun˜oz, J. C., J. Cristo´bal, J. A. Sobrino, G. So´ria, M. Ninyerola, and Pons X. 2009. Revision of the single-channel algorithm for land surface temperature retrieval from Landsat thermal-infrared data. *IEEE Transactions on Geoscience and Remote Sensing*, 47: 229–349. https://doi.org/10.1109/TGRS.2008.2007125.

Kalnay, E. and Cai, M. 2003. Impact of urbanization and land-use change on climate. *Nature* 423: 528–553.

Kanga, S., Kumar, S. and Singh, S.K. 2017b. Climate induced variation in forest fire using remote sensing and GIS in Bilaspur district of Himachal Pradesh. *International Journal of Engineering and Computer Science* 6(6): 21695–21702.

Kanga, S., Rather, M.A., Farooq, M. and Singh, S.K. 2021. GIS based forest fire vulnerability assessment and its validation using field and MODIS data: a case study of Bhaderwah forest division, Jammu and Kashmir (India). *Indian Forester* 147(2), 120–136.

Kanga, S., Singh S.K. and Sudhanshu 2017a. Delineation of urban built-up and change detection analysis using multi-temporal satellite images. *International Journal of Recent Research Aspects* 4(3): 1–9.

Kleerekoper, L., van Esch, M. and Salcedo, T.B. 2012. How to make a city climate-proof, addressing the urban heat island effect. *Resource Conservation Recycling* 64: 30–38.

Kuang, W., Liu, Y., Dou, Y., Chi, W., Chen, G., Gao, C., Yang, T., Liu, J. and Zhang, R. 2015. What are hot and what are not in an urban landscape: Quantifying and explaining the land surface temperature pattern in Beijing, China. *Landscape Ecology* 30: 357–373.

Li, J., Song, C., Cao, L., Meng, X. and Wu, J. 2011. Impacts of landscape structure on surface urban heat islands: A case study of Shanghai, China. *Remote Sensing of Environment* 115: 3249–3263.

Li, J., Wang, Y., Shen, X. and Song, Y. 2004. Landscape pattern analysis along an urban–rural gradient in the Shanghai metropolitan region. *Acta Ecologica Sinica* 24: 1973–1980.

Lopez, J.M.R., Heider, K. and Scheffran, J. 2017. Frontiers of urbanization: Identifying and explaining urbanization hot spots in the south of Mexico City using human and remote sensing. *Applied Geography* 79: 1–10.

Ma, Q., Wu, J. and He, C. 2016. A hierarchical analysis of the relationship between urban impervious surfaces and land surface temperatures: Spatial scale dependence, temporal variations, and bioclimatic modulation. *Landscape Ecology* 31: 1139–1153.

Ma, Y, Kuang, Y. and Huang, N. 2010. Coupling urbanization analyses for studying urban thermal environment and its interplay with biophysical parameters based on TM/ETM+ imagery. *International Journal of Applied Earth Observation and Geoinformation* 12: 110–118.

Mathew, A., Sreekumar, S., Khandelwal, S., Kaul, N. and Kumar, R. 2016. Prediction of surface temperatures for the assessment of urban heat island effect over Ahmedabad city using linear time series model. *Energy and Building* 128: 605–616.

McFeeters, S.K. 1996. The use of the normalized difference water index (NDWI) in the delineation of open water features. *International Journal of Remote Sensing* 17(7): 1425–1432.

Meraj, G. and Singh, S. K. 2021. Economics of the natural capital and the way forward. Preprints 2021, 2021010083. https://doi.org/10.20944/preprints202101.0083.v1.

Meraj, G., Singh, S. K., Kanga, S., & Islam, M. N. 2021a. Modeling on comparison of ecosystem services concepts, tools, methods and their ecological-economic implications: A review. *Modeling Earth Systems and Environment*, 1–20. https://doi.org/10.1007/s40808-021-01131-6

Meraj, G., Kanga, S., Kranjčić, N., Đurin, B., & Singh, S. K. 2021b. Role of natural capital economics for sustainable management of earth resources. *Earth*, 2(3): 622–634.

Mushore, T.D., Mutanga, O., Odindi, J. and Dube, T. 2016. Assessing the potential of integrated Landsat-8 thermal bands, with the traditional reflective bands and derived vegetation indices in classifying urban landscapes. *Geocarto International* 32(8): 886–899.

Nie, Q., Man, W., Li, Z. and Huang, Y. 2016. Spatiotemporal impact of urban impervious surface on land surface temperature in Shanghai, China. *Canadian Journal of Remote Sensing* 42(6): 680–689. http://doi.org/10.1080/07038992.2016.1217484.

Oke, T.R. 1976. The distinction between canopy and boundary layer heat islands. *Atmosphere* 14: 268–277.

Oke, T.R. 1982. The energetic basis of the urban heat island. *Quarterly Journal of the Royal Meteorological Society* 108: 1–24.

Oke, T.R. 1997. Urban climates and global change. In *Applied climatology: Principles and practices*, eds. A. Perry and R. Thompson, 273–287. London: Routledge.

Pearsall, H. 2017. Staying cool in the compact city: Vacant land and urban heating in Philadelphia, Pennsylvania. *Applied Geography* 79: 84–92.

Peng, J., Xie, P., Liu, Y. and Ma, J. 2016. Urban thermal environment dynamics and associated landscape pattern factors: A case study in the Beijing metropolitan region. *Remote Sensing of Environment* 173: 145–155.

Qin, Z., Karnieli, A. and Berliner, P. 2001. A mono-window algorithm for retrieving land surface temperature from Landsat TM data and its application to the Israel–Egypt border region. *International Journal of Remote Sensing* 22 (18): 3719–3746.

Rather, M. A., et al. 2018. Remote sensing and GIS based forest fire vulnerability assessment in Dachigam National park, North Western Himalaya. *Asian Journal of Applied Sciences* 11(2): 98–114.

Ren, Y., Dengm L.Y., Zuo S.D., Song, X.D., Liao, Y.L., Xu X.D., Chen, Q., Hua, L.Z. and Li, Z.W. 2016. Quantifying the influences of various ecological factors on land surface temperature of urban forests. *Environmental Pollution* 216: 519–529.

Rizwan, A.M., Dennis, L.Y.C. and Liu, C. 2008. A review on the generation, determination and mitigation of urban heat island. *Journal of Environment Science* 20: 120–128.

Sobrino, J.A., Munoz, J.C. and Paolini, L. 2004. Land surface temperature retrieval from Landsat TM5. *Remote Sensing of Environment* 9: 434–440.

Song, J., Du, S., Feng, X. and Guo, L. 2014. The relationships between landscape compositions and land surface temperature: Quantifying their resolution sensitivity with spatial regression models. *Landscape and Urban Planning* 123: 145–157.

Stone, B. Jr. 2007. Urban sprawl and air quality in large US cities. *Journal of Environmental Management* 86: 688–698.

Tran, D.X., Pla, F., Carmona, P.L., Myint, S.W., Caetano, M. and Kieua, P.V. 2017. Characterizing the relationship between land use land cover change and land surface temperature. *ISPRS Journal of Photogrammetry and Remote Sensing* 124: 119–132. http://doi.org/10.1016/j.isprsjprs.2017.01.001.

Weng, Q. 2001. A remote sensing-GIS evaluation of urban expansion and its impact on surface temperature in Zhujiang Delta, China. *International Journal of Remote Sensing* 22 (10): 1999–2014.

Weng, Q. and Yang, S. 2004. Managing the adverse thermal effects of urban development in a densely populated Chinese city. *Journal of Environmental Management* 70 (2): 145–156.

Xu, H. 2006. Modification of normalized difference water index (NDWI) to enhance open water features in remotely sensed imagery. *International Journal of Remote Sensing* 27 (14): 3025–3033.

Zha, Y., Gao, J. and Ni, S. 2003. Use of normalized difference built-up index in automatically mapping urban areas from TM imagery. *International Journal of Remote Sensing* 24(3): 583–594.

Zhang, Z., He, G., Wanga, M., Long, T., Wang, G., Zhang, X. and Jiao, W. 2016. Towards an operational method for land surface temperature retrieval from Landsat-8 data. *Remote Sensing Letters* 7(3): 279–288. http://doi.org/10.1080/2150704X.2015.1130877.

Zhou, W., Qian, Y., Li, X., Li, W. and Han, L. 2014. Relationships between land cover and the surface urban heat island: Seasonal variability and effects of spatial and thematic resolution of land cover data on predicting land surface temperatures. *Landscape Ecology* 29: 153–167.

Part D

---

*Geospatial Modeling in
Change Dynamics Studies*

14 Assessment of the Visual Disaster of Land Degradation and Desertification Using TGSI, SAVI, and NDVI Techniques

B. Pradeep Kumar, K. Raghu Babu,
M. Rajasekhar, and M. Ramachandra
Yogi Vemana University

CONTENTS

14.1 Introduction 261
14.2 Study Area 262
    14.2.1 Geology 263
    14.2.2 Geomorphology 264
14.3 Materials and Methods 265
    14.3.1 Processing of Images 266
        14.3.1.1 SAVI 266
        14.3.1.2 TGSI 267
        14.3.1.3 Classification of Land Cover Using NDVI 268
14.4 Results and Discussion 269
    14.4.1 SAVI and TGSI 269
    14.4.2 Land Cover Change by NDVI 269
14.5 Conclusion 276
References 276

14.1 INTRODUCTION

Land is an essential building block of civilization; it is necessary for raising food. Its power is maintained and valued in an unambiguously different and sometimes inappropriate manner for our excellence in life. Land depletion causes desertification, contributing to ecosystem imbalances (Thomas and Middleton 1994; Noin and

DOI: 10.1201/9781003147107-18

Clarke 1998; Eswaran et al. 2001; Meraj et al. 2021; Symeonakis and Drake 2004). In its 1995 declaration, the UNCCD described desertification as "land degradation in drylands due to different factors, including climatic variations and human activity." The concept is also associated with images of deserts moving through landscapes, invading farmland and starving vulnerable populations (Lal 1990; Reynolds et al. 2007; Cui et al. 2011; Lamchin et al. 2016; Masoudi et al. 2018; Ahmad and Pandey 2018; Kumar et al. 2018, 2019b, 2020a).

Land degradation and desertification are continuous processes. The Nairobi conference of UNCOD (United Nations Conference on Desertification) describes desertification as the diminution of the biological potential of the land, which can lead to desert-like conditions (UNCED 1992). Drought and desertification are the common features of arid and semi-arid regions. Land degradation occurs all over the world. The impact of land degradation is much related to environment (Hassanin et al. 2020; Tomar et al. 2021; Meraj and Kumar 2021; Chandel et al. 2021; Bera et al. 2021). Degradation seriously affects the ecosystem. Desertification happens due to many anthropogenic reasons like deforestation, nonsustainable land use practices, improper agricultural management, road constructions, day-by-day increase in urbanization, and other causes disturbing the natural ecosystems (Kanga et al. 2017a, b; Rather et al., 2018; Hassanin et al. 2020; Kanga et al. 2021). In addition to this extreme weather conditions, climate change and frequent draughts are some natural causes of land degradation desertification (Nachtergaele et al. 2010; Arya et al. 2009; Lamchin et al. 2016; Kumar et al. 2019a; Gujree et al. 2017; Gujree et al. 2020).

Top soil texture is closely linked to land loss and desertification, and varying levels of desertification result in different textures of the top soil, the more extreme desertification, and the composition of the top soil grain coarse (Lamchin et al. 2016). Topsoil grain size index (TGSI), soil-adjusted vegetation index (SAVI), and normalized difference vegetation index (NDVI) are indexed for soil degradation and desertification monitoring and assessment in semi-arid regions using geospatial technology (Zhu et al. 1989; Lillesand and Kiefer 1994; Zhang et al. 2015; Rawat et al. 2013; Farooq and Muslim 2014; Nathawat et al. 2010; Kumar et al. 2018; Joy et al 2019).

We made an attempt in the current study to evaluate the degradation and desertification rate in the semi-arid area of Anantapur. Desertification is caused by the wind or Aeolian action in the current study area; it is the complex phenomenon and governed by several factors such as rainfall, vegetation cover, wind direction, and its velocity (Kumar et al. 2020a). SAVI and TGSI combined are used to assess the soil texture-based degradation, based on pixel values (Anees et al. 2014; Masoudi et al. 2018).

14.2   STUDY AREA

The field of research is in the Survey of India (SOI) toposheet nos. 57B/13, 57B/14, 57F/1, 57F/2 on a scale of 1:50,000 and is stuck in northern longitudes between 77°0′ and 77°10′ and in eastern latitudes between 14°40′ and 14°55′. The study area falls in the south-central portion of the Anantapur district of AP, India, in the rain shadow

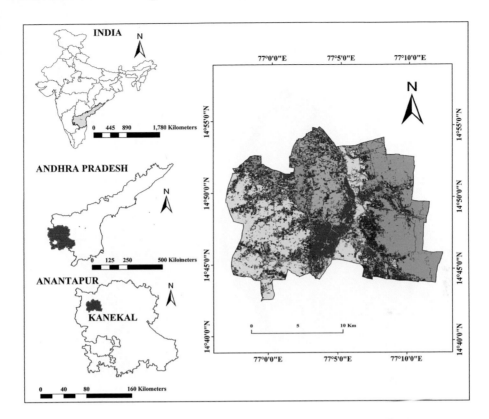

FIGURE 14.1 Location map.

zone. The study area demonstrates a humid to subhumid climate; the annual precipitation of the study area is 520 mm, after the Jaisalmer, which is the second lowest precipitation area in the world. Drought and desertification are therefore widely faced in the study area (Kumar et al. 2020a; Rajasekhar et al. 2018). During winter, the main precipitation period and the time from February to May, when the relative humidity is 50%–60% in the morning and 25%–35% in the afternoon, is the driest part of the year. During the southwestern monsoon and retreating monsoon seasons, it goes up (Figure 14.1).

14.2.1 GEOLOGY

The geology of Anantapur district has been compiled from the GIS reports and other available data from the district Gazetteer, and the geological materializations in the Anantapur district can broadly be alienated into distinct and well-marked groups; an older group of metamorphic rocks has its place to the Archean and younger group of sedimentary rocks belonging to the Proterozoic age. The study is being mainly composed of hornblende biotite gneiss, biotite gneiss, and migmatites. Some part in the south western side is composed of grey granite or pink granite (Figure 14.2).

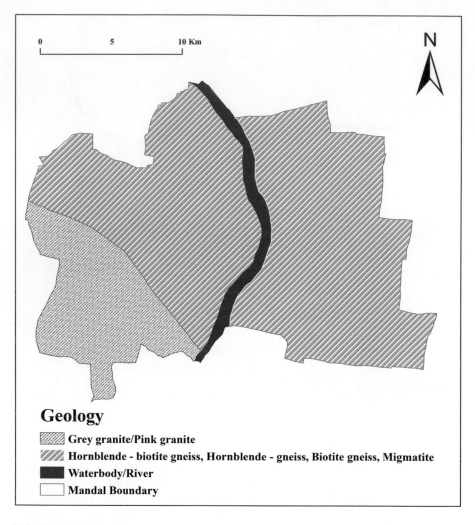

FIGURE 14.2   Geology.

14.2.2 GEOMORPHOLOGY

Geomorphology, which is the study of landforms and related physical processes, is vital to environmental management. The improved methods of mapping of geomorphological characteristics have helped in assessing the potential of land for development both in the urban and rural environments. Geomorphological processes are part of a whole system of interacting phenomena, and their significance to environmental problems has to be assessed in the light of the social, economic, and cultural conditions of the people in that area. Landforms, with their naturally associated soil and vegetation, constitute the basic and fundamental elements in the planning for development in various sectors. In the study area, Aeolian plain or sand and sand dunes have been present in the side of the Hagari river, which flows through the center of

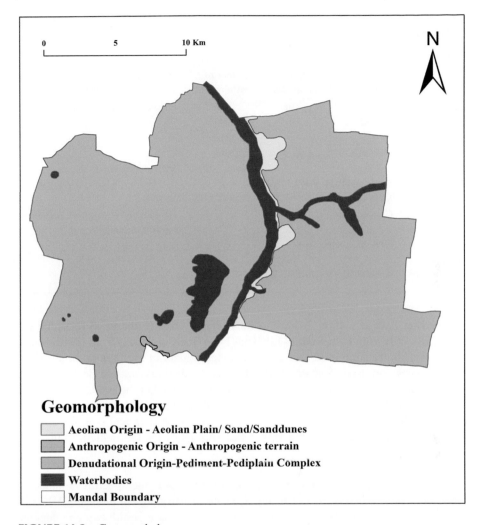

0   5   10 Km

N

Geomorphology

Aeolian Origin - Aeolian Plain/ Sand/Sanddunes

Anthropogenic Origin - Anthropogenic terrain

Denudational Origin-Pediment-Pediplain Complex

Waterbodies

Mandal Boundary

FIGURE 14.3 Geomorphology.

the study area. These sand and sand dunes get migrated because of the southwest monsoon winds (Kumar et al. 2020b; Pandey et al. 2013; Singh and Pandey 2014; Bhatt et al. 2017; Kumar et al., 2021). Denudational origin of pediment and Pedi plain complex is dominant in the study area. Anthropogenic terrain is also seen in some places of the study region (Figure 14.3).

14.3 MATERIALS AND METHODS

"In this research, the progress of desertification in the study area was evaluated using four periods of satellite remote sensing images between 1990 and 2019". Thematic mapper Landsat 4–5 data for the year 1990 image recorded on March 13th, 1990 and for the year 2000 image recorded on March 16th, 2000, and for the year 2010 image

TABLE 14.1

Data Used and Its Sources

	Data Used	Spatial Resolution	Year of Attainment	Source
	Landsat 4-5	30 m	1990	http://earthexplorer.usgs.gov/
	Landsat 7	30 m	2000	
			2010	
Satellite Data	Landsat 8	15 m	2018	
Auxiliary data	SOI maps- 57B/13, 57B/14, 57F/1,57F/2		Scale-1: 50,000	Survey of India
	Meteorological data		2000–2018	MOSDAC
Collateral data	Climate, soils, etc.			Groundwater department, Anantapur district, Andhra Pradesh
Equipment used	GPS, and anemometer			
Software used	Arc GIS 10.3 and ERDAS IMAGINE 2014			

recorded on February 16th, 2010, and for the year 2019 (Landsat 8) recorded on April 27th, 2019 with Path/Row 144/050 has been freely collected from USGS Earth Explorer. This research used four cycles of satellite data in multitemporal shifts of two variables and used 29 years of data to classify degradation and desertification map (Table 14.1). Atmospheric correction was used to predict the reflectance (P_p) of the objects, and surface reflectance was measured at the top of the atmosphere. This knowledge provides the model with a minimum of input data and combined functions. It is possible to estimate the TOA reflectance using the following formula

$$P_p = \pi L \lambda d^2 / (\text{ESUN})_\lambda \cos\theta_s$$

Where, P_p is the planetary reflectance, π is 3.14159, $L\lambda$ is the spectral radiance at the sensor's aperture, d is the Earth-Sun distance in astronomical units, $(\text{ESUN})\lambda$ is the mean solar exoatmospheric irradiance, and θ_s is the solar zenith angle in degree.

14.3.1 PROCESSING OF IMAGES

Superficial changes to land conditions are meticulously linked to desertification. The characteristics of desertification are soil erosion, decreasing soil cover and plant biomass, and micrometeorological changes. The two SAVI and TGSI metrics in this analysis measure the components of desertification.

14.3.1.1 SAVI

In the present study, the computed SAVI is used to evaluate the variations in surface porousness in among 1990 and 2020. Usually, SAVI specifies water body, vegetation, and cultivation exposure and health with respect to saturation, soil color, and moisture and therefore accounts for the high inconsistency areas. The impact of soil brightness in NDVI is correspondingly regulated by SAVI and thus minimizes soil brightness-related noise in vegetation coverage estimation. Since vegetation coverage, luminosity, and health are closely linked to surface permeability, SAVI provides a significant proxy for the identification of impermeable surfaces. SAVI calculations could be performed via the following formula.

$$SAVI = \left(\left(NIR - RED \right) / \left(NIR + RED + L \right) \right) * \left(1 + L \right)$$

where RED is the reflectance of the band 3 and NIR is the reflectance value of the near infrared band (band 4). L is the soil brightness correction factor. For dense vegetation and highly permeable surface areas, $L = 0$, and for vegetation scarce and impermeable surface areas, $L = 1$. Figure 14.4 represents the TGSI values in their distribution from 1990 to 2019.

14.3.1.2 TGSI

"TGSI was developed based on the field survey of soil surface spectral reflectance and laboratory analysis of grain comparison. Sudden or one rain fall can significantly increase the vegetation cover, and thus, NDVI and SAVI misinterpret the actual degree of desertification. To overcome this problem, Xiao et al. proposed a new index, TGSI, which is associated with the mechanical composition of topsoil" (Lamchin et al. 2016). It indicates the coarsening of topsoil grain size, which has a positive correlation with fine sand content of surface soil texture, as a manifestation of undergoing desertification. "The more severe the land degradation and desertification, the coarser the topsoil grain size comparison, which refers to the mean or

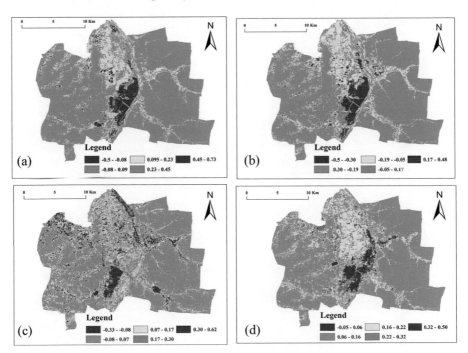

FIGURE 14.4   SAVI. (a) 1990. (b) 2000. (c) 2010. (d) 2019.

effective diameter of individual mineral grains or particles. A high TGSI value corresponds to an area with high content of fine sand in the topsoil or low content of clay-silt grains". The TGSI can be calculated as follows:

$$TGSI = (Redband(R) - Blueband(B)/(Redband(R) + Blueband(B)$$
$$+ Greenband(G)$$
$$TGSI = (R - B)/(R + B + G)$$

For Landsat 8

$$TGSI = (R_{b4} - B_{b2})/(R_{b4} + B_{b2} + G_{b3})$$

For Landsat 7 and Landsat 4–5

$$TGSI = (R_{b3} - B_{b1})/(R_{b3} + B_{b1} + G_{b3})$$

TGSI is an index used to detect the texture, or grain size, of the topsoil layer. Negative or near-0 values represent a zone or water body, and near-0.20 values indicate a high fine sand content. From 1990 to 2019, Figure 14.5 represents the TGSI values in their distribution.

14.3.1.3 Classification of Land Cover Using NDVI

NDVI is the proportion of vegetation area within a certain surface in the study area. Two methods that have been adopted for the detection of vegetation cover over the

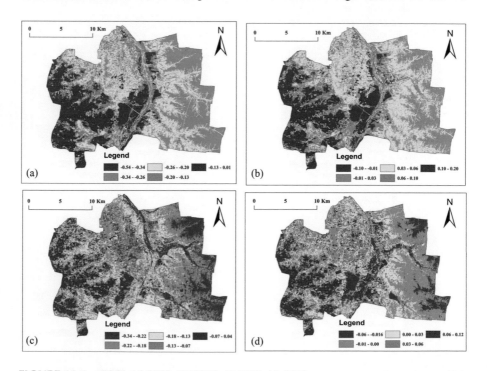

FIGURE 14.5 TGSI. (a) 1990. (b) 2000. (c) 2010. (d) 2019.

study area are field survey and RS information. According to the following equation, the NDVI could be determined by the rate between subtraction and summation of the near-infrared and RED wavelength range values,

$$NDVI = (NIR - RED)/(NIR + RED)$$

where 0 or negative values indicate no greenery or no vegetation, and +1 indicates the high density of green leaves. Low positive values of NDVI, where little difference exists in RED and NIR wavelengths, are typical of low-density vegetation such as grassland or sandy area/desert. In the semiarid areas, NDVI has been adopted for assessing the degraded/desertified land. Here, negative anomalies of NDVI indicate drought or desert conditions in the study area, and the positive anomalies indicate vegetation. Regions with lower levels of vegetation triggered by drought or degradation may be undergoing desertification because of loss of protection to the soil afforded by vegetation.

To analyze land cover change in the study region, we used four images, which were divided into two categories: those taken before 2000 (1990, 2000) and those taken after 2000 (2010, 2019), and these two were classified into four classes in the NDVI.

NDVI has been classified into three classes, namely, unaffected, degraded, and desertified.

14.4   RESULTS AND DISCUSSION

14.4.1   SAVI AND TGSI

SAVI and TGSI images are compared in Table 14.2 in terms of minimum, maximum, mean, and standard time series image variance over the past 29 years. SAVI value ranges have fluctuated in 1990 (min −0.58, max 0.73, mean 0.184, and st. dev 0.438); in 2000 (min = 0.5, max = 0.48, mean = −0.018, and st. dev 0.33); in 2010 (min = −0.33, max = 0.62, mean = 0.166, and st. dev = 0.309), and in 2019 (min = −0.05, max = 0.5, mean = 0.23, and st. dev = 0.181). The SAVI (Figure 14.4) vales have shown a decrease in tendency in their minimum values from −0.58 to −0.05, maximum values from 0.73 to 0.5, and mean values from 0.184 to 0.23, and the st. dev values continuous decrease from 0.438 to 0.181. The observed decreasing tendency of SAVI with negative anomalies (−0.58 to −0.05) will evidently reveal the increase of degradation and desertification in the study region.

TGSI (Figure 14.5) value ranges have been changed in 1990 (Min = −0.54, max = 0.01, mean = −0.224, and st. dev = 0.181), in the year 2000 (min = −0.1, max = 0.01, min = −0.224, and st. dev = 0.181), in the year 2010 (min = 0.34, max = 0.04, and st. dev = 0.125), and in the year 2019 (min = −0.06, max = 0.12, mean = 0.03, and st. dev = 0.125). Decreasing patterns or negative values or values near 0.20 have been observed, suggesting a high content of fine sand, degraded land, and desertified land.

14.4.2   LAND COVER CHANGE BY NDVI

In the analysis, the NDVI land cover changes have been divided into two categories, one is NDVI classification before 2000 (Figure 14.6a), and another one is after

TABLE 14.2
Distribution of SAVI and TGSI Changes

Variable	Year	Min	Max	Mean	Standard Deviation
SAVI	1990	−0.58	0.73	0.184	0.438
	2000	−0.5	0.48	−0.018	0.33
	2010	−0.33	0.62	0.166	0.309
	2019	−0.05	0.5	0.23	0.181
TGSI	1990	−0.54	0.01	−0.224	0.181
	2000	−0.1	0.2	0.058	0.097
	2010	0.34	0.04	−0.136	0.125
	2019	−0.06	0.12	0.03	0.06

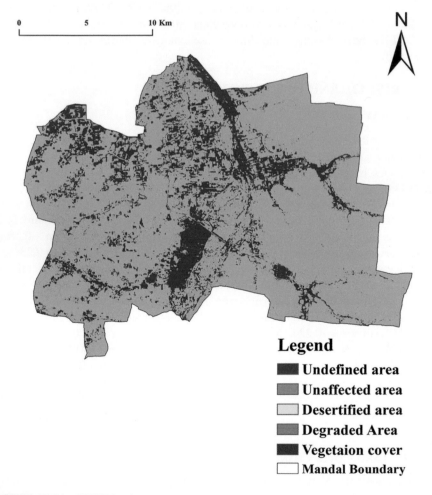

FIGURE 14.6A NDVI land cover changes before 2000.

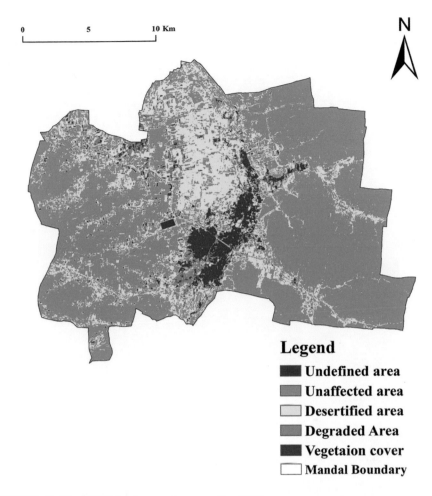

FIGURE 14.6B   NDVI land cover changes after 2000.

2000 (Figure 14.6b). We approximated the desertification process by sensing the variations of land coverage among images of these before and after 2000-year NDVI images. The change index of NDVI has been categorized into five classes, namely, undefined area, unaffected area, degraded area, desertified area, and vegetation cover of which undefined area is to be taken for error matrix which is negligible in the field check.

NDVI classification indicators have been converted into desertification degrees in LEVELs, for both two categories, i.e., before 2000 and after 2000 (Figure 14.7a and b). Table 14.3 represents the land cover category into assigned desertification degree level. It is classified into four levels. Level 1 can be classified form the unaffected area, Level 2 has been classified from vegetation cover, Level 3 has been classified form degraded area, and Level 4 has been classified from desertification area.

Table 14.4 shows the analysis of two images taken before 2000 and after 2000; the unaffected area has been decreased from 271.11 km² (67.11%) to

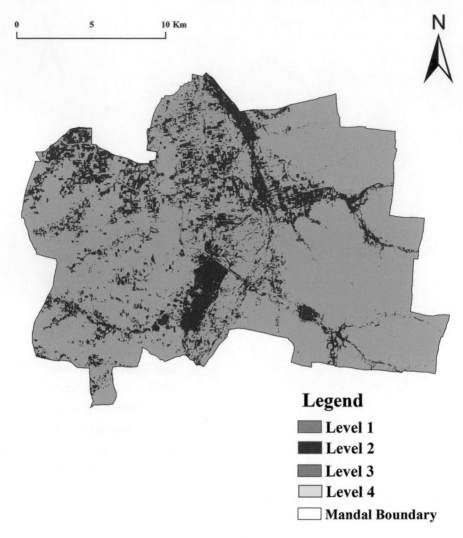

FIGURE 14.7A Desertification degree before 2000.

TABLE 14.3

Classifiction List of Land Cover into Desertification Degree Levels

Classification of Land Cover	Desertification Levels
Unaffected area	Level 1
Vegetation Cover	Level 2
Degraded area	Level 3
Desertified area	Level 4

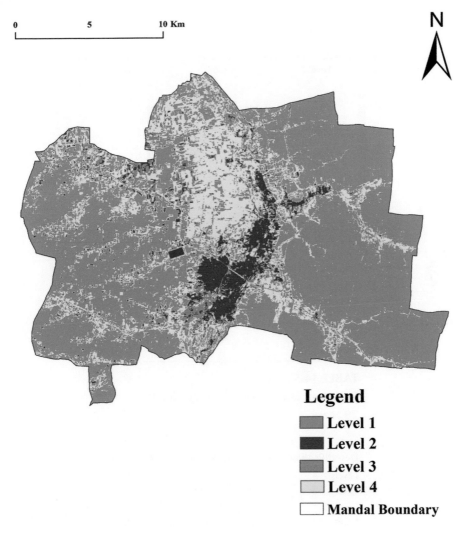

FIGURE 14.7B   Desertification degree after 2000.

TABLE 14.4
Changes of Classification in Study Region

Classification of land	Desertification Levels	Before 2000		After 2000	
		Area (km²)	Area (%)	Area (km²)	Area (%)
Unaffected area	Level 1	271.11	67.11	230.6	57.08
Vegetation area	Level 2	16.68	4.13	17.82	4.41
Degraded land	Level 3	67.89	16.81	79.56	19.7
Desertified land	Level 4	48.27	11.95	75.97	18.81
Total		403.95	100	403.95	100

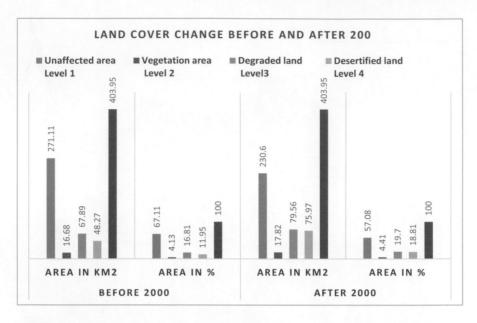

FIGURE 14.8 Changes for before and after 2000.

TABLE 14.5
Resultant Land Cover Changes for Past 29 years

	Changes	
Desertification Level	**Area (km²)**	**Area (%)**
Levels 1–4	−40.51	−10.03
Levels 2–4	1.14	0.28
Levels 3–4	11.67	2.89
Levels 4–4	27.7	6.86

230.60 km² (57.08%), vegetation cover is slightly increased from 16.68 km² (4.13%) to 17.82 km² (4.41%), degraded land is increased from 67.89 km² (16.81%) to 79.56 km² (19.70%), and desertified area is increased from 48.27 km² (11.95%) to 75.97 km² (18.81%). Figure 14.8 can show the change detection range in bar graph.

Table 14.5 reveals the resultant changes for the past 29 years (from 1990 to 2019) of Desertified levels in the study area. Figure 14.9 shows the resultant changes in the desertification levels. Levels 1–4 are decreased to 40.51 km² (−10.3%), Levels 2–4 are increased to 1.14 km² (0.28%), Levels 3 and 4 are increased to 11.67 km² (2.89%), and Levels 4–4 are increased to 27.7 km² (6.86%). Figure 14.10 shows graphical representation of land cover change detection analysis for the past 29 years.

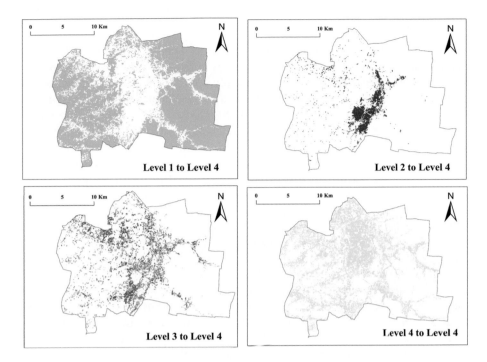

FIGURE 14.9 Expansion of Level 4 area (land cover changed from other levels).

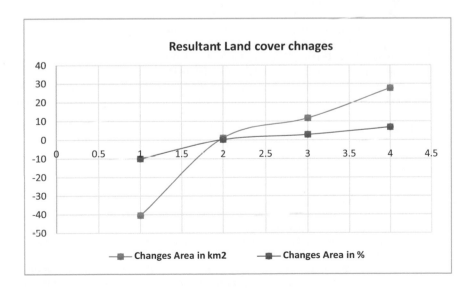

FIGURE 14.10 Land cover change detection analysis for 29 years.

14.5 CONCLUSION

In this study, we are attempting to use remotely sensed Landsat images to evaluate and map the degradation and desertification changes in the semi-arid area of Anantapur district between 1990 and 2019. Using frequently used dark object subtraction, all the satellite images were atmospherically corrected using GIS and ERDAS Imagine software. Based on the pixel values, we selected SAVI and TGSI as indicators to reflect land surface conditions based on vegetation, landscape pattern, and surface soil grain textures. Based on SAVI pixel values, we find that $L=0$ for dense vegetation and highly permeable surface areas, and $L=1$ for scarce and impermeable surface areas of vegetation (L is the correction factor for soil brightness). Based on TGSI pixel values, negative values or those near to "0" represents zone or waterbodies, and values near 0.20 indicate high contents of fine sand, and we found -0.06 is the minimum, and 0.2 is the maximum value. For assessing the land cover change in the study area, we used NDVI, and it is divided into two classes, desertification status or land cover status before the year 2000 and after the year 2000. NDVI is categorized to unaffected, vegetation area, degradation area, and desertification area, and again, those are converted into four levels. For the past 29 years from 1990 to 2019, land cover change clearly reveals that Level 1 (unaffected area) is decreased to 10010.24 acres or -10.03%, Level 2 (vegetation area) is increased to 281.7 acres or 0.28%, Level 3 (degradation area) is increased to 2883.72 acres or 2.89%, and Level 4 (desertification) is increased to 6844.82 acres or 6.86%. The expansion tendency of desertification was relatively high in the eastern part of the study region and neighboring farming land areas.

REFERENCES

Ahmad, N., & Pandey, P. (2018). Assessment and monitoring of land degradation using geospatial technology in Bathinda district, Punjab, India. *Solid Earth*, 9(1), 75–90.

Anees, M. T., Javed, A., & Khanday, M. Y. (2014). Spatio-temporal land cover analysis in Makhawan watershed (MP), India through remote sensing and GIS techniques. *Journal of Geographic Information System*, 6, 298–306.

Arya, A. S., Dhinwa, P. S., Pathan, S. K., & Raj, K. G. (2009). Desertification/land degradation status mapping of India. *Current Science*, 97, 1478–1483.

Arya, V. S., Singh, H., Hooda, R. S., & Arya, A. S. (2014). Desertification change analysis in Siwalik hills of Haryana using geo-informatics. *The International Archives of the Photogrammetry, Remote Sensing and Spatial Information Sciences*, 8, 73–81.

Bera, A., Taloor, A. K., Meraj, G., Kanga, S., Singh, S. K., Durin, B., & Anand, S. (2021). Climate vulnerability and economic determinants: linkages and risk reduction in Sagar Island, India; A geospatial approach. *Quaternary Science Advances*, 100038. doi: 10.1016/j.qsa.2021.100038.

Bhatt, C. M., Rao, G. S., Farooq, M., Manjusree, P., Shukla, A., Sharma, S. V. S. P., Kulkarni, S. S., Begum, A., Bhanumurthy, V., Diwakar, P. G., & Dadhwal, V. K. (2017). Satellite-based assessment of the catastrophic Jhelum floods of September 2014, Jammu & Kashmir, India. *Geomatics, Natural Hazards and Risk*, 8(2), 309–327.

Chandel, R. S., Kanga, S., & Singhc, S. K. (2021). Impact of COVID-19 on tourism sector: a case study of Rajasthan, India. *AIMS Geosciences*, 7(2), 224–242.

Cui, G., Lee, W. K., Kwak, D. A., Choi, S., Park, T., & Lee, J. (2011). Desertification monitoring by LANDSAT TM satellite imagery. *Forest Science and Technology*, 7(3), 110–116.

Eswaran, H., Lal, R., & Reich, P. F. (2001). Land degradation: an overview. In: Bridges, E. M., I. D. Hannam, L. R. Oldeman, F. W. T. Pening de Vries, S. J. Scherr, and S. Sompatpanit (eds.). *Responses to land degradation.* In *Proceedings of the 2nd International Conference on Land Degradation and Desertification*, Khon Kaen, Thailand. Oxford Press, New Delhi, India.

Farooq, M., & Muslim, M. (2014). Dynamics and forecasting of population growth and urban expansion in Srinagar city – a geospatial approach. *The International Archives of Photogrammetry, Remote Sensing and Spatial Information Sciences*, 40(8), 709.

Gujree, I., et al. (2017). Evaluating the variability and trends in extreme climate events in the Kashmir Valley using PRECIS RCM simulations. *Modeling Earth Systems and Environment*, 3(4), 1647–1662.

Gujree, I., Arshad, A., Reshi, A., & Bangroo, Z. (2020). Comprehensive spatial planning of Sonamarg resort in Kashmir Valley, India for sustainable tourism. *Pacific International Journal*, 3(2), 71–99.

Hassanin, M., Kanga, S., Farooq, M., & Singh, S. K. (2020). Mapping of Trees outside Forest (ToF) from Sentinel-2 MSI satellite data using object-based image analysis. *Gujarat Agricultural Universities Research Journal*, 207–213.

Joy, J., Kanga, S., and Singh, S. K. (2019). Kerala flood 2018: flood mapping by participatory GIS approach, Meloor Panchayat. *International Journal on Emerging*, 10(1), 197–205.

Kanga, S., Kumar, S., and Singh, S. K. (2017b). Climate induced variation in forest fire using Remote Sensing and GIS in Bilaspur District of Himachal Pradesh. *International Journal of Engineering and Computer Science*, 6(6), 21695–21702.

Kanga, S., Rather, M. A., Farooq, M., & Singh, S. K. (2021). GIS based forest fire vulnerability assessment and its validation using field and MODIS Data: a case study of Bhaderwah forest division, Jammu and Kashmir (India). *Indian Forester*, 147(2), 120–136.

Kanga, S., Singh, S. K., & Sudhanshu. (2017a). Delineation of urban built-up and change detection analysis using multi-temporal satellite images. *International Journal of Recent Research Aspects*, 4(3), 1–9.

Kumar, B. P., Babu, K. R., Rajasekhar, M., Narayana Swamy, B., & Ramachandra, M. (2019a). Landuse/landcover changes and geo-environmental impacts on Beluguppa Mandal of Anantapur district of Andhra Pradesh, India, using remote sensing and GIS Modelling. *Research & Reviews: Journal of Space Science & Technology*, 8(2), 6–15.

Kumar, B. P., Babu, K. R., Rajasekhar, M., & Ramachandra, M. (2019b). Assessment of land degradation and desertification due to migration of sand and sand dunes in Beluguppa Mandal of Anantapur district (AP, India), using remote sensing and GIS techniques. *Journal of Indian Geophysical Union*, 23(2), 173–180.

Kumar, B. P., Babu, K. R., Ramachandra, M., Krupavathi, C., Swamy, B. N., Sreenivasulu, Y., & Rajasekhar, M. (2020a). Data on identification of desertified regions in Anantapur district, Southern India by NDVI approach using remote sensing and GIS. *Data in Brief*, 30, 105560.

Kumar, B. P., et al. (2020b). Identification of land degradation hotspots in semiarid region of Anantapur district, South India, using geospatial modeling approaches. *Modeling Earth Systems and Environment*, 6, 1841–1852. doi: 10.1007/s40808-020-00794-x

Kumar, S., Kanga, S., Singh, S. K. & Mahendra, R. S. (2018). Delineation of Shoreline Change along Chilika Lagoon (Odisha), East Coast of India using Geospatial technique. *International Journal of African and Asian Studies*, 41, 9–22.

Lal, R, (1990). Soil erosion and land degradation: the global risks. In Lal, R., & B. A. Stewart (eds.). *Advances in soil science* (pp. 129–172). Springer, New York.

Lamchin, M., Lee, J. Y., Lee, W. K., Lee, E. J., Kim, M., Lim, C. H., Choi, H. A., & Kim, S. R. (2016). Assessment of land cover change and desertification using remote sensing technology in a local region of Mongolia. *Advances in Space Research*, 57(1), 64–77.

Lillesand, T. M., & Kiefer, R. W. (1994). *Remote sensing and image interpretation*, 3rd edn. John Wiley and Sons, New York, 161–163.

Masoudi, M., Jokar, P., & Pradhan, B. (2018). A new approach for land degradation and desertification assessment using geospatial techniques. *Natural Hazards and Earth System Sciences*, 18, 1133–1140.

Meraj, G., & Kumar, S. (2021). Economics of the natural capital and the way forward. Preprints 2021, 2021010083. doi: 10.20944/preprints 202101.0083.v1.

Meraj, G., Singh, S. K., Kanga, S., & Islam, M. N. (2021). Modeling on comparison of ecosystem services concepts, tools, methods and their ecological-economic implications: a review. *Modeling Earth Systems and Environment*, 1–20. https://doi.org/10.1007/s40808-021-01131-6

Nachtergaele, F., Petri, M., Biancalani, R., Van Lynden, G., Van Velthuizen, H., & Bloise, M. (2010). Global land degradation information system (GLADIS). Beta version. An information database for land degradation assessment at global level. Land degradation assessment in drylands. Technical report, 17.

Nathawat, M. S., et al. (2010). Monitoring & analysis of wastelands and its dynamics using multiresolution and temporal satellite data in part of Indian state of Bihar. *International Journal of Geomatics and Geosciences*, 1(3), 297–307.

Noin, D., & Clarke, J. I. (1998). Population and environment in arid regions of the world. *Man and the Biosphere Series*, 19, 1–20.

Pandey, A. C., Singh, S. K., & Nathawat, M. S. (2010). Waterlogging and flood hazards vulnerability and risk assessment in Indo Gangetic plain. *Natural Hazards*, 55(2), 273–289.

Pandey, A. C., Singh, S. K., Nathawat, M. S., & Saha, D. (2013). Assessment of surface and subsurface waterlogging, water level fluctuations, and lithological variations for evaluating groundwater resources in Ganga Plains. *International Journal of Digital Earth*, 6(3), 276–296.

Rajasekhar, M., Raju, G. S., Raju, R. S., Ramachandra, M., & Kumar, B. P. (2018). Data in brief data on comparative studies of lineaments extraction from ASTER DEM, SRTM, and Carto sat for Jilledubanderu River basin, Anantapur district, AP, India by using remote sensing and GIS. *Data in Brief*, 20, 1676–1682.

Rather, M. A., et al. (2018). Remote sensing and GIS based forest fire vulnerability assessment in Dachigam National park, North Western Himalaya. *Asian Journal of Applied Sciences*, 11(2), 98–114.

Rawat, J. S., Biswas, V., & Kumar, M. (2013). Changes in land use/cover using geospatial techniques: a case study of Ramnagar town area, district Nainital, Uttarakhand, India. *The Egyptian Journal of Remote Sensing and Space Science*, 16(1), 111–117.

Reynolds, J. F., Smith, D. M. S., Lambin, E. F., Turner, B. L., Mortimore, M., Batterbury, S. P., Downing, T. E., Dowlatabadi, H., Fernández, R. J., Herrick, J. E., & Huber-Sannwald, E. (2007). Global desertification: building a science for dryland development. *Science*, 316(5826), 847–851.

Kumar, A., Kanga, S., Taloor, A. K., Singh, S. K. & Ðurin, B., 2021. Surface runoff estimation of Sind river basin using integrated SCS-CN and GIS techniques. *HydroResearch*, 4, 61–74.

Singh, S. K., & Pandey, A. C. (2014). Geomorphology and the controls of geohydrology on waterlogging in Gangetic Plains, North Bihar, India. *Environmental Earth Sciences*, 71(4), 1561–1579.

Thomas, D. S., & Middleton, N. J. (1994). *Desertification: exploding the myth*. John Wiley and Sons, Chichester.

Tomar, J. S., Kranjčić, N., Ðurin, B., Kanga, S., & Singh, S. K. (2021). Forest fire hazards vulnerability and risk assessment in Sirmaur district forest of Himachal Pradesh (India): a geospatial approach. *ISPRS International Journal of Geo-Information*, 10(7), 447.

UNCED (1992). *Earth summit agenda 21: programme of action for sustainable development.* United Nations Department of Public Information, Rio de Janeiro.

Zhang, R., Feng, Q., Guo, J., Shang, Z., & Liang, T. (2015). Spatio-temporal changes of NDVI and climatic factors of Grassland in Northern China from 2000 to 2012. *Journal of Desert Research*, 35, 1403–1412.

Zhu, Z., Liu, S., & Di, X. (1989). *Desertification and its rehabilitation in China.* China Science Press, Beijing.

15 Dynamics of Forest Cover Changes in Hindu Kush-Himalayan Mountains Using Remote Sensing
A Case Study of Indus Kohistan Palas Valley, Pakistan

Noor ul Haq, Fazlur Rahman, and Iffat Tabassum
University of Peshawar

CONTENTS

15.1 Introduction 282
    15.1.1 Overview of Forest Cover Changes 283
15.2 Study Area 287
15.3 Materials and Methods 292
    15.3.1 Primary Data 292
    15.3.2 Focused Group Discussions 292
    15.3.3 Secondary Data 293
    15.3.4 Accuracy Assessment 293
15.4 Results and Discussion 295
    15.4.1 Trend of Forest Cover Change (1980–2017) 295
    15.4.2 Spatial and Temporal Change in Forest Cover (1980–2017) 295
    15.4.3 Forest Cover Change within 1 and 3 km of the Selected
        Settlements (1980–2017) 297
    15.4.4 Forest Cover Change within 1 and 3 km of the Roads
        (1980–2017) 300
    15.4.5 Local Committee System for Management 301
15.4 Conclusion 302
Acknowledgment 303
References 303

15.1   INTRODUCTION

Many socioeconomic and biophysical processes are responsible for forest cover changes, and a single factor cannot be blamed for deforestation and forest degradation (Geist & Lambin, 2001; Meraj et al., 2021a and b; Haq et al., 2021 a and b). Worldwide, several factors have been studied to find the causes of deforestation and forest cover changes. The primary forest cover, population growth, agricultural extension, per capita income, forest governance, mismanagement practices, food, poverty, administrative boundaries and ambiguity of area identification among forest officers are treated as drivers of change in forest cover (Zeb, 2019; Van Khuc et al., 2018; Qamer et al., 2012; Ali et al., 2006). According to Geist and Lumbin (2001), issues related to population growth (58%) and expansion in road network (32%) are reportedly related to forest cover changes in Asia.

In the Hindu Kush-Himalayas (HKH) region, forest cover change is primarily caused by population growth that has been considered as a leading driver for forest degradation (Ahmad & Nizami, 2015; Poudel & Shaw, 2015; Shehzad et al., 2014; Tsering et al., 2010; Jan et al., 2011; Qamer et al., 2012; Qasim et al., 2011; Farooq & Muslim, 2014). Many empirical studies revealed that demographic development, accessibility, and agriculture expansion are the common causes of deforestation (Haq et al., 2021b; Hussain et al., 2018; Haq et al., 2018; Rahman et al., 2014; Qasim et al., 2011; Deng et al., 2011; Southworth & Tucker, 2001). Moreover, household dynamics and the expansion of road networks have also played a key role in forest cover change (Haq et al., 2018, 2021a; Ullah et al., 2016; Robinson et al., 2014; Newman et al., 2014; Rahman et al., 2014; Diniz et al., 2013). The growing population, nearness to settlements and roads and improvements in road networks paved the way to timber harvesting in bulks as well as to overexploitation and for domestic uses, leading to forest cover changes (Sharma et al., 2016; Das, Behera, & Murthy, 2017).

The spatiotemporal patterns of forest cover changes in HKH region of Pakistan were observed by Qamer et al. (2016) and revealed that a significant decline in forest cover has occurred between 1990 and 2010, with a reduction of 1,000 hectare (ha)/year. However, this decrease in forest cover is not uniform and considerably varies in different localities Qamer et al. (2012). The forest cover change was high near the administrative boundaries at tehsil level, because of the ambiguity of area identification among forest officers. In Kurram agency of north-western Pakistan, the forest cover has been cleared at an alarming rate as a result of the increasing population, and its growing energy demands lead to deforestation and forest degradation (Hussain et al., 2018). In Roghani valley of the Hindu Raj Mountains of northern Pakistan, changes in forest cover have been attributed to changing forest ownership (Rahman et al., 2014; Haq et al., 2018).

Increasing demand of fuel wood and timber at local and regional levels exerts considerable pressure on forest resources. Disintegration of joint and extended families and increasing trend of nuclear families has multiple impacts on forest cover. Due to increasing number of households, the need of timber, firewood as well as fodder for livestock also increases. In such cases, areas adjacent to the settlement units are cleared first (Ali et al., 2006). Moreover, to feed the increasing population, more and more area has been brought under plough, and in many cases, agriculture encroached on forest land (Shehzad et al., 2014; Rodriguez & Pérez, 2013; Ullah et al., 2016; Haq, 2019; Abdullah, 2001).The timber extraction, roads, and settlement projects are considered as bases of forest cover change as reported by Potapov et al. (2015) and Diniz et al. (2013).

Generally, accessibility and improvement in road network have been considered as a main factor in forest cover change. According to Ali et al. (2005) more than 50% of the forest cover in Basho valley has been converted into other land uses because of improvement in accessibility. The link road has provided a route for legal and illegal commercial harvesting of wood till 1987. Similar to other parts of HKH region, forest resources of the study area are under high stress from demographic growth, unchecked and ruthless cutting of trees for fuelwood, timber and fodder. Specifically, forest cover changes in and around the settlements and along the link roads leading to subvalleys in the study area are considerable (Haq et al., 2021b; Kanga et al., 2017a and b; Rather et al., 2018; Hassanin et al., 2020; Kanga et al., 2021). The present study is conducted in one of the tehsils of Indus Kohistan i.e. Palas valley of HKH Mountains where the natural forest is continuously shrinking. This study investigates forest cover dynamics for about four decades. The main objective of this study is to investigate forest cover changes in and around the settlements and along the link road to highlight the impact of settlement expansion and road extension on forest cover.

The valley is one among those remote areas of the HKH region whose geography and natural resources, especially forest cover change dynamics, have little been studied and lack forest inventory. The valley has a considerable amount of forest resources, which are under stress from increasing human and livestock population, accessibility and illegal cutting of tress. Forest resources of the valley are distributed among the people of different villages and kept under communal ownership. Usually, the dwellers have equal access and withdrawal rights of forest resources. Here, the communities have established a local institution for management of these resources. For comprehensive study and to detect forest cover changes, a case study approach is adopted to highlight the drivers of forest cover changes and to quantify forests loss in the study area. Moreover, it is adopted in order to develop an approach for management and sustainable use of the mountainous forest and to understand the ongoing changes in this remote mountainous belt and its impact on the ecological characteristics. In addition, this study will be helpful for policy maker and infrastructure development authorities for future development and to establish the monitoring system for forest and other land-use/cover changes in the study area.

15.1.1 OVERVIEW OF FOREST COVER CHANGES

Mountain forest covers over 9 million square kilometers of the Earth's surface, and this represents almost 23% of the Earth's forest cover. In mountainous areas, forests play a key role, providing goods and services essential to the livelihood of both highland and lowland communities. The biodiversity protected in the healthy mountain forests provides more than a few products, such as wood, fuel, medicinal and fragrant plants, forage, and a wide range of food stuff that ensure the good fortune of local dwellers. These forests also have a central position in terms of climate change, in lieu of vital ecosystems for the well-being of the planet earth. As a matter of reality, they protect the Earth and contribute a shielding effect to the environment from carbon dioxide emissions. Also, mountains covered with green, dense and healthy forests are undeniably one of the utmost outstanding visions provided to the planet Earth. Human mysticism, custom along with tourism usually attract these landscapes. However, these attractive and profitable landscapes are under threat, and deforestation has been extensively practiced for short-term benefits, without paying due attention to long-term influences. Population increase and progress in extensive farming have forced smallholding farmers to move closer to peripheral areas or steep slopes and consequently induced the clearing of forest areas. Moreover, the conservation of nutritious forests frequently may not be the primary priority for personal commercial enterprise. Vitally, mountain forests perform a defensive function in natural hazards; hence, after the forest cover loss, the land is left unprotected, and there is increase runoff and soil erosion, infuriating landslides, avalanches and floods, to the destruction of villages, transport systems, human infrastructures and of the food security of susceptible population (Haq et al., 2021a and b; Hassanin et al., 2020; Tomar et al., 2021; Meraj et al., 2021; Chandel et al., 2021; Bera et al., 2021).

The population of Pakistan has reached 220 million in 2020 at a consistent growth rate of 1.98. Density of population, on average, raised from 200 to 257 people/km^2 (tabular data source: population division of the UN department of economic and Social Affairs). The National Population Vision, 2030 predicts that Pakistan will rank fifth globally by 2030, with a population ranging between 230 and 260 million human beings, 60% of whom will stay in cities. Moreover, there is lack of consistency between supply and demand due to population increase as well as incompatible forest resource base. Supply has step-by-step improved from smallholding trees but stress on natural forests has remained to increase extensively. In contrast, only population size is not responsible for deforestation and forest cover change, but other factors also, like population density, labor force, house hold income, age structure of the house hold and life cycle (Sydenstricker-Neto, 2012). Demography, accessibility, land use policy and biophysical factors have a significant impact on forest cover in accessible places (Vanonckelen & Van Rompaey, 2015). Similarly, proximate drivers like agriculture expansion, timber extraction, logging, fuel wood, charcoal production and livestock grazing cause forest cover changes (Hussain et al., 2018). In Hindu Raj Mountains of Pakistan, the socioeconomic factors, mainly population growth and household dynamics, resulted a high rate of deforestation (Haq et al., 2018). The impacts of roads, growing human and livestock population accelerated forest cover changes in different localities (Suleri, 2002; Rodriguez & Pérez, 2013; Newman et al., 2014; Ullah et al., 2016). Similarly, system of tenure changing, pattern of ownership, household dynamics and population increase, accessibility, proximity to the settlement are closely connected with deforestation and degradation (Rahman et al., 2014; Shehzad et al., 2014).

According to some studies, forest cover has been declined in the HKH mountain region of Pakistan. Qamer et al. (2016) revealed that a significant reduction in forest cover has been occurred between 1990 and 2010, and the forest cover decreased from 95,000 to 75,000 ha in two decades. Qamer et al. (2012) reported that the rate of forest cover change was 1,268 ha/year in Swat and Shangla districts of Khyber Pakhtunkhwa from 2001 to 2009. Change in forest cover was more near the administrative boundaries at tehsil level, because of the ambiguity of area identi-fication among the forest officers. Furthermore, the inhabitants of the valley extract fuelwood, timber and forage for animals from these forests. Like other marginalized rural areas (Ahmad & Nizami, 2015; Poudel & Shaw, 2015; Shehzad et al., 2014; Qamer et al., 2012; Qasim et al., 2011), the forest cover in Palas valley has been also decreased due to mismanagement. The transformation of forest cover to agricul-ture land and shrubs bushes indicates the increasing population and food demands. Similar to our study, other research studies also reported the conversion of forest cover to agriculture land due to growing population (Haq et al., 2021a; Hussain et al., 2018; Haq et al., 2018; Rahman et al., 2014; Qasim et al., 2011). Growing population needs more food and space for living, and to meet these demands, more land has been brought under cultivation and house construction. In this regard, forest areas close to human settlements suffered substantially as reported by (Munawar & Udelhoven, 2020; Bhatta et al., 2019; Das, Behera, & Murthy, 2017; Sharma et al., 2016) which eventually resulted in forest cover change in the study area.

The study of Hussain et al. (2018) revealed that forest cover has been cleared at an alarming rate due to increasing population and their growing energy demands in Kurram agency of north-western Pakistan. Haq et al. (2018), Robinson et al. (2014), Newman et al. (2014), Diniz et al. (2013), Rahman et al. (2014) and Ali et al. (2006) reported several factors behind the increasing rate of forest cover degradation among which population growth, household dynamics and the expansion of road networks have played a key role. Haq et al. (2021b) and Abdullah (2001) reported a strong rela-tionship between transport accessibility, socioeconomic conditions and deforestation and that constructing a link road in forest area resulted in causeless deforestation which is a significant tool for forest and other land use/cover changes. The shrinkage of forest cover in the investigated area is settlement expansion as a result of growing population, accessibility and other socio-economic drivers. Palas valley is a moun-tainous area and has severe extended winter season. Similar to other mountainous localities of the HKH region (Ullah et al., 2016; Newman et al., 2014; Rahman et al., 2014; Diniz et al., 2013), in the absence of alternate fuel sources, the inhabitants totally depend on forest for fuelwood. Deforestation is a complex phenomenon that involves many socioeconomic and biophysical driving forces and cannot be attrib-uted to a single driving force (Zeb, 2019; Wester et al., 2019; Geist & Lambin, 2001); however, population growth and accessibility usually cause and intensify changes in the forest cover. Ali et al. (2006), Shehzad et al. (2014), Rodriguez and Pérez (2013), Qamer et al. (2012), Van Khuc et al. (2018) and Zeb (2019) reported that population density, accessibility, demands for firewood, timber, forage, cultivable land and prox-imity to settlement and roads result in forest cover changes and are considered as the foremost drivers of deforestation. Brinkmann et al. (2014) have studied the deforesta-tion in South-Western Madagascar, using remote sensing and GIS, and found 45% loss in forest cover during the last four decades. The forest loss was not constant spatially and temporally. It was high in remote areas and near to the small settle-ment, which were away from and poorly connected to market and other settled areas. Qasim et al. (2011) have taken a notice of the contrasting accounts on the state of for-est for Pakistan generally and district Swat particularly. Many public sources of the country have claimed for increase in forest cover, while other scientific studies and international statistics claim tremendous decrease. Regardless of claims that forest has increased in Swat district, it decreased considerably.

The research of Ellis and Porter-Bolland (2008) revealed that local community forest management has linkage with the forest cover changes in Mexico. They evaluated contrasting annual rate of deforestation and concluded that forest preservation or maintenance was affected by local community forest institutions and landscape zonation by resident population. The results showed that community-based forest administration can play a crucial role in forest management. It was recommended for local land use management system as a conservation approach in which local people were considered central actors to control deforestation. Cubbage et al. (2007) have highlighted the significance of sustainable forestry of the present day. Nowadays, forest management sustainability is taken as cross-cutting theme in forestry. Forests provide diverse kind of advantages that are social, environmental and cost-effective in nature so the forest management policies ought to be multidimensional. Aspects having impact on forest policy have been investigated in traditional as well as present-day tools for policy-making. In addition, Bruns (2005) stresses the significance of community participation in forest management practices in this way, "As government and other organizations are trying to enhance the management of natural resources, participatory and community-based strategies have assured valuable benefit and as a result received increasing support in the policy of national and international agencies". Furthermore, Rai has the same idea while saying "In northeast India, community-based natural resource management has a long history. Indigenous understanding and adaptation are the combined information, with development from generation to generation. The expectancy under community management and indigenous understanding on biodiversity play a key role in management of natural resource through traditional practices".

Rasolofoson et al. (2015) have carried out a study in Madagascar to evaluate the effectiveness of community-based forest management and deforestation. According to them, strong evidence is not guaranteed for conservation success of community forest management and utilization of forest resources for commercial purposes. Their findings suggest that differentiating among types of community-based forest management is crucial while evaluating effectiveness. They obviously estimated influences of uncertainty on the type of community-based forest management, and scholars ought to shed light on those aspects that encourage effective community-based forest management. Shahbaz et al. (2007) have analyzed the forest polices of Pakistan and observed that most of the policy initiatives, till currently, are designed for forest conservation and pay no attention to the livelihood necessities for local communities, though even the management elements of these policies were not executed efficiently. Furthermore, people's involvement in plantation and forests management no way give enough consideration, so social and cultural factors of forest management had been ignored. The origins of this method may possibly trace back to colonial era. They depict that the earlier forest strategies (1955, 1962, 1975 and 1980) have been related more or less by changing the governments to fulfill the government's political objectives. But, the rules of 1991 and 2001 are claimed to be participatory, but the local stakeholders blame these to be "donor-driven" policies, which take no notice of the ground level realities and requirements of the local inhabitants. As a result, the stakeholders frequently found themselves in a situation in which country policies again do not assist or have risky consequences on their living strategies. In this situation, the policies do not meet the expectations of individuals who in turn are compelled to use the natural resources unsustainably to secure their livelihoods. Hence, neither the developmental nor the conservational targets are met resulting in forest cover changes and degradation as observed in Palas valley.

15.2   STUDY AREA

Kohistan district of Khyber Pakhtunkhwa province of Pakistan, covering an area of 7,492 km^2 (2,893 sq. m). Geographically, it is located between 34° 54′ to 35° 52′ North latitudes and 72° 43′ to 73° 57′ East longitudes. At north and northeast, it is surrounded by Diamer districts of Gilgit Baltistan, on the southeast by Mansehra District, while it shares its borders with Kagan valley of Mansehra District in the east. It is bordered in south by Battagram district and on the west by Shangla and Swat districts (Figure 15.1). Moreover, this district is bordered by two massive mountain ranges of the world. The twigs of the Himalayas are on the eastern bank, while the branches of the Karakorum are on the western bank. The Hindukush Mountain ranges extent up to Kandia valley from the North. In this way, three massive mountain ranges of the world coverage are on Kohistan making it one of the unique mountain ecosystems of the world. The highest peak in the area is around 5,500 m, while numerous pinnacles exist up to 2,100 m (Rafiq, 1996a). It is situated on such a global place where it works as a natural boundary for environmental regions in the chains of HKH Mountains. Moreover, it extends in south easterly direction across the Indus River from Shandur ranges to the Kagan valley in the northern Hazara in a continuous mountainous area about 160 km wide. These mountains system uniqueness results in rich flora and fauna in the region (Barth, 1956; Jan & Tahirkheli, 1969).

Kohistan, the place of mountains, the land of rebels and the state of free people, was called, "Yaghistan" during the colonial period. In reality, there is hardly any plain area; it is all mountainous, gigantic and monstrous (Jan & Tahirkheli, 1969; Rafiq, 1996b; GoP, 2000). The beauty of the district is unmatched with the name given "paradise of earth". The green pastures, forests, unique species, and majestic mountains make it literally heaven. This study is conducted in Palas valley located in Indus Kohistan (Figure 15.2). It covers an area of about 1,400 km^2 and elevation ranging from 700 to 5,200 m above mean sea level. The valley extends from 34.73°N to 35.21°N latitude and 72.90°E to 73.58°E longitude containing the most widespread and best protected areas of natural forest in the HKH mountains system of Pakistan (Hellquist, 2005). From the north side, the valley is confined by Jalkot, at the east by Naran, from south by Alai and from the west by Pattan.

Climatically, the study area experiences dry subtropical and temperate type of climate with sharp local variation with altitude and aspect. Pattan with an altitude of 739 m above mean sea level, the summer mean maximum temperature of June–July is about 38°C and mean minimum 22°C, while the winter mean maximum temperature remains above 0°C but does not exceed 15°C. However, the mean minimum temperature falls up to 6°C in January in the lower parts, while it remains below freezing point for more than 3 months at 3,000 m or above. Most of the precipitation

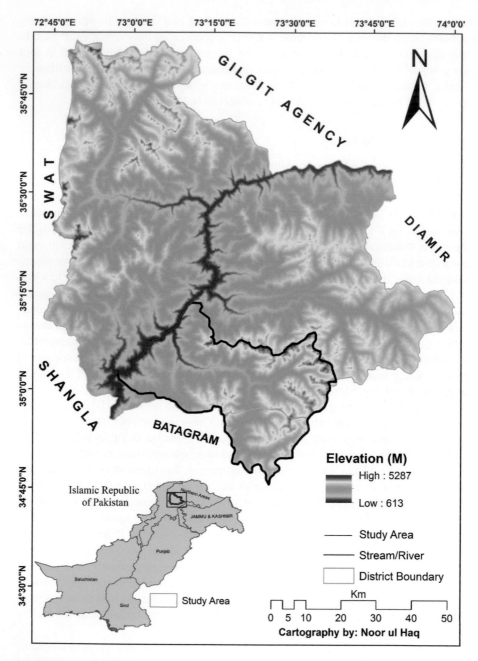

FIGURE 15.1 Location and physiography of Kohistan district. (DEM taken from USGS website and prepared in Arc GIS 10.2.1.)

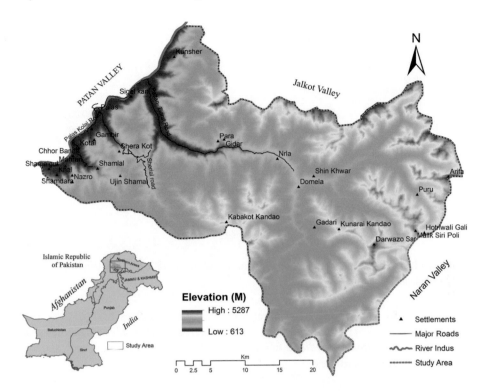

FIGURE 15.2   Location map of Palas valley. (DEM taken from USGS website and prepared in Arc GIS 10.2.1.)

occurs in the form of snow on high peaks (Dichter, 1967; Barth, 1956; Gujree et al., 2017). The average precipitation of 10 years from 2005 to 2015 recorded at Pattan Meteorological Station was 10 mm (Figure 15.3) in the valley bottom; however, at higher elevation, it might be 30–40 mm.

The study area has natural forests officially declared as Reserved and Guzara forests. Based on prevailing diverse and irregular climatic conditions, these forests are categorized as scrub to coniferous. These forests are famous in Asia with respect to their kind, density, and wild life. In the upper hilly areas, Deodar with intermixture of Blue Pine, Kail and Fir spruce forests are available in huge amount, while oak trees are abundant in the lower areas. The high mountains are covered with thick beautiful forests of excellent quality which contain cedar, juniper, pine, fir, Olea Erruinea, chilgoza, oak, walnut and birch trees. The low-lying areas around the Indus River are characterized by scrub and thorny forests of Palosa (a thorny, flowering tree), while the rest of the area, up to 3,000 m altitude, supports dense deodar and pine forests wherever the slope is not too steep. Furthermore, above 3,000 m altitude, a few pine trees, and occasional groves of birch, may be seen, though at such altitudes, up to the snowline (4–4,875 m), the scant area is mostly covered by thick, short grass, moss and innumerable flowers in the short summer season. Vegetation is sparse, common plants up to 1,500 m elevation are broad-leaved (Barth, 1956; Perveen, Khan, & Shah, 2013). The most common species are *Juniperus communis*

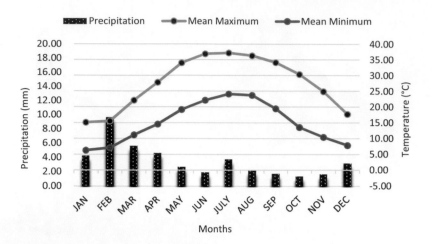

FIGURE 15.3   Pattan mean monthly temperature and precipitation (2005–2011). (Pakistan Meteorological Department.)

(juniper), *Pinus wallichiana* (blue pine), *Abies webbiana* (silver fir), *Quercus iiex* (oak), *Aesculus indica* (bankhor), Pinus species, *Cedrus deodara* (deodar), *Abies pindrow* (palunder), *Pinus gerardiana* (chalghoza), *Juglans regia* (walnut), etc., and are distinctive of higher altitudes.

According to the census of 2017, the total population of the district is 784,711 with a contribution of 275,461 people from Palas valley. It was declared as tehsil in 1998 with a population of 165,613 persons in 1998 (95 persons/km²). According to the population survey of Himalayan Jungle Project (HJP) in 1992, the inhabitants in the valley were 60,524 individuals. According to recent census, the population of the district was 784,711 persons with a male population of 434,956 and a female population 349,746 persons. Sex ratio was 124.3 male per 100 female and density was 104 person/sq. km with an average annual increase rate of 2.70%. There is considerable migration to the lowers valleys and population growth rate was only 0.09% from 1981 to 1998. In 1998 out of the total population in Palas, 54.5% were male and 45.5% were female. The share of total population in Kuz Palas is higher (72%) than Bar Palas (28%). The average household size is 6.4 persons for the whole valley and was higher (6.81) for Bar Palas (GoP, 2017). The whole district has been divided into three tehsils, and the population of each tehsil is given below.

According to census 1981, the population of district was 465,237 persons with male population 265,312 and female population 199,925 persons with a sex ratio 132.7 males per 100 females, and the population density was 61.4 persons/sq. km. The population according to 1998 census was 472,570 persons with a male population of 261,942 and a female population 210,628 persons. Sex ratio was 124.4 male per 100 females and the density was 63 persons per square km, an average annual growth rate of 0.09%. According to 2017 census, the population was 784,711 persons (Table 15.1) with a male population of 434,956 and a female population 349,746 persons. Sex ratio was 124.3 males per 100 females, and the density was 104 persons/sq. km with an average annual growth rate of 2.70%. Population density of the district

TABLE 15.1
Population Statistics of Subdivision of Kohistan District

Years	Name	Population	Male	Female	Sex Ratio	Average Household Size	Growth Rate
1981	Dassu	180,124	103,237	76,887	134.2	7.5	7.69
	Pattan	285,113	162,075	123,038	131.7	10.2	12.09
1998	Dassu	184,746	102,866	81,880	125.6	6.8	0.15
	Palas	165,613	92,330	73,283	126.0	6.4	1.73
	Pattan	122,211	66,746	55,465	120.3	6.0	−1.62
2017	Dassu	222,282	–	–	–	7.6	1
	Palas	275,461	–	–	–	7.7	3.4
	Pattan	202,913	–	–	–	7.6	3.4

Source: GoP (1983, 2000, 2017).

TABLE 15.2
Study Area: Union Council-Wise Population, Number of Households, Household Size and Population

Union Council	Number of Households	Population	Average Household Size
Bar Sherial	1,629	11,428	7
Kuz Paro	3,834	27,245	7
Sharid	4,198	27,470	6
Shalaken Abad	2,847	19,116	7
Kolai	5,257	50,784	10
Shara Kot	1,578	12,853	8
Mada Khel	4,448	37,335	8
Beach Bela	3,175	20,458	6
Batera	896	5,927	7
Hiran	3,358	25,302	7
Khota Kot	1,463	16,334	11
Kun Sher	1,531	10,961	7
Kuz Sherial	1,503	10,248	7
Total	35,717	275,461	7 (Average)

Source: GoP (2017).

was 104 persons/sq. km and declared rural population. The average household size of district has decreased from 9 persons in 1981 to 6.4 persons in 1998 which has again increased to 7.6 persons in 2017 (GoP, 1983, 2000, 2017).

There are a total of 35,717 households (Table 15.2) in Palas valley. A total of 275,461 people inhabit the valley with an average of seven persons per household. Kolai is the most populated union council which inhabits 50,784 people, with household 5,357, almost one fourth of the total population of the Palas valley. It is roughest on the basis of topography, but the high population density is due to richness in natural forest and pastures. Batera is the most thinly populated union council even having a large size and thick forest cover but remoteness in accessibility and improvement in other basic facilities like drinking water, agriculture land repel population. Kuz Paro, Sharid, Bar Sherial and Mada Khel are also large in size and rich in natural forests and hence densely populated. Household size is rather large in the lower portion of the valley which is 10–11 in different union councils. While in the upper union councils, the household size ranges from 5 to 7 persons per household.

15.3   MATERIALS AND METHODS

Mixed method was used to collect data from various sources. However, the main focus of the study was based on secondary data. The following procedure was followed for data collection.

15.3.1   PRIMARY DATA

Focused Group Discussions (FGDs) were conducted to acquire data about forest ownership, communal and customary laws, forest property rights and utilization system. During these discussions, unstructured questions were asked about forest management, utilization and measures to combat deforestation. Furthermore, information was collected through personal-observation and individual interviews from the knowledgeable and elders of the communities. Global Positioning System (GPS) survey was conducted for field verification.

15.3.2   FOCUSED GROUP DISCUSSIONS

The focused discussion were conducted with the elders of different villages in the valley in order to get information regarding forest ownership system, communal and customary laws for the management and utilization of forest resources. The FGDs were conducted in September and October 2017 and 2018 in different villages of the study area. Ten FDGs were conducted in the whole valley out of which two were from the forest department officials of Palas. The main focus of discussion was on the system of forest tenure, management and utilization system, local committee system for the management of forest resources and measures to control deforestation. These discussions provided essential information, which became the base for discussion on forest tenure/*wesh* system, locally established intuitions for management and utilization of forest resources in the study area

Another technique of data collection was individual interviews from the knowledgeable persons of the community. However, ten well-informed elders of each village were interviewed. Unstructured questions were asked regarding the forest management and utilization system, locally established institution for the management of forest resources, traditional and customary laws and their current situation. The interviews of the respondent were recorded in an audio recorder with their permission in order to utilize it in future.

15.3.3   SECONDARY DATA

Secondary data including already georeferenced and rectified SENTINEL 2A and LANDSAT satellite imageries along with Digital Elevation Model (DEM) were obtained from United State Geological Survey (USGS), Global Land Cover Facility (GLCF) websites and Pakistan Space and Upper Atmosphere Research Commission (SUPARCO). The selected data sets were free from cloud cover and acquired in Autumn season, in order to help in separating different land use/cover classes particularly agriculture land from the forest area as a large number of trees are present on the boundaries of agriculture land which make it difficult to differentiate between forest cover and agriculture land. Moreover, the trees shed their leaves in autumn season, and all shrubs and grasses become dried, and therefore, the forest cover can easily be distinguished from other land use/cover.

15.3.4   ACCURACY ASSESSMENT

Accuracy assessment of the satellite data sets of 1980, 2000, 2010 and 2017 was done utilizing the Landsat and Sentinel information of the corresponding years. More than 100 points were taken using stratified random sampling method for each class in all the data sets. For the creation of ground truth or reference points, shape file of the point features was created. The sample data of the year 1980, 2000, 2010 and 2017 were converted to Kmz file format and plotted on Google earth based on very high resolution satellite (VHRS) images.

The sample points were compared with the Google earth images, and maximum data were correctly placed. These points were again imported in Arc GIS (ArcMap) and the data were exported for creating Shapefile again. The point data were converted to raster pixels of 30 meter resolution for the first three data sets and 10 m resolution for the last one using point to raster operation in conversion tools. The classified images and pixels for each data set were combined separately and for the creation of confusion matrix, pivot table was used for attribute data. For further analysis, the pivot table was exported and opened in MS. Excel. User and Producer accuracy were calculated for each class separately by using the following formulas.

User accuracy = Total no. of correctly placed pixels of a class/Total no. of the referenced pixels of that class row * 100

Producer accuracy = Total no. of correctly placed pixels of a class/Total no. of the referenced pixels of that class column * 100

The commission and Omission errors were also calculated in percentage from the confusion matrix using the following formulas.

Commission errors = Total no. of incorrectly placed pixels of a class column/ Total no. of the referenced pixels of that class column * 100

Omission errors = Total no. of incorrectly placed pixels of a class row/Total no. of
the referenced pixels of that class row * 100

The overall accuracy was calculated in percentage with the help of the following formula

Overall accuracy = Total no. of correctly placed pixels of all classes/Total no. of
referenced pixels * 100

Kappa coefficient was calculated separately for all the data sets; it is used to measure the agreement between the model prediction of the classified image and ground reality. The value of the kappa coefficient comes in between 0 and 1. 0 indicates the complete randomness, while 1 shows perfect agreement between the models used and ground reality. So the value nearer to 1 shows significant agreement and nearer to 0 indicates randomness of agreement. The kappa coefficient was calculated with the assistance of the following formula.

$$K = N \sum_{i}^{r} = 1 \ X_{ii} - \sum_{i}^{r} = 1 \left(x_{i+} * x_{+i} \right) / N^2 - \sum_{i}^{r} = 1 \left(x_{i+} * x_{+i} \right)$$

N is the total quantity of reference pixels in confusion matrix.
    r is the entire quantity of rows in a matrix.
    X_{ii} is the total number of accurate pixels in row i and column i.
    x_{+i} is the total for row i and x_{i+} is the total for column i.
    Simply the formula can be written as

   K = (Total pixels * Sum of correct)– (Sum of all the (Row total * Column total) /
(Square of total pixels)) – (Sum of all the (Row total * Column total)).

The forest management data were collected from Pakistan Forest Institute (PFI) and Forest Management Center (FMC) Peshawar. The property tenure data were collected from Kohistan forest divisions. Population data were collected from Population Census Reports of Kohistan. Road network data were obtained from communication and works department Kohistan and Google earth.

Cartographic and statistical methods were applied to analyze the data. ArcGIS 10.2.2 and Erdas Imagine 2014 software were used for image processing, analysis and map making. Satellite datasets was selected based on the same season and minimum cloud cover. Spectral bands of the same spatial resolution were stacked. Histogram equalization and color composite techniques were applied for image enhancement to give better visibility as used in the study of Hussain et al., 2018; Shehzad et al., 2014; Qamer et al., 2012. Supervised classification (Maximum Livelihood Classifier) was used to delineate different classes, using ground checkpoints. The images were classified into different LULC classes to develop land cover maps of the study area. These classes were physically verified on ground using GPS. Signature files were created first in ArcGIS and then in ERDAS for the six land use/cover classes which include forest cover, agriculture land, bare soil/rocks, snow cover/glaciers and water bodies. In order to create signature file in ArcGIS polygon shape files were developed for sample data, and then signature files were developed. Forest cover changes were assessed through change detection techniques. Digital LULC classes were prepared and the area of these classes was figured utilizing "calculate geometry". Finally, LULC maps were prepared. GPS points and Google earth images were used for accuracy assessment. Impact of population growth and accessibility were assessed through Euclidean distance, buffer analysis and statistical correlation. Locally established management system was correlated with the management system of forest department. Database for the secondary data, population data and road networks data was developed in MS. Excel 2013. Furthermore, data were analyzed and tabulated, and various charts and diagrams were prepared in MS. Excel and interpreted in MS Word 2013. Geospatial techniques like Euclidean distance and buffer analysis were used to analyze the impact of population growth and accessibility on forest cover in Arc GIS.

15.4   RESULTS AND DISCUSSION

15.4.1   Trend of Forest Cover Change (1980–2017)

In Hindu Kush-Himalayas Mountain, the northeastern sides of the study area were covered with thick forest 26.5% in 1980 which decreased gradually to 24.8% in 2000. Then onward, the rate of forest cover change increases sharply to 18.88% till 2010. In the last period, the change again became gradual with shrinkage of 14.22% out of the total area (Table 15.3) with an annual decrease rate of 0.33%. On contrary, agriculture land and shrubs/bushes gained persistent growth throughout the study period. The bare soil/rocks class continuously increased except for the last period (2010–2017) as snow cover was converted to it, for instance, see change detection map Figure 15.4 and Table 15.3.

In the study area, the trend of forest cover was negative in all the study period. In the first period (1980–2000), the trend was low with a decreasing rate of 115.5 ha per annum, and overall decrease of 2311 ha (6.2%) was recoded. From 2000, the decreasing rate was maximum, and annual negative trend was 822 ha, more than the first period. The overall decrease in period 2 (2000–2010) was 8,226 ha (2.3%). Similarly, in the third period (2010–2017), the negative trend was again increased, and the annual shrinkage was 934 ha/year (Table 15.4), and overall negative trend recorded was 6,538 ha (2.5%). The total forest cover loss from 1980 to 2017 was 17,076 ha (12.22%) out of the total area.

15.4.2   Spatial and Temporal Change in Forest Cover (1980–2017)

Despite considerable changes in the vicinity of settlements and existing roads, the generalized picture of the whole study area is quite different. Due to remoteness, more than half (52.5%) of the study area remained unchanged. The main changes were observed in forest cover (6.8%) and shrub/bushes (12.3%) which were transformed to

TABLE 15.3

Forest and Other Land Use/Cover Statistics in hectare (ha) and percentage (%) from 1980 to 2017

Land Use/Cover	1980		2000		2010		2017		1980–2017	
	ha	%	ha	%	Ha	%	ha	%	ha	%
Forest cover	36,942	26.45	34631.37	24.79	26375.92	18.88	19866.17	14.22	–17075.58	–12.22
Agriculture land	1,323	0.95	1742.04	1.25	2552.17	1.83	3139.9	2.25	1816.9	1.30
Shrubs/bushes	35,400	25.34	38321.53	27.44	46310	33.15	56803.17	40.67	21403.17	15.32
Bare soil/rock	50,614	36.24	52940.45	37.90	56901.11	40.74	48675.21	34.85	–1939	–1.39
Glaciers/snow	15400	11.03	12021.63	8.61	7550.9	5.41	11301.69	8.09	–4098.31	–2.93
Total	139679	100	139679	100	139679	100	139679	100.		
Over all accuracy	94.6 %		95.5%		94.4%		93%			
Kappa coefficient	93.41%		94.39%		93.2%		91.4 %			

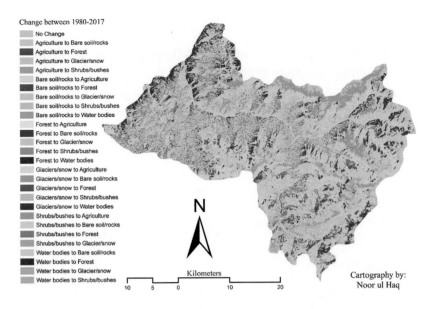

FIGURE 15.4 Change detection of forest cover and other land use changes in Palas valley, Kohistan (1980–2017).

TABLE 15.4
Periodical Forest Cover Change in Palas Valley

	Period 1 (1980–2000)		Period 2 (2000–2010)		Period 3 (2010–2017)	
Forest Cover	Area (ha)	%	Area (ha)	%	Area (ha)	%
Change	2311.11	6.2	8226.45	2.3	6538.75	2.5
Annual	115.5	0.31	822.6	0.23	934.0	0.35

bare soil/rocks; on the other hand, during the same time, some of the bare soil/rocks were converted into forest cover (1.8%) and shrub/bushes (2.1%). Furthermore, during the study period, 8% and 0.9% of forest cover was converted to shrub/bushes and agriculture land, while 1.5% of shrub/bushes were converted back to forest. These results indicated that major transformation in forest cover is due to the conversion of forest land to shrub/bushes followed by bare soil/rocks. However, during the study period from 1980 to 2017, one of the major changes occurred in forest cover and that was a decrease by 12.2% (Figure 15.4 and Table 15.5).

15.4.3 Forest Cover Change within 1 and 3 km of the Selected Settlements (1980–2017)

The forest cover within 1 km of the settlements in year 1980 was 1,271 ha (50.54%) which decreased to 997 ha (39.9%) in 2000. It again reduced to 769 ha (30.63%) in 2017. A decrease of 19.95% has been recorded. Agriculture land increased from

TABLE 15.5

Change Detection of Forest and Other Land-Use/Cover Changes in hectare (ha) and percentage (%) from 1980 to 2017

Land Use/Cover Change	Area (ha)	%	Land Use/Cover Change	Area (ha)	%
No change	73317.8	52.5	Water bodies to glaciers/snow	5.7	0.0
Forest to bare soil	9470.3	6.8	Bare soil to forest	2521.3	1.8
Shrubs/bushes to Bare soil	17162.8	12.3	Shrubs/bushes to forest	2055.6	1.5
Glacier/snow to Bare soil	4184.3	3.0	Glacier/snow to forest	216.9	0.2
Agriculture to Bare soil	45.3	0.0	Agriculture to forest	1.0	0.0
Water bodies to Bare soil	12.8	0.0	Water bodies to forest	0.4	0.0
Forest to Shrub/bushes	11128.2	8.0	Forest to water bodies	3.3	0.0
Bare soil to shrubs/BUSHES	2991.4	2.1	Bare soil to water bodies	49.6	0.0
Glacier/snow to Shrubs/bushes	1222.0	0.9	Shrubs/bushes to water bodies	0.05	0.0
Agriculture to Shrubs/bushes	2116.9	1.5	Glacier/snow to water bodies	51.3	0.0
Forest to glaciers/snow	174.8	0.1	Forest to agriculture	1258.9	0.9
Bare soil to glaciers/snow	9717.3	7.0	Bare soil to agriculture	332.0	0.2
Shrubs/bushes to Glaciers/snow	764.1	0.5	Shrubs/bushes to Agriculture	849.33	0.6
Agriculture to Glaciers/snow	0.2	0.0	Glacier/snow to Agriculture	12.8	0.0
			Total	139666.4	100

TABLE 15.6

Forest and Other Land Use/Cover within 1 km of the Settlement (1980–2017)

Land Use/Cover Classes	Year 1980		Year 2000		Year 2017		Change 1980–2017	
	Area (ha)	%	Area (ha)	%	Area (ha)	%	Area (ha)	%
Forest cover	1271.16	50.54	997.84	39.68%	769.53	30.63	−501.63	−19.95
Agriculture	10.44	0.42	54.2	2.16%	67.9	2.61	57.46	2.28
Shrubs/bushes	634.68	25.24	1120.86	44.57%	1255.59	49.97	620.91	24.69
Bare soil/rock	586.8	23.33	340.74	13.55%	392.22	15.61	−194.59	−7.74
Glacier/snow	11.88	0.47	1.26	0.05%	29.72	1.18	17.84	0.71

0.42% to 2.16% in year 2000 and 2.61% in 2017 with an increasing rate of 2.28% in whole study period. Similarly, shrubs/bushes were recoded 25.24% in 1980 increased to 44.57% in year 2000 and 49.97% in 2017 (Table 15.6). The share of bare soil/rocks was 23.33% in 1980 decreased to 13.55% in 2000 and increased to 15.61% in 2017. A decrease of 7.75% has been recorded from 1980 to 2017.

Similarly, in 1980, within 3 km of the settlements, the forest cover was 9,511 ha (44.8%) which decreased to 7,822 ha (36.9%) in 2000 and again decreased to 6004.5 ha (28%) of the total area in 2017. A decrease of (16.53%) has been recorded from 1980 to 2017. The total area covered by agriculture land in 1980 was 150 ha (0.71%) which increased to 395 ha (1.86%) in 2000 and 590.46 ha (2.78%) in 2017. An increase of almost (2.1%) has been recorded in study period. In the same way,

TABLE 15.7

Forest and Other Land Use/Cover within 3 km of the Settlement (1980–2017)

Land Use/Cover Classes	Year 1980		Year 2000		Year 2017		Change 1980–2017	
	Area (ha)	%	Area (ha)	%	Area (ha)	%	Area (ha)	%
Forest cover	9513.53	44.81	7822.8	36.86	6004.58	28.28	−3508.95	−16.53
Agriculture	150.49	0.71	395	1.86	590.46	2.78	439.97	2.07
Shrubs/bushes	6054.96	28.51	8832.12	41.61	10495.82	49.44	4440.86	20.92
Bare soil/rock	4944.99	23.30	4105.1	19.34	3846.88	18.12	−1098.11	−5.17
Glacier/snow	566.64	2.67	70.22	0.33	292.87	1.38	−273.77	−1.29

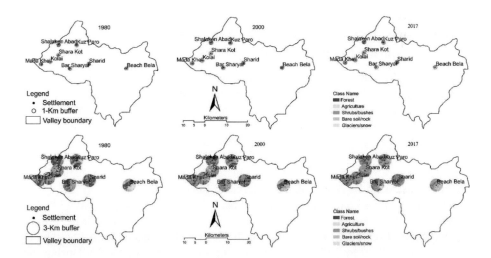

FIGURE 15.5 Forest and other land use/cover changes within 1 and 3 km of the settlement (1980–2017).

shrubs/bushes were 6,051 ha (28.51%) in 1980 increased to 8,832 ha (41.61%) in 2000 and 10495.9 (49.5%) in 2017 (Table 15.7 and Figure 15.5). The share of bare soil/rocks was 23.30% in 1980 which decreased to 19.34% in 2000 and 18.12% in 2017. A decrease of 5.17% has been recorded from 1980 to 2017. The results indicate that within 3 km of the settlements, 3,509 ha (16.6%) of the forest cover has been reduced in the study period.

Forest cover change is more prominent in surrounding areas of the settlements as people use them for firewood and other purposes and deforestation declines as we move far from the settlement/villages. The analysis reveals that changes in forest cover are higher within 1 km of the settlement as compared to 3 km. Agricultural land and shrubs/bushes have been increased, while other land use/cover classes show a gradual decrease. This increase in agricultural land and shrubs/bushes may be the

TABLE 15.8

Forest and Other Use/Cover Change within 1 km of the Roads (1980 and 2017)

Land Use/Cover Classes	Year 1980		Year 2017		Change	
	Area (ha)	%	Area ha	%	Area ha	%
Forest cover	4625.28	36.03	4337.79	33.79	−287.49	−2.24
Agriculture	191.88	1.49	768.92	5.99	577.04	4.50
Shrubs/bushes	3283	25.58	2678.61	20.87	−604.39	−4.71
Bare soil/rock	4736.52	36.90	4356.35	33.94	−380.17	−2.96

result of growing population as more land has been cleared for agriculture and house construction when the number of people increased.

15.4.4 FOREST COVER CHANGE WITHIN 1 AND 3 KM OF THE ROADS (1980–2017)

The results indicated that within 1 km of the road networks, forest cover and shrubs/bushes decreased while other land use/cover increased. Within 1 km of these roads, forest cover was 4625.28 ha (36%) which decreased to 4337.79 ha (33.79%) in 2017. A decrease of 288 ha (2.24%) has been recorded in the study period. Similarly, the shrubs/bushes decreased from 3,283 ha (25.9%) to 2,678 ha (20.8) with a decrease of 4.71% (Table 15.8 and Figure 15.6). During the same period, agriculture land increased from 1.5% to 6% in 2017 with an increasing rate of 4.5%. The share of bare soil/rock was 36.9% in 1980 which decreased to 33.9% in 2017.

Similarly, within 3 km radius of the road networks, the forest cover was 13,847 ha (41%), which reduced to 13,519 ha (40.3%) in 2017. A shrinkage of 328 ha, almost 1%, has been observed. Similarly, the shrubs/bushes also shrunk from 10,691 ha to 10,572 ha from 1980 to 2017, with a decreasing rate of 118 ha (0.35%). During the same period, the agriculture land increased from 405 ha (1.2%) to 1,515 ha (4.5%) (Table 15.9), and an increase of 1,110 ha (3.3%) has been observed in 2017.

The forest cover and shrubs/bushes have been converted to agriculture land and man-made features like houses because of population growth. Another cause of decrease in vegetation cover near roads is the preference of people to live close to these roads for better accessibility. The reduction in forest cover with increase in the accessibility indicates the impact of road on the forest cover. The extension in road networks accelerates forest cover change for road construction, and these roads increase deforestation for timber transportation and fuel wood extraction.

We have evaluated the relationship of forest cover with the proximity of the settlement and road networks. Our results highlights that within 1 and 3 km radius of the settlements and road networks, reduction has been observed in forest cover while expansion in agriculture land. These changes in the forest and other land use/cover are the impact of growing population and extension in ease of access to the inaccessible areas. The change detection map and Table 15.5 illustrate this condition in

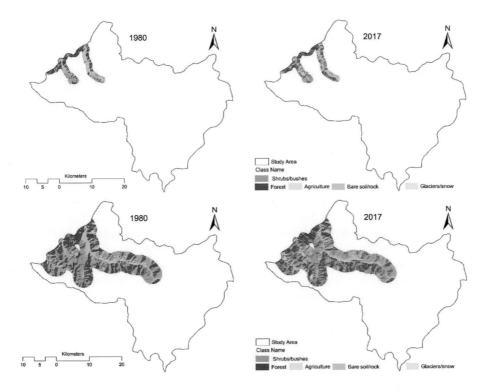

FIGURE 15.6 Forest and other land use/cover change within 1 and 3 km of the road network (1980–2017).

TABLE 15.9
Forest and Other Use/Cover Changes within 3 km of the Roads (1980–2017)

Land Use/Cover Classes	Year 1980		Year 2017		Change	
	Area (ha)	**%**	**Area (ha)**	**%**	**Area (ha)**	**%**
Forest cover	13847.4	41.29	13519.37	40.31	−328.03	−0.98
Agriculture	405	1.21	1515.25	4.52	1110.25	3.31
Shrubs/bushes	10691.64	31.88	10572.83	31.53	−118.81	−0.35
Bare soil/rock	8593.56	25.62	7929.57	23.64	−663.99	−1.98

pictorial and statistical form and reveal that large areas of forest have been transformed to other land use/cover.

15.4.5 LOCAL COMMITTEE SYSTEM FOR MANAGEMENT

In Palas valley, the management of forest is communal, and the local community members have established a committee for forest management and its utilization. The management system of the forest resources is conceived by the local community.

For the suitable management and utilization of forest assets, the committee is constituted in each community. In Koli Palas, one elder of the village from each tribe and clan is selected as the member of the committee. There are 24 tribes, and each has one member in the committee which is locally called "*zaituon*", the plural of which is *zaituo*. *Zaituon* is responsible to combine all the residents together for everything that needs cooperative effort. Great care is needed to make sure about the selection of *zaituon* which is strict and hard work. They keep check and balance on trees in the forests and look after the crops and animals. If the *zaituon* catches someone logging unauthorized wood, they charge them about 1,000 INR for each tree. If somebody has been punished by the *zaituon* and refuses to give the fine, the *zaituon* takes valuable things from the offender's house and keeps it till the offender pays the penalty. In addition to the fine, the *zaituon* also makes offenders to replant trees that were cut down. The responsibility of the *zaituon* committee is to manage the forest affairs like its cutting, its safety, management and conservation and distributions of money at the time of selling. 80% share is distributed in each clan on the basis of individuals of the household. 20% share is going to the government in the selling of forest. The committee of *zaituon* takes a decision in a meeting in consultation with the elders of the community, about the tract/portion of the forest to be opened for cutting. They also decide the date and duration of the tree cutting.

According to focused group discussion with the elders of the community, the forest department appointed a forest guard to check and control illegal cutting of forests. If someone has cut the trees in the forest, the guard takes action against illegal harvester and then takes penalties from the wrong doers with the consultation of the elders of the community. In few villages of the study area, forest guards are not local, which creates problems for the local committee as they do not know the terms and conditions of the local committee. Furthermore, in some villages like Shalkan Abad, Bar Paro, Kuz Paro and Sharid, there is no proper management system at local level. Nevertheless, committees have been constituted to monitor cutting of and distribution of money when forests are sold to a contractor.

15.4　CONCLUSION

The main conclusions in this chapter are that forest cover has been changed throughout the study period. The transformation of forest cover to agriculture land indicates the increasing population and food demands. To meet these demands, more land has been brought under cultivation and house construction which ultimately resulted in forest cover change. This study found a strong inverse relationship between population growth, accessibility and forest cover. The improvement in accessibility in the form of road networks leads to changes in forest and other use/cover. Before 1980, most of the valley was inaccessible, and thick forests were found on the mountains. After 1980, road network extension caused deforestation in two ways; the forest was cleared in those areas from where the transportation networks passed. In addition to accessibility, growing population demands for fuelwood and construction purposes including space heating and kitchen also increase, as a large number of people use maximum amount of fuelwood for room heating in winter. It is also observed that within 1 km radius of the settlements, deforestation is high as compared to 3 km radius. The logic is that people cut trees near the villages/settlements for fuelwood and house construction, etc. In contrast, due to lack of roads in the precipitous mountains, the forest cover increases as we move far away from the settlements. It is further concluded that the forest management is communal, and the local community members established a committee for the management and its proper utilization in each community. They are accountable for uniting all the inhabitants for whatever that involves collective effort and managing the forest affairs like its cutting, safety, management and conservation, distributions of money at the time of selling of forest and looking after the crops and animals. Every individual of the study area is bound to follow the rules and regulations. The committee of *zaituon* is nominated annually or seasonally; however, an individual possibly will serve persistently as long as the inhabitants are contented with his performance. Although growing population and road network extension are the main driving forces of change in forest cover, deforestation is a complex phenomenon involving many socioeconomic and biophysical driving forces other than the abovementioned ones. Thus, this study is an attempt to investigate a few causes of change in forest cover. However, further research is required to find out other grounds of forest cover changes in the study area. We recommend that forest management alternatives for fuelwood, construction wood and forage should be provided in low altitude oak forest to assist burden on forest resources adjacent to human settlements. Furthermore, this type of research can be convenient to generate a baseline for a long-term monitoring system based on GIS and RS skills.

ACKNOWLEDGMENT

This chapter is part of the M.Phil. thesis submitted to the Department of Geography, University of Peshawar, Pakistan. This study is part of the National Research Program for Universities (NRPU), Project No. 20.2396/NRPU/R&D/HEC 2014/163, funded by the Higher Education Commission of Pakistan. The authors gratefully acknowledge this as well as the help, cooperation and hospitality of the local inhabitants rendered during the field survey.

REFERENCES

Abdullah, S. M. S. (2001). Transport accessibility and deforestation: Empirical evidence from the Klang-Langat watershed study. *Proceedings of the Eastern Asia Society for Transportation Studies, 3*(3), 249–256.

Ahmad, A., & Nizami, S. M. (2015). Carbon stocks of different land uses in the Kumrat valley, Hindu Kush Region of Pakistan. *Journal of Forestry Research, 26*(1), 57–64. https://doi.org/10.1007/s11676-014-0008-6

Ali, J., & Benjaminsen, T. A. (2004). Fuelwood, timber and deforestation in the Himalayas: The case of Basho Valley, Baltistan Region, Pakistan. *Mountain Research and Development, 24*(4), 312–318.

Ali, T., Shahbaz, B., & Suleri, A. (2006). Analysis of myths and realities of deforestation in Northwest Pakistan: Implications for forestry extension. *International Journal of Agriculture and Biology, 8*(1), 107–110.

Barth, F. (1956). *Indus and Swat Kohistan: An Ethnographic Survey.* Vol. II. Studies Honouring the Centennial of Universitetets Etnografiske Museum, Oslo.

Bera, A., Taloor, A. K., Meraj, G., Kanga, S., Singh, S. K., Durin, B., & Anand, S. (2021). Climate vulnerability and economic determinants: Linkages and risk reduction in Sagar Island, India; A geospatial approach. *Quaternary Science Advances*, 100038. https://doi.org/10.1016/j.qsa.2021.100038

Bhatta, L. D., Shrestha, A., Neupane, N., Jodha, N. S., & Wu, N. (2019). Shifting dynamics of nature, society and agriculture in the Hindu Kush Himalayas: Perspectives for future mountain development. *Journal of Mountain Science*, *16*(5), 1133–1149. https://doi.org/10.1007/s11629-018-5146-4

Brinkmann, K., Noromiarilanto, F., Ratovonamana, R. Y., & Buerkert, A. (2014). Deforestation processes in south-western Madagascar over the past 40 years: What can we learn from settlement characteristics? *Agriculture, Ecosystems & Environment*, *195*, 231–243.

Bruns, B. (2005). Community-based principles for negotiating water rights: Some conjectures on assumptions and priorities. In *International Workshop on African Water Laws: Plural Legislative Frameworks for Rural Water Management in Africa*, Johannesburg, South Africa.

Chandel, R. S., Kanga, S., & Singh, S. K. (2021). Impact of COVID-19 on tourism sector: a case study of Rajasthan, India. *AIMS Geosciences*, *7*(2), 224–242.

Cubbage, F., Harou, P., & Sills, E. (2007). Policy instruments to enhance multi-functional forest management. *Forest Policy and Economics*, *9*(7), 833–851.

Das, P., Behera, M. D., & Murthy, M. S. R. (2017). Forest fragmentation and human population varies logarithmically along elevation gradient in Hindu Kush Himalaya-utility of geospatial tools and free data set. *Journal of Mountain Science*, *14*(12), 2432–2447. https://doi.org/10.1007/s11629-016-4159-0

Deng, X., Huang, J., Uchida, E., Rozelle, S., & Gibson, J. (2011). Pressure cookers or pressure valves: do roads lead to deforestation in China?. *Journal of Environmental Economics and Management*, *61*(1), 79–94.

Dichter, D. (1967). *The North-West Frontier of West Pakistan: A Study in Regional Geography*. Clarendon Press, Oxford.

Diniz, F. H., Kok, K., Hott, M. C., Hoogstra-Klein, M. A., & Arts, B. (2013). From space and from the ground: Determining forest dynamics in settlement projects in the Brazilian Amazon. *International Forestry Review*, *15*(4), 442–455. https://doi.org/10.1505/146554813809025658

Ellis, E. A., & Porter-Bolland, L. (2008). Is community-based forest management more effective than protected areas?: A comparison of land use/land cover change in two neighboring study areas of the Central Yucatan Peninsula, Mexico. *Forest Ecology and Management*, *256*(11), 1971–1983.

Farooq, M., & Muslim, M. (2014). Dynamics and forecasting of population growth and urban expansion in Srinagar city – A geospatial approach. *The International Archives of Photogrammetry, Remote Sensing and Spatial Information Sciences*, *40*(8), 709.

Geist, H. J., & Lambin, E. F. (2001). *What Drives Tropical Deforestation? A Meta-Analysis of Proximate and Underlying Causes of Deforestation Based on Subnational Case Study Evidence*. LUCC International Project Office, University of Louvain, Department of Geography, Belgium.

GoP (1983). *Census Report of District Kohistan*. Population Census Organization Statistics Division Government of Pakistan, Islamabad.

GoP (2000). *Census Report of District Kohistan*. Population Census Organization Statistics Division Government of Pakistan, Islamabad.

GoP (2017). *Census Report of District Kohistan*. Population Census Organization Statistics Division Government of Pakistan. https://www.pbscensus

Gujree, Ishfaq, et al. (2017). Evaluating the variability and trends in extreme climate events in the Kashmir Valley using PRECIS RCM simulations. *Modeling Earth Systems and Environment*, *3*(4), 1647–1662.

Haq, N. (2019). Dynamics of forest cover changes in Hindu Kush-Himalayas Mountain using remote sensing (1980–2017): A case study of Palas valley, Pakistan. M.Phil. Thesis submitted to University of Peshawar, Pakistan

Haq, F., Rahman, F., Tabassum, I., Ullah, I., & Sher, A. (2018). Forest Dilemma in the Hindu Raj Mountains Northern Pakistan: impact of population growth and household dynamics. *Small-scale Forestry*, *17*(3), 323–341. https://doi.org/10.1007/s11842-018-9390-9

Haq, N., et al. (2021a). Extension of roads towards forest in Palas Valley Indus Kohistan, Hindu Kush-Himalayan Mountains, Pakistan. *GeoJournal*. https://doi.org/10.1007/s10708-021-10437-y

Haq, N., et al. (2021b). Forest cover dynamics in Palas Valley Kohistan, Hindu Kush-Himalayan Mountains, Pakistan. *Journal of Mountain Science*, 18(2), 416–426. https://doi.org/10.1007/s11629-020-6093-4

Hassanin, M., Kanga, S., Farooq, M., & Singh, S. K. (2020). Mapping of Trees outside Forest (ToF) from Sentinel-2 MSI satellite data using object-based image analysis. *Gujarat Agricultural Universities Research Journal*, 207–213.

Hellquist, A. (2005). Are divergent preferences between benefactors and beneficiaries an obstacle to community-based conservation? A case study of the Palas Valley, northern Pakistan. http://lup.lub.lu.se/student-papers/record/1334494

Hussain, K., Haq, F., & Rahman, F. (2018). Shrinking greenery: Land use and land cover changes in Kurram Agency, Kohi Safid Mountains of north-western Pakistan. *Journal of Mountain Science*, 15(2), 296–306. https://doi.org/10.1007/s11629-017-4451-7

Jan, G., Khan, M. A., Farhatullah, J. F., Ahmad, M., Jan, M., & Zafar, M. (2011). Ethnobotanical studies on some useful plants of Dir Kohistan valleys, KPK, Pakistan. *Pakistan Journal of Botany*, 43(4), 1849–1852.

Jan, M. Q., & Tahirkheli, R. A. (1969). The geology of the lower part of Indus Kohistan (Swat), West Pakistan. *Journal of Himalayan Earth Sciences*, 4.

Kanga, S., Kumar, S., & Singh, S. K. (2017b). Climate induced variation in forest fire using remote sensing and GIS in Bilaspur district of Himachal Pradesh. *International Journal of Engineering and Computer Science*, 6(6), 21695–21702.

Kanga, S., Rather, M. A., Farooq, M., & Singh, S. K. (2021). GIS based forest fire vulnerability assessment and its validation using field and MODIS data: A case study of Bhaderwah forest division, Jammu and Kashmir (India). *Indian Forester*, 147(2), 120–136.

Kanga, S., Singh, S. K., & Sudhanshu (2017a). Delineation of urban built-up and change detection analysis using multi-temporal satellite images. *International Journal of Recent Research Aspects*, 4(3), 1–9.

Meraj, G., Singh, S.K., Kanga, S. et al. (2021a). Modeling on comparison of ecosystem services concepts, tools, methods and their ecological-economic implications: a review. *Modeling Earth Systems and Environment*. https://doi.org/10.1007/s40808-021-01131-6

Meraj, G., Kanga, S., Kranjčić, N., Đurin, B. & Singh, S.K. (2021b). Role of Natural Capital Economics for Sustainable Management of Earth Resources. *Earth*, 2(3), 622–634.

Meraj, G., & Kumar, S. (2021). Economics of the natural capital and the way forward. Preprints 2021, 2021010083. https://doi.org/10.20944/preprints202101.0083.v1

Meraj, G., Singh, S. K., Kanga, S., & Islam, M. N. (2021). Modeling on comparison of ecosystem services concepts, tools, methods and their ecological-economic implications: A review. *Modeling Earth Systems and Environment*, 1–20

Munawar, S., & Udelhoven, T. (2020). Land change syndromes identification in temperate forests of Hindukush Himalaya Karakorum (HHK) mountain ranges. *International Journal of Remote Sensing*, 41(20), 7735–7756. https://doi.org/10.1080/01431161.2020.1763509

Newman, M. E., McLaren, K. P., & Wilson, B. S. (2014). Assessing deforestation and fragmentation in a tropical moist forest over 68 years; the impact of roads and legal protection in the Cockpit Country, Jamaica. *Forest Ecology and Management*, *315*, 138–152. https://doi.org/10.1016/j.foreco.2013.12.033

Perveen, F., Khan, A., & Shah, A. H. (2013). Hunting and trapping pressure on the Himalayan Grey Goral Naemorhedus goral (Hardwicke) (Artiodactyla Bovidae) in Kohistan, Pakistan. *American Journal of Zoology Reserved*, *1*, 5–11.

Potapov, P. V., Turubanova, S. A., Tyukavina, A., Krylov, A. M., McCarty, J. L., Radeloff, V. C., & Hansen, M. C. (2015). Eastern Europe's forest cover dynamics from 1985 to 2012 quantified from the full Landsat archive. *Remote Sensing of Environment*, *159*, 28–43. https://doi.org/10.1016/j.rse.2014.11.027

Poudel, S., & Shaw, R. (2015). Demographic changes, economic changes and livelihood changes in the HKH Region. In: Nibanupudi, H., & Shaw, R. (eds) *Mountain Hazards and Disaster Risk Reduction. Disaster Risk Reduction (Methods, Approaches and Practices)* (pp. 105–123). Springer, Tokyo. https://doi.org/10.1007/978-4-431-55242-0_6

Qamer, F. M., Abbas, S., Saleem, R., Shehzad, K., Ali, H., & Gilani, H. (2012). Forest cover change assessment in conflict-affected areas of northwest Pakistan: The case of Swat and Shangla districts. *Journal of Mountain Science*, *9*(3), 297–306. https://doi.org/10.1007/s11629-009-2319-1

Qamer, F. M., Shehzad, K., Abbas, S., Murthy, M. S. R., Xi, C., Gilani, H., & Bajracharya, B. (2016). Mapping deforestation and forest degradation patterns in western Himalaya, Pakistan. *Remote Sensing*, *8*(5), 385. https://doi.org/10.3390/rs8050385

Qasim, M., Hubacek, K., Termansen, M., & Khan, A. (2011). Spatial and temporal dynamics of land use pattern in District Swat, Hindu Kush Himalayan region of Pakistan. *Applied Geography*, *31*(2), 820–828. https://doi.org/10.1016/j.apgeog.2010.08.008

Rafiq, R. A. (1996a). *Taxonomical, Chorological and Phytosociological Studies on the Vegetation of Palas Valley*. Pakistan Agriculture Research Centre, Islamabad.

Rafiq, R.A. (1996b). Three new species from Palas Valley, district Kohistan, North West Frontier Province, Pakistan. *Novon*, *6*(3), 295–297.

Rahman, F., Haq, F., Tabassum, I., & Ullah, I. (2014). Socio-economic drivers of deforestation in Roghani Valley, Hindu-Raj Mountains, Northern Pakistan. *Journal of Mountain Science*, *11*(1), 167–179. https://doi.org/10.1007/s11629-013-2770-x

Rasolofoson, R. A., Ferraro, P. J., Jenkins, C. N., & Jones, J. P. (2015). Effectiveness of community forest management at reducing deforestation in Madagascar. *Biological Conservation*, *184*, 271–277.

Rather, M. A., et al. (2018). Remote sensing and GIS based forest fire vulnerability assessment in Dachigam National park, North Western Himalaya. *Asian Journal of Applied Sciences*, *11*(2), 98–114.

Robinson, B. E., Holland, M. B., & Naughton-Treves, L. (2014). Does secure land tenure save forests? A meta-analysis of the relationship between land tenure and tropical deforestation. *Global Environmental Change*, *29*, 281–293. https://doi.org/10.1016/j.gloenvcha.2013.05.012

Rodriguez, L. G., & Pérez, M. R. (2013). Recent changes in Chinese forestry seen through the lens of Forest Transition theory. *International Forestry Review*, *15*(4), 456–470.

Shahbaz, B., Ali, T., & Suleri, A. Q. (2007). A critical analysis of forest policies of Pakistan: implications for sustainable livelihoods. *Mitigation and Adaptation Strategies for Global Change*, *12*(4), 441–453.

Sharma, M., Areendran, G., Raj, K., Sharma, A., & Joshi, P. K. (2016). Multitemporal analysis of forest fragmentation in Hindu Kush Himalaya – A case study from Khangchendzonga Biosphere Reserve, Sikkim, India. *Environmental Monitoring and Assessment*, *188*(-10), 596. https://doi.org/10.1007/s10661-016-5577-8

Shehzad, K., Qamer, F. M., Murthy, M. S. R., Abbas, S., & Bhatta, L. D. (2014). Deforestation trends and spatial modeling of its drivers in the dry temperate forests of northern Pakistan: A case study of Chitral. *Journal of Mountain Science*, *11*(5), 1192–1207. https://doi.org/10.1007/s11629-013-2932-x

Southworth, J., & Tucker, C. (2001). The influence of accessibility, local institutions, and socioeconomic factors on forest cover change in the mountains of western Honduras. *Mountain Research and Development*, *21*(3), 276–283. https://doi.org/10.1659/0276-4741(2001)021[0276:TIOALI]2.0.CO;2

Suleri, A. Q. (2002). *The State of Forests in Pakistan through a Pressure-State-Response Framework. International Forestry Review.* Sustainable Development Policy Institute, Islamabad.

Sydenstricker-Neto, J. (2012). Population and deforestation in the Brazilian Amazon: A mediating perspective and a mixed-method analysis. *Population and Environment*, *34*(1), 86–112.

Tomar, J. S., Kranjčić, N., Đurin, B., Kanga, S., & Singh, S. K. (2021). Forest fire hazards vulnerability and risk assessment in Sirmaur district forest of Himachal Pradesh (India): A geospatial approach. *ISPRS International Journal of Geo-Information*, 10(7), 447.

Tsering, K., Sharma, E., Chettri, N., & Shrestha, A. (eds) (2010) *Climate Change Vulnerability of Mountain Ecosystems in the Eastern Himalayas; Climate Change Impact an Vulnerability in the Eastern Himalayas – Synthesis Report.* ICIMOD, Kathmandu.

Ullah, S., Farooq, M., Shafique, M., Siyab, M. A., Kareem, F., & Dees, M. (2016). Spatial assessment of forest cover and land-use changes in the Hindu-Kush mountain ranges of northern Pakistan. *Journal of Mountain Science*, *13*(7), 1229–1237. https://doi.org/10.1007/s11629-015-3456-3

Van Khuc, Q., Tran, B. Q., Meyfroidt, P., & Paschke, M. W. (2018). Drivers of deforestation and forest degradation in Vietnam: An exploratory analysis at the national level. *Forest policy and economics*, *90*, 128–141. https://doi.org/10.1016/j.forpol.2018.02.004

Vanonckelen, S., & Van Rompaey, A. (2015). Spatiotemporal analysis of the controlling factors of forest cover change in the Romanian Carpathian Mountains. https://www.nccr-north-south.unibe.ch

Wester, P., Mishra, A., Mukherji, A., & Shrestha, A. B. (eds) (2019). *The Hindu Kush Himalaya Assessment: Mountains, Climate Change, Sustainability and People* (p. 627). Springer Nature, Cham.

Zeb, A. (2019). Spatial and temporal trends of forest cover as a response to policy interventions in the district Chitral, Pakistan. *Applied Geography*, *102*, 39–46. https://doi.org/10.1016/j.apgeog.2018.12.002

16 Remote Sensing and Geographic Information System for Evaluating the Changes in Earth System Dynamics
A Review

Asraful Alam
University of Calcutta

Nilanjana Ghosal
Jadavpur University

CONTENTS

16.1   Introduction 310
16.2   Remote Sensing and GIS as Tools.. 311
          16.2.1   Remote Sensing 311
          16.2.2   GIS 311
16.3   Role of GIS and Remote Sensing in Assessing Different Changes in
          Earth's System Dynamics... 312
          16.3.1   Application in the Field of Land Use and Land Cover................ 312
                      16.3.1.1   Natural Resource Management.................................... 312
                      16.3.1.2   Application in the Field of Agriculture....................... 313
                      16.3.1.3   Application in the Field of Forestry............................ 313
                      16.3.1.4   Application in the Field of Geology............................ 314
                      16.3.1.5   Application in the Field of Hydrology 314
                      16.3.1.6   Application in the Field of Sea Ice.............................. 315
                      16.3.1.7   Application in the Field of Ocean Monitoring............ 315
                      16.3.1.8   Application in the Field of Coastal Management 315
                      16.3.1.9   Application in the Field of Environmental Monitoring ...316
                      16.3.1.10 Application in the Field of Ecology and
                                     Biodiversity 316
16.4   Conclusion 317
References... 317

DOI: 10.1201/9781003147107-20

16.1   INTRODUCTION

Remote sensing and geographical information system (GIS) are powerful tools that are very important in every area of human endeavor over which geographers are very proud. They showcase some of the significant ways technology has resulted in an immense benefit to humanity. Remote sensing and GIS have significantly contributed to over 80% accurate and real-time geo-information data and support for sustainable development efforts (Longato et al., 2019). Nowadays, the remote sensing and GIS field has become exciting and desirable with rapidly expanding opportunities. It provides vital tools that can be applied in the various levels leading to decision-making toward sustainable socio-economic development and conservation of natural resources. Remote sensing and GIS technology and its applications in different fields have experienced a successful outcome in recent decades. Remote sensing technology in present times has proved to be of great importance in acquiring data for adequate resource management and hence could also be applied to coastal environment monitoring and management. On the other hand, a GIS aids in analyzing trends and estimating changes that have occurred in many themes, which aids in the management of the decision-making process. In particular, GIS and remote sensing technologies provide the ability to collect data quickly, process and integrate data and information, and present outcomes in geographically referenced maps and reports (Wilkinson, 1996).

The three elements of geo-informatics technology are remote sensing (RS), GISs, and global positioning systems (Ranjan et al., 2016). Remote sensing is the collecting of data on a small or large scale about an object or phenomenon (for example, earth surface features) recorded via real-time devices such as airplanes, spaceships, satellites, boats, or ships that are not in direct touch with the thing (Solomon and Ghebreab, 2011). It is frequently utilized in geology, ecology, geography, hydrology, land surveying, and natural resource management, among other fields. It is also used for military, intelligence, commercial, economic, planning, and humanitarian purposes (Goyal et al., 2020). Similarly, a GIS is a set of tools for managing and analyzing spatial data to help users organize, store, edit, analyze, and display positional and attribute information about geographic data (Solomon and Ghebreab, 2011).

In recent years, the rapid growth of spatial technology has made new tools and capacities available to extension services and consumers for geographical data management (Milla et al., 2005). RS and GIS are two technologies that are pretty useful (Bernd et al., 2017). During the last 30 years, GIS and RS have become essential tools for extracting data and information from the ground (Ranjan et al., 2016; Singh et al., 2016). These days, spatial, temporal, and spectral resolved satellite data are widely available, and their uses in the GIS context have expanded for research purposes (Ranjan et al., 2016).

This chapter mainly focuses on applying RS and GIS in assessing various dynamics related to Earth in today's time. Remote sensing and GIS technologies are well-established tools (Raup et al., 2007). They often work hand-in-hand in mapping, analyzing, and disseminating spatial information. The subject of correctly combining both technologies and assessing both systems has been the focus of research in recent years (Geels, 2004). Several satellite data are currently being used to form digital elevation models and monitor the global and regional area for sustainable development and protection of the environment (Coskun et al., 2008). Hence, with the advancement of computer technology, the present-day GIS and RS software have become more powerful. It has opened the door for immense large-scale mapping opportunities, updating existing maps, project planning, and decision-making. It has linked several discrete technologies into a whole for proper map overlay analysis and modeling that are beyond the capability of manual methods (Reddy and Reddy, 2008).

16.2 REMOTE SENSING AND GIS AS TOOLS

16.2.1 REMOTE SENSING

Remote sensing is defined as the science and technology by which the characteristics of objects of interest can be identified, measured, or analyzed without coming into direct contact. In its broadest definition, remote sensing involves detecting and collecting electromagnetic waves from target places inside the sensor instrument's range of view. Remote sensing, according to the United Nations, is the process of sensing the Earth's surface from space using the characteristics of electromagnetic waves generated, reflected, or diffracted by the detected objects to improve natural resource management, land use, and environmental protection (Yadav et al., 2013). RS of Earth has come a long way from nineteenth-century aerial photography to the latest UAV remote sensing. In a general sense, RS presently means satellite RS, which started with the launch of Landsat-1 in 1972 for civilian applications (Roy et al., 2017). Remote sensing includes the recording of reflected or emitted energy (i.e., EMR). Based on the source of EMR, remote sensing is classified into "active" and "passive". It has been used in different fields such as geology, ecology, geography, hydrology, land surveying, and natural resource management. Remote sensing plays a vital role in military, intelligence, commercial, economic, planning, and humanitarian applications (Goyal et al., 2020).

16.2.2 GIS

GIS is "a system of hardware and software that supports the capture, management, manipulation, analysis, and display of geographic information" (Al-Ramadan, 2005). It can be used to analyze the spatial characteristics of the data over various digital layers. Even though GIS has been there since the 1960s, its applications grew in the 1990s. Many software systems have been created to cover a wide range of fields, including earth and environmental sciences, management of natural resources, terrain modeling, agriculture, forest management, construction engineering, land use development and planning control, population distribution, settlement, transportation, education, and health planning (Ondieki and Murimi, 1997). It is an expanding information technology for creating databases with spatial information (Akeem et al., 1986). GIS also combines technical and human resources with a set of organizing procedures to produce information in support of decision-making (Ondieki and Murimi, 1997; Akeem et al., 1986).

16.3 ROLE OF GIS AND REMOTE SENSING IN ASSESSING DIFFERENT CHANGES IN EARTH'S SYSTEM DYNAMICS

The integration of RS and GIS has received considerable attention. The availability of remotely sensed data from different sensors of various platforms with a wide range of spatiotemporal, radiometric, and spectral resolutions has made remote sensing, perhaps, the best source of data for large-scale applications (Melesse et al., 2007). Similarly, GIS technology can capture all kinds of geographical reference data and digitally manipulate images from the Earth's surface and present the data in a three-dimensional format. It provides the facilities for data capture, data management, and data manipulation and has superb abilities that distinguish GIS from the other information systems. It is valuable to a wide range of public and private enterprises for explaining the events, predicting the outcomes, and planning strategies. Therefore, the unification of RS and GISs is believed to be an efficient and powerful technique that is used in the assessment, exploration, evaluation, analysis, monitoring, and management of different data. The application of remote sensing along with GIS in assessing changes in the Earth's system has been used in various dynamic fields, which have been discussed as follows.

16.3.1 APPLICATION IN THE FIELD OF LAND USE AND LAND COVER

Land use and land cover (LULC) change are critical drivers of global change and have significant implications for many international policy issues (Olokeogun et al., 2014). Although the phrases land cover and land uses are frequently used interchangeably, they have very different meanings. The physical cover on the ground is referred to as land cover, whereas the purpose of the land is referred to as land use (Thakur et al., 2017). LULC are both dynamic, and they provide a comprehensive picture of anthropogenic activities and their interactions with the environment (Chavare, 2015). Information on what change happens, when and where they occur, the rate at which they occur, as well as the social and physical forces that cause those changes is essential to understand how LULC change would affect and interact with global earth systems (Olokeogun et. al., 2014). Thereby, recent technologies like RS and GIS help us assess by giving a quicker and cost-effective analysis on various LULC mapping (Chavare, 2015). The application of RS and GIS in multiple aspects of LULC is discussed as follows.

16.3.1.1 Natural Resource Management
- Wildlife Habitat Protection (McDermid et al., 2009)
- Baseline Mapping for GIS Input
- Urban Expansion/Encroachment
- Land Cover Area Estimation
- Routing and Logistics Planning for Seismic/Exploration/Resource Extraction Activities
- Damage Delineation (Tornadoes, Flooding, Volcanic, Seismic, Fire)
- Legal Boundaries for Tax and Property Evaluation
- Wildlife habitat protection

- Baseline mapping for GIS input
- Land cover area estimation
- Urban expansion
- Routing and logistics planning for seismic/exploration/resource extraction activities (Yi and Özdamar, 2007)
- Damage delineation (tornadoes, flooding, volcanic, seismic, fire) (Eloff, 2009)
- Legal boundaries for tax and property evaluation (Dimopoulos et al., 2014)

16.3.1.2 Application in the Field of Agriculture

Even though the terms land cover and land uses are commonly interchanged, they have very various connotations. Land cover refers to the physical covering of the Earth, whereas land use refers to the function of the land (Thakur et al., 2017). LULC are both dynamic, providing a comprehensive picture of human activities and their surrounding environment (Chavare, 2015). Understanding how LULC change affects and interacts with global earth systems requires knowledge of what changes occur, where and when they appear, the rate during which they occur, and the social and physical causes that produce those changes (Olokeogun et al., 2014). The various agricultural applications with the help of RS and GIS are discussed as follows:

- Crop Type Classification (Orynbaikyzy et al., 2019)
- Classification of Agriculture Areas
- Crop Condition Assessment
- Crop Yield Estimation (Prasad et al., 2006)
- Mapping of Soil Characteristics
- Crop Area Estimation
- Compliance Monitoring (Farming Practices)
- Crop Monitoring

16.3.1.3 Application in the Field of Forestry

Forests are essential renewable natural resources and significantly preserve an environment suitable for human life (Sonti, 2015). It plays a vital role in balancing the Earth's CO_2 supply and exchange, acting as a critical link between the atmosphere, geosphere, and hydrosphere (Jitendra et al., 2016). Forest resource management in today's ever-changing world is becoming more complex and demanding to forest managers. Thereby, the application of GIS and related technologies has become necessary as they provide foresters with powerful tools for record-keeping, analysis, and decision-making (Sonti, 2015). Application of remote sensing and GIS in the field of forestry is discussed as follows:

- Reconnaissance mapping: forest cover monitoring, depletion, surveillance, and measuring biophysical characteristics of forest stands are some of the objectives that national forest/environmental agencies must meet (Yemshanov et al., 2012).
- Commercial forestry enterprises and resource management organizations rely on inventory and mapping programs to collect harvest data, update

inventory for timber supply, wide forest type, agricultural processes, and biomass measurements.
- Environmental monitoring: forest conservationists are concerned with the amount, health, and diversity of the world's forests (Conrad and Hilchey, 2011).

16.3.1.4 Application in the Field of Geology

Geological mapping methods have been undergoing continuous change along with technological and scientific advances in other relevant fields. Along with RS techniques, GIS is also increasingly used to prepare geological maps and obtain the basic geological information on which further detailed work is based (Bhan and Krishnanunni, 1983). The various applications of RS and GIS in geology are discussed as follows:

- Surficial deposit/bedrock mapping (Stack et al., 2016)
  - Lithological mapping
  - Structural mapping
- Surficial deposit
- Lithological mapping
- Sand and gravel exploration
- Mineral exploration
- Land resource assessment
- Geo-hazard mapping
- Planetary mapping (Nass et al., 2011)

16.3.1.5 Application in the Field of Hydrology

The endless recirculation of water in the atmosphere-hydrosphere-lithosphere is known as the hydrologic cycle. This cycle is studied according to the particular scale of reference, i.e., global scale, basin-scale, etc. The importance of integrating RS and GIS to study hydrologic processes has been realized by water resource workers in recent times. The use of computers and remotely sensed data has given us a huge opportunity to merge data sets and update the information by a low-cost operation. In particular, the widely used technique known as a GIS provides means for merging spatial and attribute data into the computerized database system allowing input, storage, retrieval, overlay, analysis, and tabulation of the geographically referenced data (Shih, 1996). The applications of RS and GIS in various fields of hydrology are discussed as follows:

- Wetland mapping and monitoring (Lang et al., 2008)
- Soil moisture estimation
- River and lake ice monitoring (Chaouch et al., 2014)
- Glacier dynamics monitoring
- River or delta change detection
- Drainage basin mapping and watershed mapping (Butt et al., 2015)
- Irrigation scheduling
- Mapping of groundwater recharge zone (Yeh et al., 2009)

16.3.1.6   Application in the Field of Sea Ice

Sea ice covers about one-tenth of the world's ocean surface and plays a vital role in the global climate system through interaction with the atmosphere and ocean (Toyota and Jedlovec, 2009). Sea ice also acts as a significant factor in commercial shipping and fishing industries, Coast Guard and construction operations, and global climate change studies. Examples of sea ice information and application include:

- Ice concentration (Spreen et al., 2008)
- Glacier mass balance determination
- Iceberg detection and tracking (Churchill et al., 2004)
- Surface topography
- Historical ice and iceberg conditions and dynamics for planning purposes (Eik, 2008)
- Meteorological or global change research (Zhang and Jia, 2013).

16.3.1.7   Application in the Field of Ocean Monitoring

The ocean is the vast and complex part of the Earth's surface. As a result, observation, monitoring, and studies of oceanic processes and aspects are critical. Understanding the oceans is vital for pure scientific curiosity. Global climate change issues, fisheries, aquaculture conservation, minerals resource (both renewable and nonrenewable) exploitation, coastal zone management, transportation/recreation, marine pollution hazards, submarine communication, and more require acoustic propagation for strategic planning (Hafiz, 2019). Therefore, RS and GIS integration is necessary as these technological tools are significant ways to monitor large areas of the oceans and provide an integrated framework for the detection and localization of various issues related to the oceanic region (Fustes et al., 2012). The major fields of the ocean where GIS and remote sensing are applied are mentioned as follows:

- Ocean pattern identification
- Storm forecasting
- Fish stock and marine mammal assessment (Carretta et al., 2018)
- Oil spill and more (Fingas and Brown, 2014).

16.3.1.8   Application in the Field of Coastal Management

Coastal locations all over the world are under severe stress for a variety of reasons. High population expansion, urbanization, and the construction of fish ponds have contributed to major environmental issues. Over the last several centuries, the effects have been terrible. An estimated 30% of the coastal land area has been altered or destroyed, and 70% of all beaches have been eroded. Due to such inefficient governance around the globe, what is needed is integrated coastal zone management, along with the use of the most advanced means of securing information for management purposes (Twumasi et al., 2016). It has been seen that satellite remote sensing enables us to locate and monitor various aspects like shoreline changes, bathymetry, wetland management, suspended sediment dynamics, pollution due to oil slicks, and chlorophyll content. RS and GIS can be effectively used in mapping and monitoring coastal resources, detecting shoreline changes, studying coastal landforms, and more. Hence, integration of GIS and remote sensing data is required. It is expected that coupling these two techniques can play an essential role in minimizing the danger posed by the flooding and sea-level rise (Twumasi et al., 2016). Some of the spheres of coastal management where techniques of RS and GIS are applied are discussed as follows:

- Mangrove and coral reef mapping (Chauvaud et al., 1998).
- Coastal resource survey and management (Brock and Purkis, 2009)
- Coastal change monitoring and analysis
- Modeling coastal process (Bhardwaj, 2007)

16.3.1.9   Application in the Field of Environmental Monitoring

RS and GISs play a pivotal role in environmental mapping, mineral exploration, agriculture, forestry, infrastructure planning and management, disaster mitigation, and management. For the past few decades, RS and GISs have increased in importance as a primary instrument for gathering data on practically every element of the world (Kumar et al., 2013). In environmental planning and management, the depiction of spatial data on maps has become an increasingly valuable tool. Thus, RS and GIS capabilities make them excellent tools in quickly presenting decision-makers with large-area maps of target features and assisting environmental practitioners in having access to comprehensive data, methods, and tools through the usage of these techniques. Some of the fields of environmental monitoring where GIS and RS are used are discussed as follows:

- Environmental impact assessment (Vorovencii, 2011)
- Wildlife habitat analysis
- Analysis of earthquake
- Flood management (Plate, 2002)
- Mining environment
- Urban and wasteland environment (Kumar et al., 2013)

16.3.1.10   Application in the Field of Ecology and Biodiversity

Conservation of biodiversity has been put to the highest priority through the Convention on Biological Diversity (CBD); understanding patterns of biodiversity distribution is essential to conservation strategies. Conservation of biodiversity has been put to highest priority through CBD. Understanding patterns of biodiversity distribution is necessary for conservation practices. The conventional methods for the collection of information related to biodiversity are time-consuming. Hence, there is a need for technology for the holistic management of biodiversity. As a result, the modern technology of RS and GIS can be applied as it is believed that these technological tools will help to analyze the data spatially, offering various options to the researchers and environmentalists (Reddy, 2010). Some of the spheres of biodiversity where techniques of RS and GIS are applied are discussed as follows:

- Biodiversity assessment
- Landscape analysis
- Habitat mapping (McDermid et al., 2005)

- Biodiversity modeling (Turner et al., 2003)
- Biodiversity mapping
- Biodiversity monitoring and assessment (Reddy, 2010)

In an era of unparalleled data proliferation from different sources, the need for efficient approaches that can optimize data combinations while also solving increasingly complex application challenges has never been greater. GIS and RS integration delves into this enormous potential, focusing on the careful and painstaking components of analytical data matching and thematic compatibility (Mesev, 2008). Hence, from the above discussion, it is believed that recent advancements in the fields of RS, GIS, and a higher level of computation will not only help to provide and handle a range of data simultaneously but also in a cost-efficient manner for the applicants for various fields when required (Singh et al., 2016).

16.4   CONCLUSION

Remote sensing and GIS are integral to each other, and virtually, RS is of no use without the development of GIS and vice versa. While RS has the capability of providing a large amount of data of the whole Earth and very frequently, GIS has the capabilities of analyzing a large amount of data within no time (Thakur et al., 2017). Most importantly, geographic information processing deals with the query, analyses, reporting, and output. Remote sensing has always provided a primary source of geographic data to these systems (Archibald, 2008). The technological tools of RS and GISs arose separately but are now inextricably linked. Most GIS features are collected by satellite imagery or aerial photogrammetry (digital photogrammetry), and GIS is the application where this imagery is most generally visualized. Our definition of integration includes using each technology to benefit the other and applying both technologies for modeling and decision support. Thus, through this, the necessity of integration of RS and GIS can be justified (Zurmotal, 2016). Nowadays, the field of RS and GIS has become glamorous with rapidly expanding opportunities (Thakur et al., 2017). Both tools are equally efficient as well as powerful. They are used in the assessment, exploration, evaluation, analysis, monitoring, and management of different dynamic changes that are taking place on the Earth's surface. Various researches and studies have also shown the positive influence of space tools and techniques (Jitendra et al., 2016). Therefore, we can conclude by saying that both RS and GIS have become an integral part of assessing the dynamic changes occurring on the Earth's surface. Hence, applying these two joint approaches is required to evaluate and solve the problems related to various fields of our biosphere (Jitendra et al., 2016).

REFERENCES

Akeem, A. B., Martins, A. O., & Gbola, K. A. (1986). Relevance of remote sensing and GIS in sustainability of the built-up environment. *6th National conference of Environmental Studies on The Nigeria economy and sustainable built environment: The way forward organised by the Federal Polytechnic Ilaro Ogun State*, May 8th–11th 2017. Ogun State, Southwest Nigeria.

Al-Ramadan, B. (2005). Introduction to GIS technology and its applications. *GIS@ Development Middle East*, *1*(1), 26–29.

Bernd, A., Braun, D., Ortmann, A., Ulloa Torrealba, Y. Z., Wohlfart, C., & Bell, A. (2017). More than counting pixels–perspectives on the importance of remote sensing training in ecology and conservation. *Remote Sensing in Ecology and Conservation*, *3*(1), 38–47.

Bhan, S. K., & Krishnanunni, K. (1983) Applications of remote sensing techniques to geology. *Proceedings of the Indian Academy of Sciences Section C: Engineering Sciences*, *6*(4), 297–311.

Bhardwaj, R. K. (2007). Application of GIS Technology for coastal zone management: Hydrographer perspective. In Ranade, P. S. (Ed.), *Management of Coastal Resources: An Introduction*, pp. 168–181. The Icafi Univesity Press, Hyderabad.

Brock, J. C., & Purkis, S. J. (2009). The emerging role of lidar remote sensing in coastal research and resource management. *Journal of Coastal Research*, *53*, 1–5.

Butt, A., Shabbir, R., Ahmad, S. S., & Aziz, N. (2015). Land use change mapping and analysis using remote sensing and GIS: A case study of Simly watershed, Islamabad, Pakistan. *The Egyptian Journal of Remote Sensing and Space Science*, *18*(2), 251–259.

Carretta, J. V., Forney, K. A., Oleson, E. M., Weller, D. W., Lang, A. R., Baker, J., Muto, M., Hanson, B., Orr, A. J., Huber, H. R., & Lowry, M. S. (2018). US Pacific marine mammal stock assessments: 2017. *NOAA Technical Memorandum NMFS-SWFSC*, *602*, 161.

Chaouch, N., Temimi, M., Romanov, P., Cabrera, R., McKillop, G., & Khanbilvardi, R. (2014). An automated algorithm for river ice monitoring over the Susquehanna River using the MODIS data. *Hydrological Processes*, *28*(1), 62–73.

Chauvaud, S., Bouchon, C., & Maniere, R. (1998). Remote sensing techniques adapted to high resolution mapping of tropical coastal marine ecosystems (coral reefs, seagrass beds and mangrove). *International Journal of Remote Sensing*, *19*(18), 3625–3639.

Chavare, S. (2015). Application of remote sensing and GIS in landuse and land cover mapping of sub-watershed of Wardha River Basin. *Proceedings of National Conference on Development & Planning for Drought Prone Areas*. doi: 10.14358/PERS.69.4.369.

Churchill, S., Randell, C., Power, D., & Gill, E. (2004, September). Data fusion: Remote Sensing for target detection and tracking. In *IGARSS 2004. 2004 IEEE International Geoscience and Remote Sensing Symposium* (Vol. 1). IEEE, Anchorage, AK.

Conrad, C. C., & Hilchey, K. G. (2011). A review of citizen science and community-based environmental monitoring: Issues and opportunities. *Environmental Monitoring and Assessment*, *176*(1–4), 273–291.

Dimopoulos, T., Labropoulos, T., & Hadjimitsis, D. G. (2014, August). Comparative analysis of property taxation policies within Greece and Cyprus evaluating the use of GIS, CAMA, and remote sensing techniques. In *Second International Conference on Remote Sensing and Geoinformation of the Environment (RSCy2014)* (Vol. 9229, p. 92290O). International Society for Optics and Photonics, Paphos, Cyprus.

Eik, K. (2008). Review of experiences within ice and iceberg management. *The Journal of Navigation*, *61*(4), 557.

Eloff, C. (2009). Spatial technology with an emphasis on remote sensing applications for safety and security for macro level analysis. *Acta Criminologica: African Journal of Criminology & Victimology*, *22*(1), 25–36.

Fingas, M., & Brown, C. (2014). Review of oil spill remote sensing. *Marine Pollution Bulletin*, *83*(1), 9–23.

Fustes, D., Cantorna, D., Dafonte, C., Iglesias, A., & Arcay, B. (2012). Applications of cloud computing and gis for ocean monitoring through remote sensing. In Mukhopadhyay, S. (Ed.), *Smart Sensing Technology for Agriculture and Environmental Monitoring*, pp. 303–321. Springer, Berlin, Heidelberg.

Geels, F. W. (2004). From sectoral systems of innovation to socio-technical systems: Insights about dynamics and change from sociology and institutional theory. *Research policy*, *33*(6–7), 897–920.

Goyal, M. K., Sharma, A., & Surampalli, R. Y. (2020). Remote sensing and GIS applications in sustainability. *Sustainability: Fundamentals and Applications*, 605–626. https://doi.org/10.1002/9781119434016.ch28

Hafiz, A. (2019). Application of remote sensing in Oceanographic Research. *International Journal of Oceanography and Aquaculture*, *3*(1). doi: 10.23880/ijoac-16000159.

Jitendra, S., Krishna Kumar, Y., Neha, G., & Vinit, K. (2016). *Remote Sensing and Geographical Information System (GIS) and Its Applications in Various Fields*, pp. 158–178. India: Rakesh Sohal.

Kingra, P., Majumder, D., & Singh, S. P. (2016). Application of remote sensing and GIS in agriculture and natural resource management under changing climatic conditions. *Agricultural Research Journal*, *53*, 295–302.

Kumar, S., Arivazhagan, S., & Rangarajan, N. (2013). Remote sensing and GIS applications in environmental sciences – A review. *Journal of Environmental Nanotechnology*, *2*(2), 92–101.

Lang, M. W., Kasischke, E. S., Prince, S. D., & Pittman, K. W. (2008). Assessment of C-band synthetic aperture radar data for mapping and monitoring Coastal Plain forested wetlands in the Mid-Atlantic Region, USA. *Remote Sensing of Environment*, *112*(11), 4120–4130.

Longato, D., Gaglio, M., Boschetti, M., & Gissi, E. (2019). Bioenergy and ecosystem services trade-offs and synergies in marginal agricultural lands: A remote-sensing-based assessment method. *Journal of Cleaner Production*, *237*, 117672.

McDermid, G. J., Franklin, S. E., & LeDrew, E. F. (2005). Remote sensing for large-area habitat mapping. *Progress in Physical Geography*, *29*(4), 449–474.

McDermid, G. J., Hall, R. J., Sanchez-Azofeifa, G. A., Franklin, S. E., Stenhouse, G. B., Kobliuk, T., & LeDrew, E. F. (2009). Remote sensing and forest inventory for wildlife habitat assessment. *Forest Ecology and Management*, *257*(11), 2262–2269.

Melesse, A. M., Weng, Q., Thenkabail, P. S., & Senay, G. B. (2007). Remote sensing sensors and applications in environmental resources mapping and modelling. *Sensors*, *7*(12), 3209–3241.

Mesev, V. (2008). *Integration of GIS and Remote Sensing* (Vol. 19). John Wiley & Sons, New York.

Milla, K. A., Lorenzo, A., and Brown, C. (2005). GIS, GPS, and remote sensing technologies in extension services: Where to start, what to know. *Journal of Extension*, *43*(3), 16.

Nass, A., van Gasselt, S., Jaumann, R., & Asche, H. (2011). Implementation of cartographic symbols for planetary mapping in geographic information systems. *Planetary and Space Science*, *59*(11–12), 1255–1264.

Olokeogun, O. S., Lyiola, O. F., & Lyiola, K. (2014). Application of remote sensing and GIS in land use/land cover mapping and change detection in Shasha forest reserve, Nigeria. *International Archives of the Photogrammetry, Remote Sensing and Spatial Information Sciences*, *40*, 613–616.

Ondieki, C. M., & Murimi, S. (1997). Applications of geographic information systems. *Environmental Monitoring*, *11*, 314–340.

Orynbaikyzy, A., Gessner, U., & Conrad, C. (2019). Crop type classification using a combination of optical and radar remote sensing data: A review. *International Journal of Remote Sensing*, *40*(17), 6553–6595.

Plate, E. J. (2002). Flood risk and flood management. *Journal of Hydrology*, *267*(1–2), 2–11.

Prasad, A. K., Chai, L., Singh, R. P., & Kafatos, M. (2006). Crop yield estimation model for Iowa using remote sensing and surface parameters. *International Journal of Applied Earth Observation and Geoinformation*, *8*(1), 26–33.

Ranjan, A. K., Vallisree, S., & Singh, R. K. (2016). Role of geographic information system and remote sensing in monitoring and management of urban and watershed environment: Overview. *Journal of Remote Sensing & GIS*, *7*(2), 60–73.

Raup, B., Kääb, A., Kargel, J. S., Bishop, M. P., Hamilton, G., Lee, E., Paul, F., Rau, F., Soltesz, D., Khalsa, S. J. S., & Beedle, M. (2007). Remote sensing and GIS technology in the Global Land Ice Measurements from Space (GLIMS) project. *Computers & Geosciences*, *33*(1), 104–125.

Reddy, C. S. (2010). *Remote Sensing and GIS Applications in Biodiversity Studies.* Training Programme on Biodiversity Assessment, Institute of Bioresources and Sustainable Development (IBSD), Imphal.

Reddy, M. A., & Reddy, A. (2008). *Textbook of Remote Sensing and Geographical Information Systems* (p. 4). BS Publications, Hyderabad.

Roy, P. S., Behera, M. D., & Srivastav, S. K. (2017). Satellite remote sensing: Sensors, applications and techniques. *Proceedings of the National Academy of Sciences, India Section A: Physical Sciences*, *87*, 465–472.

Shih, S. F. (1996). Integration of remote sensing and GIS for hydrologic studies. In Singh, V. & Fiorentino, M. (Eds.), *Geographical Information Systems in Hydrology*, pp. 15–42. Springer, Dordrecht.

Solomon, S., & Ghebreab, W. (2011). Remote sensing and GIS techniques for tectonic studies. In Gupta, H. K. (Ed.), *Encyclopedia of Solid Earth Geophysics. Encyclopedia of Earth Sciences Series*, pp. 1030–1034. Springer, Dordrecht.

Sonti, S. H. (2015). Application of geographic information system (GIS) in forest management. *Journal of Geography & Natural Disasters*, *5*(3), 1000145.

Spreen, G., Kaleschke, L., & Heygster, G. (2008). Sea ice remote sensing using AMSR-E 89-GHz channels. *Journal of Geophysical Research: Oceans*, *113*(C2), 1–14.

Stack, K. M., Edwards, C. S., Grotzinger, J. P., Gupta, S., Sumner, D. Y., Calef III, F. J., Edgar, L. A., Edgett, K. S., Fraeman, A. A., Jacob, S. R., & Le Deit, L. (2016). Comparing orbiter and rover image-based mapping of an ancient sedimentary environment, Aeolis Palus, Gale crater, Mars. *Icarus*, *280*, 3–21.

Thakur, J. K., Singh, S. K., & Ekanthalu, V. S. (2017). Integrating remote sensing, geographic information systems and global positioning system techniques with hydrological modeling. *Applied Water Science*, *7*(4), 1595–1608.

Toyota, T., & Jedlovec, G. (2009). Application of remote sensing to the estimation of sea ice thickness distribution. In Toyota, T. (Ed.), *Advances in Geosciences and Remote Sensing*, pp. 21–44. Intech, Vienna.

Twumasi, Y. A., Merem, E. C., & Ayala-Silva, T. (2016). Coupling GIS and remote sensing techniques for coastal zone disaster management: The case of Southern Mississippi. *Geoenvironmental Disasters*, *3*(1), 1–9.

Vorovencii, I. (2011). Satellite remote sensing in environmental impact assessment: An overview. *Bulletin of the Transilvania University of Brasov. Forestry, Wood Industry, Agricultural Food Engineering. Series II*, *4*(1), 73.

Wilkinson, G. G. (1996). A review of current issues in the integration of GIS and remote sensing data. *International Journal of Geographical Information Science*, *10*(1), 85–101.

Yadav, S. K., Shubhangi, R., & Shyam, S. R. (2013). Remote sensing technology and its applications. *Inernational Journal of Advancements in Research & Technology*, *2*(10), 25–30.

Yeh, H. F., Lee, C. H., Hsu, K. C., & Chang, P. H. (2009). GIS for the assessment of the groundwater recharge potential zone. *Environmental Geology*, *58*(1), 185–195.

Yemshanov, D., McKenney, D. W., & Pedlar, J. H. (2012). Mapping forest composition from the Canadian National Forest Inventory and land cover classification maps. *Environmental Monitoring and Assessment*, *184*(8), 4655–4669.

Yi, W., & Özdamar, L. (2007). A dynamic logistics coordination model for evacuation and support in disaster response activities. *European Journal of Operational Research*, *179*(3), 1177–1193.

Zhang, A., & Jia, G. (2013). Monitoring meteorological drought in semiarid regions using multi-sensor microwave remote sensing data. *Remote Sensing of Environment*, *134*, 12–23.

Part E

Geospatial Modeling in Policy and Decision-Making

17 Sustainable Livelihood Security Index
A Case Study in Chirrakunta Rurban Cluster

Supratim Guha
Indian Institute of Technology Ropar
Vellore Institute of Technology

Dillip Kumar Barik
Vellore Institute of Technology

Venkata Ravibabu Mandla
National Institute of Rural Development & Panchayati Raj,
Ministry of Rural Development, Government of India

CONTENTS

17.1 Introduction 326
17.2 Study Area 327
17.3 Materials and Methods 327
    17.3.1 Normalization of Indicators Using the Functional
          Relationship with SLSI 327
    17.3.2 Selection of Variables for Computing SLSI 332
    17.3.3 Ecological Security Indicators 332
    17.3.4 Economic Efficiency Indicators 333
    17.3.5 Social Equity Indicators 333
17.4 Results and Discussion 334
    17.4.1 Ecological Security Index 334
    17.4.2 Economic Equity Index 336
          17.4.2.1 Social Equity Index (SEI) 336
    17.4.3 Sustainable Livelihood Security Index (SLSI) 340
17.5 Conclusion 341
References 343

DOI: 10.1201/9781003147107-22

17.1 INTRODUCTION

In 2016, the government of India introduced "Shyama Prasad Mukherjee Rurban Mission" (SPMRM) which is a modern type of rural development project. In this particular project, a group of panchayats are considered as a single unit called *Cluster* (Guha et al., 2018). The mission is divided into three main stages, primarily at the household level followed by the panchayat level and finally at the aforesaid cluster level. The objective of the mission is to stimulate local economic development, enhance basic services, ensure healthy environment, and create sustainable Rurban Cluster. Accordingly, the project focusses on development ensuring Sustainable Livelihood. The term sustainable livelihood implies living in an economically, ecologically, and socially sustainable manner. In this regard, Chambers (1986) elaborated that sustainable livelihood possesses adequate stock, wealth, a steady supply of money, and food for social and physical well-being. Therefore, sustainable livelihood now provides a broad goal for breadline communities (Bull, 2015). Effectively, sustainable livelihood provides security against potential poverty. However, achieving the sustainable livelihood is not a rigid issue (Scoones, 2009). In addition, sustainable livelihood can be ensured by some strategies which focus to improve unsuitable rural resources, high population growth rate, a vulnerable environment, and significant social inequity specifically disparate distribution and allocation of wealth and land (Qu et al., 2011; Shaw and Kristjanson., 2014; Ouyang et al., 2014; Dai and Dien, 2013; Wu, 2004; Altaf et al., 2014; You and Zhang 2017; Gujree et al., 2020). UK Department for International Development has developed a sustainable livelihood framework to analyze the factors that affect sustainable livelihoods (Scoones, 1998). Likewise, Cooperative for Assistance and Relief Everywhere (CARE) from the USA has also introduced a program that frames the Household Livelihood Security (HLS) to understand the varying relationship between households and society (McCaston, 2005). In particular, HLS covers six security areas: food, health, economics, education, shelter, and community participation to emphasize the multidimensional dynamics of the factors causing poverty (Ghanim, 2000). So, the HLS approach can be used to measure the livelihood security of rural people at both family and community level (Lindenberg, 2002). Sustainable Livelihood Security (SLS) is a livelihood option that is ecologically secure, economically efficient and socially equitable, underscoring in dimensions of ecology, economics, and equity (Swaminathan 1991a and b; Meraj et al., 2021a and b). The concept implies the protection or assurance of means of livelihood for the masses both in the present and future. In particular, SLS reflects the concern of both intergenerational and intragenerational equity equally. The concept of SLS is implemented at both the macro and micro levels (Guha et al., 2018). The macro-level prescriptions for ensuring SLS include prohibiting population explosion, reducing distress migration thus preventing exploitation, and supporting long-term sustainable resource management (Hassanin et al., 2020; Tomar et al., 2021; Meraj et al., 2021; Chandel et al., 2021; Bera et al., 2021). On the other hand, micro and local levels, the critical ingredients of SLS, are adequate stocks and supplies of food and cash to meet basic needs and access to resources, income, and assets to offset shocks (McCracken and Pretty, 1988; Singh and Hiremath, 2010).

Saleth and Swaminathan (1993) have proposed Sustainable Livelihood Security Index (SLSI) as a means of identifying the necessary conditions for sustainable livelihood or sustainable development in a given region (Moser, 1996; Sajjad et al., 2014). For instance, SLSI has been applied to evaluate the livelihood security of farmers in highland and lowland communities of the Kali-Khola agricultural watershed in western Nepal (Arshad et al., 2019; Bhandari and Grant, 2007). The aim of this study is to identify the current state of sustainability in the intra-cluster level in Chirrakunta Rurban Cluster, (which is one of the 100 clusters selected by govt of India) Telangana, India. Consequently, the study focuses on identifying unsustainable hotspot zones and effective planning to achieve sustainable tags for the entire Chirrakunta Cluster.

Furthermore, the study showcases the potential and robustness of the SLSI at micro-level like the SPMRM project.

17.2   STUDY AREA

In December 2016, Asifabad Mandal was selected as a tribal cluster named Chirrakunta Cluster (Figure 17.1). The area coordinates under 79.056° to 79.399° East longitudes and 19.248° to 19.449° North latitudes which cover 382.28 square kilometers of area. Also, the total population of the Chirrakunta Cluster is 43,728 out of which 21.72% population is tribal. As a result, the area is ecologically strong with less human density, less built-up area, and high forest cover, socioeconomic conditions not being good enough though. Further, the average per capita per day income is a mere 123 INR which is quite less than the country's average figure. Likewise, the primary literacy rate is 28.94% in Asifabad which is again less than India's overall figures i.e. 74.04% (Census, 2011).

17.3   MATERIALS AND METHODS

Remote sensing data are suitable to extract spatial information because of its synoptic view, repetitive coverage, and real-time data acquisition (Belal and Moghanm, 2011; Shaw and Das, 2017, Ranga et al., 2020; Meraj et al., 2020, Kanga et al., 2020a and b). In this study, IRS LISS-IV satellite imageries are used to identify land use land cover (LULC) of the area from where some useful information like forest cover, waterbody, built-up area were extracted (Figure 17.2). The rest of the data were collected from the Rurban mission Telangana department (Table 17.1).

17.3.1   Normalization of Indicators Using the Functional Relationship with SLSI

There are two types of indicators based on the functional relationship with sustainability. The first one is a positive indicator. In particular, the positive indicators are those when sustainability increases with increase in the value of the indicator. On the other hand, negative indicators are those when sustainability increase with a decrease in the value of the indicator. As different indicators have different units

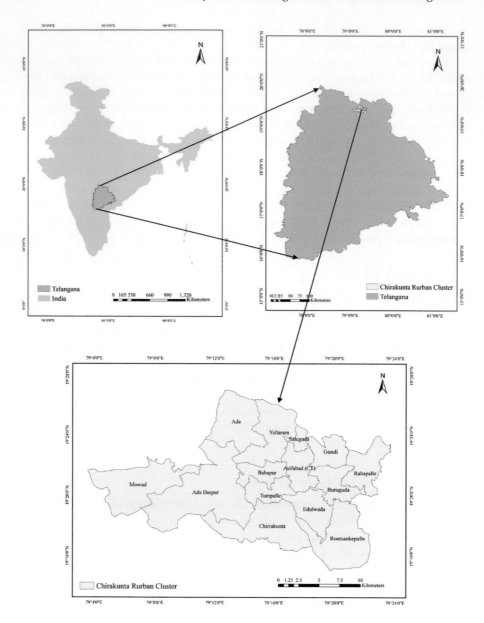

FIGURE 17.1 Location of study area.

and scales so in first step we normalized all indicators on a 0–1 unit scale. For that instance, UNDP's Human Development Index (UNDP, 2006) was used.

To follow the index, the collected data were arranged in the quadrangular matrix with rows representing panchayats and columns representing indicators for each component of sustainability. For example, let there be L ($j = 1, 2, 3,..., L$) panchayats and say K ($i = 1, 2, 3,..., K$) indicators were assigned. In effect, the table will be L rows and K columns.

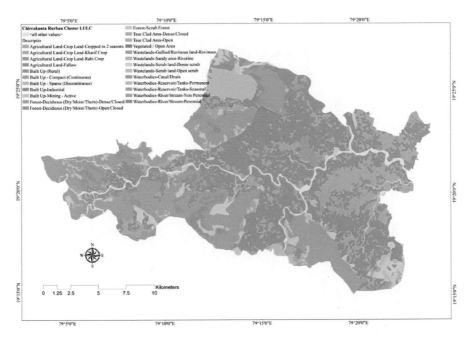

FIGURE 17.2   LULC for the Chirrakunta Rurban Cluster.

Let x_{ij} be the value of the ith indicator corresponding to jth panchayat. In the case of positive indicators, normalization can be done by using the Equation (17.1).

$$Z_{ij} = \frac{X_{ij} - \text{Min}\{X_{ij}\}}{\text{Max}\{X_{ij}\} - \text{Min}\{X_{ij}\}}$$

(17.1)

Conversely, if the variable has a negative functional relationship with sustainability, then the normalization can be done by using the Equation (17.2)

$$Z_{ij} = \frac{\text{Max}\{X_{ij}\} - X_{ij}}{\text{Max}\{X_{ij}\} - \text{Min}\{X_{ij}\}}$$

(17.2)

Where x_{ij} is the value of the ith indicator related to the jth panchayat, Max $\{X_{ij}\}$ and Min $\{X_{ij}\}$ is the maximum and minimum value of the ith indicator among all the L panchayats, respectively.

For all the indices, $j = 1, 2,\ldots, 14$ as there are 14 panchayats in the cluster; $i = 1, 2,\ldots, K$ equals 4 for ecological security index (ESI), economic efficiency index (EEI), and social equity index (SEI).

Further, ESI was calculated by using the Equation (17.3).

$$\text{ESI}_j = \sum_{1}^{i} W_i \times Z_{ij}$$

(17.3)

TABLE 17.1

Raw Data Used for the Calculation of ESI, EEI, SEI and SLSI in Chirrakunta Rurban Cluster

Name of the Panchayat	Ecological Security Indicators				Economic Efficiency Indicators				Social Equity Indicators			
	Human Density	Forest Cover (%)	Built Up Area (%)	Water Body (%)	Agricultural Production (kg/ha)	Milk Yield Per Animal (kg/day)	Home Based Industry Present	Average Income Per Capita Per Day	Primary Literacy Rate	Difference between Sex Ratio	% of Households with LPG Connectivity	% of Households with Digital Connectivity
Ada	66	18.16	0.69	11.31	1,908	2.4	8	191.91	36.25	100	46.51	0.17
Ada Dasnapur	28	44.18	0.46	1.32	1,723	2.1	4	100.98	25.11	3	45.12	0.19
Asifabad	714	5.6	3.8	0.82	1,853	2.1	76	159.81	25.26	32	6.84	0.74
Babapur	268	6.31	0.71	0.61	1,871	2.2	7	154.83	27.07	11	6.98	0.21
Buruguda	207	6.74	0.73	1.13	1,694	1.3	7	59.15	29.14	51	18.18	0.85
Chirrakunta	88	24.7	1	3.6	1,739	1.7	25	107.99	20.37	10	0.19	0.38
Edulawada	107	6.43	0.45	0.52	1,727	1.8	26	105.78	35.14	19	21.06	0.94
Gundi	127	6.92	0.52	1.09	1,705	1.4	4	91.19	29.01	46	3.77	0.6
Mowad	44	34.64	0.44	2.71	1,701	0.9	1	87.2	24.67	3	63.05	1.04
Rahapally	103	6.97	1.17	1.57	1,911	1.9	23	193.2	37.4	47	40.15	0.15
Rout Sankepalle	67	16.21	0.53	0.62	1,715	1.4	9	90.46	32.97	48	16.86	0.26
Saleguda	292	6.01	0.23	0.14	1,889	2	23	152.27	26.32	4	6.23	1.3
Tumpalle	123	7.98	0.53	1.45	1,768	1.6	13	109.02	27.21	65	14.71	0.42
Yellaram	114	14.77	0.84	0.38	1,763	1.2	12	120.2	29.25	20	12.82	0.29

Where ESI_j represents the ESI for jth panchayat and W_i represents the weight associated with the ith indicator included for ESI.

Iyengar and Sudarshan (1982) methods have assigned the weight based on the variance of the indicator across the panchayats. Here, the same postulate has been used to assign the weight i.e. the weights are varying inversely with the variance of the normalized index. The weight has been obtained by using the Equation (17.4).

$$W_i = \frac{c}{\sqrt{\mathrm{Var}\left(Z_{ij}\right)}}$$
(17.4)

Where c is a normalized constant such that

$$c = \sum_{j=1}^{k} \frac{1}{\sqrt{\mathrm{Var}\left(Z_{ij}\right)}}$$
(17.5)

The selection of weight of each index done in this way to ensure that large variation in any one of the indicators would not unduly dominate the contribution of the rest of the indicators and distort the inter panchayat comparisons. Similar ESI_j, EEI_j, and SEI_j were calculated using Equations (17.3)–(17.5). The computed sustainability indices (ESI, EEI, and SEI) lie in between 0 and 1, where, 1 indicates maximum sustainability and 0 indicates less sustainable. The summation of each sustainable index with proper weight is SLSI and it can be calculated by Equation (17.6).

$$SLSI_j = \left(W_{ESI} \times ESI_j\right) + \left(W_{EEI} \times EEI_j\right) + \left(W_{SEI} \times SEI_j\right)$$
(17.6)

Where W_{ESI}, W_{EEI}, and W_{SEI} are the weights for ESI, EEI and SEI, respectively, and W_{ESI} was calculated by using Equations (17.7) and (17.8).

$$W_{ESI} = \frac{c}{\sqrt{\mathrm{Var}\left(ESI_{ij}\right)}}$$
(17.7)

$$c = \left| \frac{1}{\sqrt{\mathrm{Var}\left(ESI_{ij}\right)}} + \frac{1}{\sqrt{\mathrm{Var}\left(EEI_{ij}\right)}} + \frac{1}{\sqrt{\mathrm{Var}\left(SEI_{ij}\right)}} \right|^{-1}$$
(17.8)

Similarly, W_{EEI} and W_{SEI} were calculated from Equations (17.7) and (17.8).

For meaningful characterization of ESI_j of different panchayats, four real numbers C_1, C_2, C_3, and C_4 were used to divide the values obtained into five linear intervals namely [0, C_1], [C_1, C_2], [C_2, C_3], [C_3, C_4], and [C_4, 1] with same probability weight of 20%, that is,

$$P[0 \leq \text{ESI}_j \leq C_1] = 0.20$$

$$P[C_1 \leq \text{ESI}_j \leq C_2] = 0.40$$

$$P[C_2 \leq \text{ESI}_j \leq C_3] = 0.60$$

$$P[C_3 \leq \text{ESI}_j \leq C_4] = 0.80$$

$$P[C_4 \leq \text{ESI}_j \leq 1] = 1.00$$

These intervals have been used in this study to characterize the various level of sustainability as given below:

1. Very low ecologically sustainable if $0 \leq \text{ESI}_j \leq C_1$
2. Low ecologically sustainable if $C_1 \leq \text{ESI}_j \leq C_2$
3. Moderately ecologically sustainable if $C_2 \leq \text{ESI}_j \leq C_3$
4. Highly ecologically sustainable if $C_3 \leq \text{ESI}_j \leq C_4$
5. Extremely ecologically sustainable if $C_4 \leq \text{ESI}_j \leq 1$

Similarly, the panchayats were also classified into five categories based on scores of EEI_j, SEI_j, and SLSI_j.

17.3.2 SELECTION OF VARIABLES FOR COMPUTING SLSI

It is important to select proper indicator to describe the status of each of the three dimensions which are not an easy task ad nauseam (Fredericks, 2012). Not surprisingly, there may be chances to get a gap or overlapped indicators due to less or more number of indicators respectively. The following sections are explained about the selection of indicators for the ecologic, economic, and social equity status of each panchayat in Chirrakunta Rurban Cluster. In this study, we have selected indicators as per the literature survey, current requirements for panchayats, data availability and in addition to that effort made by the Government to promote. Furthermore, all the above factors are combined together based on their criteria to construct the ESI, EEI, SEI, and then finally the SLSI. In this study, four indicators are considered under each category.

17.3.3 ECOLOGICAL SECURITY INDICATORS

Forest cover is one of the key indicators in ecological sustainability. For example, forest cover plays a significant role in the carbon cycle, water cycle, soil preservation, and pollution control and also gives shelter to some habitats. In addition, forests provide resources for humans not only in terms of food but also housing, agriculture, and a cluster of marketable forest products. Unfortunately, conservation of the forest resources is threatened due to tremendous growth in population, fragmentation of the land holdings, demand for the fodder, and energy resources.

A minimum forest cover is essential for a healthy environment to ensure ecological security. Variables of human density were selected to reflect the extent of stress on the overall ecological security in terms of forest loss and habitat degradation (Maikhuri et al., 2001; Arjunan et al., 2005). In particular, the increase of population in a comparatively short duration is responsible for increasing the built-up area. Gradually, the built-up areas are grabbing forests, agricultural fields, and water bodies, thus hampering the sustainability. Hence, it is a negative indicator in the determination of SLSI. Similarly, waterbodies have a strong relationship with climatic changes having great impacts on anthropogenic water resources (Li et al., 2013; Pekel et al., 2016; Tao et al., 2015; Tulbure and Broich, 2013; Tulbure et al., 2014; Zou et al., 2017). Therefore, it is very much essential to protect waterbodies for environmental as well as anthropogenic purposes.

17.3.4 ECONOMIC EFFICIENCY INDICATORS

The domestic industries or Home-Based Industries (HBI) have great potential to enhance the economic growth of a particular area which can assure the economic sustainability of the area. Agricultural Land Productivity implies the overall ratio of the "total food yield" in relation to the "area" (where grains are produced for our survival). This indicator ensures food security of the area thus making it is a positive indicator. Likewise, milk yield per animal per day is obtained by dividing total milk production by a number of milch animals (Kumar et al., 2014). Not surprisingly, this is also a positive indicator as not only it safeguards food security but also focusses on nutrition management. Finally, the income per capita which is basically dependent upon the previous indicators and some other indicators are also significant as they reveal the economic condition of the area.

17.3.5 SOCIAL EQUITY INDICATORS

The basic objective/indicator of any place or community would require identification of its level of education and the state of living. For instance, the objective of primary literacy is to help in getting a good rank which depends on the acquiring of the basic knowledge for reading, writing, and calculating basic mathematics. High literacy level of learners represents a well-enhanced economy as well as society. Furthermore, constructive literacy skills can result in better education and employment options. For example, individuals need to enhance their knowledge by adopting the new skills to keep up with the pace of change in this twenty-first century. Hence, the primary literacy rate is selected as a positive social indicator in this study.

People these days have adopted LPG because of its efficiency and ease of use than the conventional one. Furthermore, the fuel being Sulphur-free, will not produce soot, and smoke as it does not contain any unburnt carbon residue. As a result, it emits comparatively lesser greenhouse gas than other fuels. Hence, LPG connectivity is a positive indicator when it comes to SLSI measurement.

Likewise, digital connectivity has proven to be a nondetachable tool for day to day business and society's needs in this modern world. As a result, it helps various public services such as education, employment opportunities and many more by accessing internet. As such, digital connectivity has been selected as a positive indicator in the present study.

Sex ratio is the ratio of the number of males to females in a given population. A favorable sex ratio refers to an even sex ratio where the numbers of males and females are evenly matched. An imbalanced sex ratio can lead to various social problems like failure to marry or difficulties in starting families. Therefore, a favorable sex ratio might help in solving such problems. In addition, a favorable sex ratio might be a key factor in eliminating some of the societal issues that have been psychologically linked to gender imbalance like prostitution and cheating (as a reverse impact of excess in men). Thus, in this study "difference in sex ratio" has been selected as an indicator. In this case, the difference between male and female or female and male populations in every 1,000 populations has been estimated. As the difference between sex ratio is negatively related to the sustainability of society, the index "difference between sex ratio" is selected as a negative indicator.

17.4   RESULTS AND DISCUSSION

Development is a ceaseless and continuous process. However, time and funds are the major two potential factors for development process. Therefore, it is necessary to identify under-developed areas and also the underdevelopment indicator for which development is more essential. Consequently, identification of the area-specific requirements towards sustainable development can be done using SLSI tool. In addition, SLSI also can identify the priority.

17.4.1   ECOLOGICAL SECURITY INDEX

'Sustainably developed' indicates preserving the natural resources and ecological systems for the social and economic well-being and significant imperative to fulfill the future requirements of humanity (Littig and Grießler, 2005). As Per ESI Score, Panchayats Have Been Classified into Four Categories I.E. Extreme Sustainable (ES), High Sustainable, Moderate Sustainable, and Very Low Sustainable Category (VLS). As a result, ESI classification shows that western panchayats are much more secured than eastern Panchayats (Figure 17.3). Among the panchayats, one, three, nine and one panchayat are categorized in Extremely Sustainable, Highly Sustainable (HS), MS, and VLS category, respectively. Fortunately, not a single panchayat lies in the 'Low Sustainability' (LS) category. In addition, the Panchayat Ada which covers 8.57% area of the cluster secured Extremely Sustainable category (Figure 17.4). In this area, Human density is very low, with only 66 persons/sq.km. Similarly, built up area also covers a very small fraction of the total panchayat area. On the other hand, forest cover in these areas is quite good. 18.16% area of this panchayat is covered by forest and 11.31% area is covered by waterbody. Further, Ada Dasnapur, Chirrakunta and Mowad panchayats covered 41.57% of total cluster area and 17.49% of the population of the cluster. Therefore, these panchayats are categorized as High Sustainable category. In particular, the average human density is 48.14 persons/sq.km and an average of the built-up area is only 0.633% of the total panchayat area in these

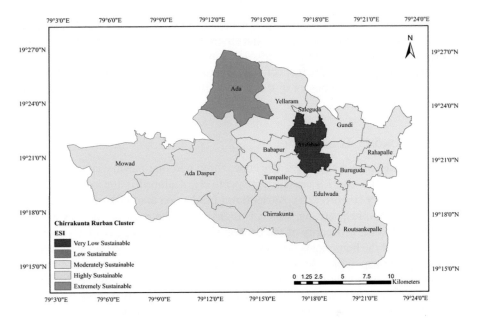

FIGURE 17.3   Spatial variation in ESI.

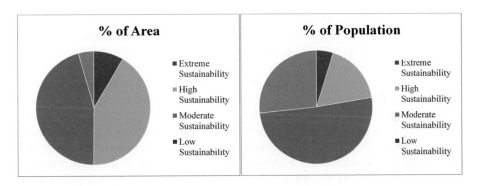

FIGURE 17.4   Area and population distribution in various degrees of sustainability classes based on ESI score.

panchayats. In addition, average of forest covers is 34.50% of this area, whereas the average water-body coverage is 2.54%. Furthermore, nine panchayats which nurture around half of the population of the cluster fall under MS category where average human density is about 174.26 persons/sq.km, which is quite higher than the other two categories. Conversely, the average built-up area is almost the same as the previous category. As for the remaining indicators, result is pretty lower than the previous category. For example, average forest cover and average water-body are 8.7% and 0.834%, respectively.

Asifabad Panchayat comes under VLS category where human density is 714 persons/sq.km. Not surprisingly, built up area also covers a very large area in this panchayat along with the fact that, forest cover and water body occupy only 5.6%

and 0.82% area of the total Asifabad Panchayat, respectively. As a result, Asifabad Panchayat exerts extreme pressure on the environment and natural resources. In the rankings, Ada, Ada Dasnapur, and Mowad stand in the first three places with ESI scores 0.808, 0.750 and 0.726, respectively, whereas Asifabad, Babapur and Saleguda occupy the bottom three places in the table with 0.016, 0.416 and 0.431 ESI score (Table 17.2).

17.4.2 ECONOMIC EQUITY INDEX

Analysis of EEI shows that four, five, one, three and one panchayats come under VLS, LS, MS, HS, and ES category, respectively. Asifabad is the only panchayat under ES category (Figure 17.5). Agricultural production and milk yield is 1853 kg/ha and 2.1 kg/day, respectively, which is higher from the most of panchayats. HBI has attributed to the prosperity of the economy in this area. In particular, more than three-fourths of the households are associated with HBI in this panchayat. Naturally, the average per capita income is 159.81 INR. For this reason, this panchayat stands on very sound economic foundations which help to ensure sustainable livelihood. In same way, Ada, Rahapally, Saleguda secures HS tag as per EEI score which covers 58.07 square kilometer area and carries 12.89% population of the Chirrakunta Cluster (Figure 17.6). For example, average agricultural production in these areas is 1902.66 kg/ha which is a bit higher than Asifabad, whereas average milk production per animal is 2.1 kg/day. Along with that, 18% of households are associated with HBI in this area. In effect, the average per capita income is higher than the most of the other panchayats, that is 179.126 INR.

Furthermore, Babapur is the only panchayat that comes under the MS category with 1,871 kg/ha agricultural production and 2.2 kg/day milk yield. However, only 7% of households are linked with HBI. Consequently, average per capita income of the Babapur Panchayat is 154.83 INR. Likewise, Ada Dasnapur, Chirrakunta, Edulawada, Tumpalle, and Yellaram come under LS category with average agricultural production 1744 kg/ha only. Likewise, average milk production is only 1.68 kg/day which is significantly low. Along with that, only 16% of the households are connected with HBI. As a result, per capita income is only 108.79 INR that is much lower than all previous categories. At last, VLS category covers almost 30% area and 22.58% population of the study area. In this category, average agricultural production is 1703.75 kg/ha which is lower than any other category. Along with that, milk yield is 34.3% lower than the LS category. In addition, only 5.25% households are associated with HBI. For that reason, average per capita income is very poor in these areas at a lowly 82 INR. In standing, Asifabad, Ada, Rahapally secured the first three places in the EEI table with 0.836, 0.728 and 0.706 scores. In contrast, Mowad, Buruguda, and Gundi stand in three bottom places in the table.

17.4.2.1 Social Equity Index (SEI)

Social equity is one of the basic pillars for sustainable development. Also, Social equity is a pre-requisite for clean development procedure (Kumar et al. 2014; Altaf et al., 2013; Meraj et al., 2015, 2016). In particular, it entails fairly accessible ways to the resources and livelihood. Along with that, social equity emphasizes the principle

TABLE 17.2
Indicates Values of the Sustainability Indicators

Name of the Panchayat	Ecological Security Indicators					Economic Efficiency Indicators					Social Equity Indicators				
	Human Density Index	Forest Cover Index	Built Up Index	Water Body Index	Ecological Security Index	Agricultural Production Index	Milk Yield Index	Home Based Industry Index	Average Income Per Capita Index	Economic Efficiency Index	Primary literacy Rate Index	Difference between Sex Ratio Index	LPG Connectivity Index	Digital Connectivity Index	Social Equity Index
Ada	0.945	0.326	0.871	1.000	0.808	0.986	1.000	0.093	1.000	0.728	0.932	0.000	0.737	0.017	0.431
Ada Dasnapur	1.000	1.000	0.936	0.106	0.750	0.134	0.800	0.040	0.315	0.324	0.278	1.000	0.715	0.035	0.516
Asifabad	0.000	0.000	0.000	0.061	0.016	0.733	0.800	1.000	0.758	0.836	0.287	0.701	0.106	0.513	0.402
Babapur	0.650	0.018	0.866	0.042	0.416	0.816	0.867	0.080	0.721	0.587	0.393	0.918	0.108	0.052	0.379
Buruguda	0.739	0.030	0.860	0.089	0.452	0.000	0.267	0.080	0.000	0.093	0.515	0.505	0.286	0.609	0.478
Chirrakunta	0.913	0.495	0.784	0.310	0.633	0.207	0.533	0.320	0.368	0.365	0.000	0.928	0.000	0.200	0.286
Edulawada	0.885	0.022	0.938	0.034	0.495	0.152	0.600	0.333	0.351	0.372	0.867	0.835	0.332	0.687	0.685
Gundi	0.856	0.034	0.919	0.085	0.498	0.051	0.333	0.040	0.241	0.168	0.507	0.557	0.057	0.391	0.382
Mowad	0.977	0.753	0.941	0.230	0.726	0.032	0.000	0.000	0.211	0.058	0.252	1.000	1.000	0.774	0.751
Rahapally	0.891	0.035	0.737	0.128	0.470	1.000	0.667	0.293	1.010	0.706	1.000	0.546	0.636	0.000	0.561
Rout Sankepalle	0.943	0.275	0.916	0.043	0.560	0.097	0.333	0.107	0.236	0.196	0.740	0.536	0.265	0.096	0.421
Saleguda	0.615	0.011	1.000	0.000	0.431	0.899	0.733	0.293	0.701	0.627	0.349	0.990	0.096	1.000	0.603
Tumpalle	0.862	0.062	0.916	0.117	0.513	0.341	0.467	0.160	0.376	0.329	0.402	0.361	0.231	0.235	0.310
Yellaram	0.875	0.238	0.829	0.021	0.506	0.318	0.200	0.147	0.460	0.271	0.521	0.825	0.201	0.122	0.428
Weightage	0.259	0.212	0.269	0.260		0.196	0.261	0.297	0.246		0.263	0.258	0.247	0.232	

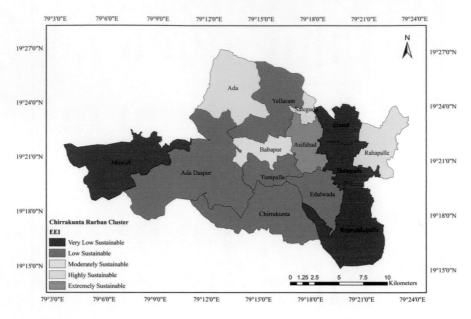

FIGURE 17.5   Spatial variation in EEI.

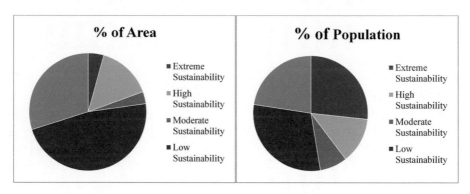

FIGURE 17.6   Area and population distribution in various degrees of sustainability classes based on EEI score.

of each citizen, independent of personal traits or economic status, deserves and possesses a right for fair treatment by the political systems, providing special observation to the requirements of vulnerable and weak populations (Chitwood, 1974; Farooq and Muslim, 2014; Nathawat et al., 2010; Kumar et al., 2018; Joy et al. 2019). As per SEI score, three, seven and four panchayats are classified in HS, MS, and LS category (Figure 17.7). As a matter of fact, not a single panchayat comes under ES category and VLS category. Although, Edulawada, Mowad, and Saleguda achieved HS level which covers only 17.58% area and 12.08% population of the cluster (Figure 17.8). Here average primary literacy rate is 28.71%. In addition, average of difference in sex ratio is only 8.66. Along with that, 30.113% and 1.093% of

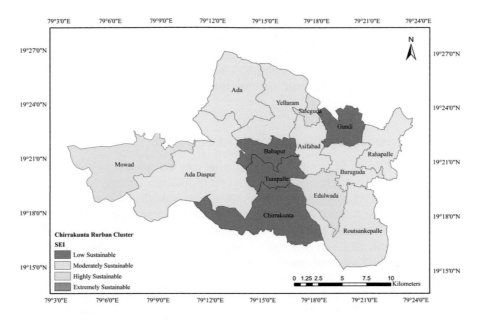

FIGURE 17.7   Spatial variation in SEI.

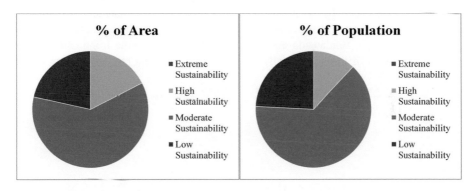

FIGURE 17.8   Area and population distribution in various degrees of sustainability classes based on SEI score.

households are using LPG connectivity and digital connectivity, respectively. On the other hand, seven panchayats come under the MS category which covers 60.91% area of the Chirrakunta Cluster and gives shelter to 63.72% population. However, average primary literacy rate in these areas is 30.76% which is higher than the previous category but average of sex ratio difference is 43, which is much below than the previous category. 26.64% and 0.3785% of households are using LPG connectivity and digital connectivity, respectively, in this area which is much lesser than the previous category. Further, 4.2% population of the Chirrakunta Cluster lives under LS category spread over only 21.51% area of the total cluster. In this area, the average literacy rate is 25.89% whereas the average difference between sex ratio is 33, better than the previous category. Figures of the remaining indicators are not impressive. Only 6.41% and 0.40% of households are using LPG and digital connectivity facility. In terms of rankings, Mowad, Edulawada, and Saleguda are placed in the top three spots of SEI result 0.751, 0.685 and 0.603, respectively. In contrast, Chirrakunta, Gundi, and Babapur are the most insecure three panchayats with 0.286, 0.310 and 0.379 SEI score.

17.4.3 SUSTAINABLE LIVELIHOOD SECURITY INDEX (SLSI)

Finally, as per the SLSI classification one, nine and four panchayats are classified in HS, MS, and LS category (Figure 17.9). Out of fourteen panchayats, none of the panchayats is placed under VLS category. Also, not a single panchayat secured ES category. However, Ada is the only panchayat which achieved HS tag. In addition, the Panchayats are placed in Extremely Sustainable, HS and moderately sustainability group as per ESI, EEI, and SEI score, respectively. Alternatively, nine panchayats spread over 60.91% area and covering 63.72% population of the cluster comes under MS category (Figure 17.10). In between these panchayats, Saleguda and Mowad are the only panchayats which secured two HS categories (Table 17.3). In particular, Saleguda secured HS category as per EEI and SEI classification but poor ESI score prevented them from achieving any of the first two categories as per SLSI classification. Mowad achieves HS category in ESI and SEI but as per EEI classification, Mowad is categorized by VLS category. Likewise, Ada Dasnapur, Chirrakunta secured HS tag as per ESI score but other two pillars of SLSI forced them to stay in the MS category. Edulawada and Rahapally follow the same trend. Edulawada secured HS tag in SEI and Rahapally secured HS tag in EEI but due to poor result

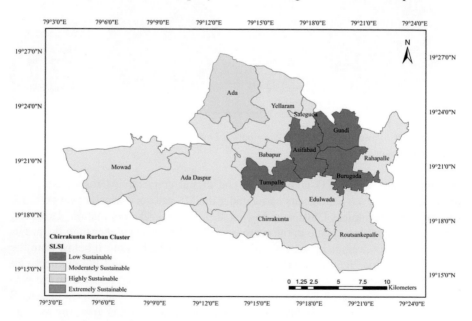

FIGURE 17.9   Spatial Variation in SLSI.

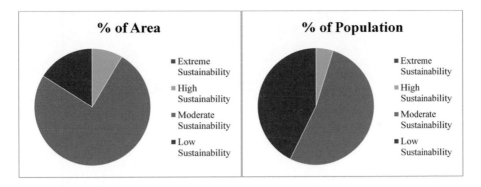

FIGURE 17.10   Area and population distribution in various degrees of sustainability classes based on SLSI score.

in the other categories, these panchayats come under MS category. On the other hand, Babapur, Rout Sankepalle, and Yellram are the only three panchayats which could not achieve HS category in any one pillar of SLSI but each of those panchayat achieves two MS tag in any two pillars of SLSI which helps them to enter MS group as per SLSI score. Unfortunately, Asifbad, Buruguda, Gundi, and Tumpalle covers 16.09% area and 42.71% population of the total cluster categorized by LS category as per SLSI classification. In between these panchayats, Asifabad achicved first place with ES tag in EEI but on the other hand, this panchayat is placed in the last place with VLS score in ESI, and also SEI score is not good enough. As a result, this panchayat is placed in the LS group in SLSI. In the ranking final ranking chart, Ada. Mowad and Rahapall secured the first three spot with 0.621, 0.576 and 0.567 SLSI score. Conversely, Gundi, Buruguda, and Tumpalle are placed in the last three spots with 0.367, 0.377 and 0.379 respectively.

17.5   CONCLUSION

SLSI is a kind of summation index, an exhaustive tool and a blueprinted to bestow an efficacious sustainable development planning. SLSI provides a foundation which depends on the relative scores, prioritizing the allotment of the development funds and the programs in addition to appropriate activities to attain sustainable development of a particular panchayat or an area. For example, panchayats which mostly comprise of tribal population viz., Ada Dasnapur, Mowad, and Chirrakunta secured minimum HS label as per ESI classification but failed to enter in any of the first three categories as per EEI score. Therefore, efforts need to be intensified from expert fields to boost the productivity in the agricultural field through advanced technology usage, eco-friendly practices in terms of agronomy with spot-particular crop diversity, along with enhancement in irrigation method with an efficient use of water. In addition, proper usage of electricity should be promoted when it comes to using agricultural machinery. Along with that, knowledge of properly trained farmers can improve the decision-making process and a healthy yield can be expected, scientific animal husbandry will increase milk production, HBI can be promoted by proper skill

TABLE 17.3
Panchayat Wise Ranking of the ESI, EEI, SEI and SLSI

Name of the Panchayat	ESI			EEI			SEI			SLSI		
	ESI Value	Degree of Sustainability	ESI Rank	EEI Value	Degree of Sustainability	EEI Rank	SEI Value	Degree of Sustainability	SEI Rank	SLSI Value	Degree of Sustainability	SLSI Rank
Ada	0.808	ES	1	0.728	HS	2	0.431	MS	7	0.621	HS	1
Ada Dasnapur	0.750	HS	2	0.324	LS	9	0.516	MS	5	0.544	MS	6
Asifabad	0.016	VLS	14	0.836	ES	1	0.402	MS	10	0.385	LS	11
Babapur	0.416	MS	13	0.587	MS	5	0.379	LS	12	0.441	MS	7
Buruguda	0.452	MS	11	0.093	VLS	13	0.478	MS	6	0.377	LS	13
Chirrakunta	0.633	HS	4	0.365	LS	7	0.286	MS	14	0.414	MS	9
Edulawada	0.495	MS	9	0.372	LS	6	0.685	HS	2	0.550	MS	5
Gundi	0.498	MS	8	0.168	VLS	12	0.382	LS	11	0.367	LS	14
Mowad	0.726	HS	3	0.058	VLS	14	0.751	HS	1	0.576	MS	2
Rahapally	0.470	MS	10	0.706	HS	3	0.561	MS	4	0.567	MS	3
Rout Sankepalle	0.560	MS	5	0.196	VLS	11	0.421	MS	9	0.410	MS	10
Saleguda	0.431	MS	12	0.627	HS	4	0.603	HS	3	0.554	MS	4
Tumpalle	0.513	MS	6	0.329	LS	8	0.310	LS	13	0.379	LS	12
Yellaram	0.506	MS	7	0.271	LS	10	0.428	MS	8	0.415	MS	8
	Weightage – 0.315			Weightage – 0.241			Weightage – 0.444					

nourishment practices. Presence of market feasibility can provide a stage to showcase industrial production and connectivity between panchayats and markets which in turn is important to bring down the transportation costs. Therefore, proper implementation of various flagship program like Deen Dayal Upadhyaya Grameen Kaushalya Yojana (DDUGKY), Rashtriya Krishi Vikas Yojna (RKVY), Pradhan Mantri Krishi Sinchai Yojna (PMKSY), Paramparagat Krishi Vikas Yojana (PKVY), Rashtriya Madhyamik Shiksha Abhiyan (RMSA) and Pradhan Mantri Gram Sadak Yojana (PMGSY) are of utmost importance for these panchayats. In the same way, Asifabad has a very poor score in ESI. Thus, prioritized efforts should be initiated to restore ecologically balanced sustainability. In particular, steps must be taken starting from the protecting the forest areas, stoppage of built up sprawls, preservation of the water bodies and addition to this promoting plantation. Therefore, Mahatma Gandhi National Rural Employment Guarantee Act (MGNREGA) is a very important scheme for Asifabad Panchyat. Under this scheme various environmental and ecological programs may be undertaken. Babapur and Chirrakunta Panchayats are categorized under Less Sustainable category as per SEI. Therefore, social structure reformation is utmost priority in these panchayats. That is why the efforts should be made toward enhancing the equity in benefit sharing through better schooling facility, protection of the female child, promoting digital connectivity, and LPG connection. National Rural Health Mission (NHM), Sarva Shiksha Abhiyan (SSA), Rashtriya Uchchatar Shiksha Abhiyan, Rajiv Gandhi Grameen LPG Vitaran Yojana (RGGLVY), Digital India are most important flagship programs for these panchayats which can be helpful to improve social equity in these panchayats. Further, for panchayats like Ada minor funding and development can be helpful for healthy livelihood. Conversely, in panchayats Gundi and Tumpalle which are not categorized by MS tag in at least two pillars of SLSI, proper planning on the basis of holistic and integrated approach with the utilization of the local resources along with appropriate management of the environment should be applied. Evidently, the results demonstrate that SLSI is a very robust and powerful tool for decentralized development, especially with respect to projects like SPMRM where panchayats have different requirements and priorities. SLSI can identify proper spatial requirements. Therefore, the current study suggests that, its replication is important for selected clusters in SPMRM project. This will help all the stakeholders with the upgraded management of natural resources by developing the balance between the economic, ecological, and social facets of sustainable development. Accordingly, the security of the current as well as future generation in terms of sustainable livelihood can be ensured.

REFERENCES

Altaf, F., Meraj, G., & Romshoo, S. A. (2013). Morphometric analysis to infer hydrological behaviour of Lidder watershed, Western Himalaya, India *Geogr. J.* 2013. doi: 10.1155/2013/178021.

Altaf, F., Meraj, G., & Romshoo, S. A. (2014). Morphometry and land cover based multi-criteria analysis for assessing the soil erosion susceptibility of the western Himalayan watershed. *Environ. Monit. Assess.* 186(12): 8391–8412.

Arjunan, M., Puyravaud, J. P., & Davidar, P. (2005). The impact of resource collection by local communities on the dry forests of the Kalakad–Mundanthurai tiger reserve. *Trop. Ecol.* 46: 135–143.

Arshad, A., Zhang, Z., Zhang, W., & Gujree, I. (2019). Long-term perspective changes in crop irrigation requirement caused by climate and agriculture land use changes in Rechna Doab, Pakistan. *Water* 11(8): 1567.

Belal, A. A., & Moghanm, F. S. (2011). Detecting urban growth using remote sensing and GIS techniques in Al Gharbiya governorate, Egypt. *Egypt. J. Remote Sens. Space Sci.* 14: 78–79.

Bera, A. A., Taloor, A.K., Meraj, G., Kanga, S., Singh, S.K., Durin, B., & Anand, S. (2021). Climate vulnerability and economic determinants: Linkages and risk reduction in Sagar Island, India; A geospatial approach. *Quat. Sci. Adv.* 100038. doi: 10.1016/j. qsa.2021.100038.

Bhandari, B. S., & Grant, M. (2007). Analysis of livelihood security: A case study in the Kali-Khola watershed of Nepal. *J. Environ. Manage.* 85: 17–26.

Bull, K. (2015). *Improving Rural Livelihoods and Environments in Developing Countries. Report.* Institute for Land, Water and Society, Charles Sturt University. http://www.csu. edu.au/research/ilws/research/sra-rurallive (Accessed on 12th January 2018).

Chambers, R. (1986). *Sustainable Livelihoods: An Opportunity for the World Commission on Environment and Development.* Institute of Development Studies. University of Sussex, Brighton. https://opendocs.ids.ac.uk/opendocs/bitstream/handle/123456789/873/rc338. pdf?sequence=1 (Accessed on 12th January 2018).

Chandel, R. S., Kanga, S., & Singhc, S. K. (2021). Impact of COVID-19 on tourism sector: A case study of Rajasthan, India. *AIMS Geosci.* 7(2): 224–242.

Chitwood, S. R. (1974). Social equity and social service productivity. *Public Adm. Rev.* 34: 29–35.

Dai, D. D., & Dien, N. T. (2013). Difficulties in transition among livelihoods under agricultural land conversion for industrialization: Perspective of human development. *Mediterranean J. Soc. Sci.* 4: 259–267.

Farooq, M., & Muslim, M. (2014). Dynamics and forecasting of population growth and urban expansion in Srinagar city – A geospatial approach. *Int. Arch. Photogramm. Remote Sens. Spat. Inf. Sci.* 40(8): 709.

Fredericks, S. E. (2012). Justice in sustainability indicators and indexes. *Int. J. Sustain. Dev. World Ecol.* 19: 490–499.

Ghanim, I. (2000). *Household Livelihood Security: Meeting Basic Needs and Fulfilment of Rights.* CARE-USA Discussion Paper, Atlanta, GA.

Guha, S., Mandla, V. R., Barik, D. K., Das, P., Rao, V. M., Pal, T., & Rao, P. K. (2018). Analysis of sustainable livelihood security: A case study of Allapur S Rurban cluster. *J. Rural Dev.* 37(2): 365–382.

Gujree, I., Arshad, A., Reshi, A., & Bangroo, Z. (2020). Comprehensive Spatial planning of Sonamarg resort in Kashmir Valley, India for sustainable tourism. *Pac. Int. J.* 3(2): 71–99.

Hassanin, M., Kanga, S., Farooq, M., & Singh, S. K. (2020). Mapping of Trees outside Forest (ToF) from Sentinel-2 MSI satellite data using object-based image analysis. *Gujarat Agric. Univ. Res. J.* 207–213.

Iyengar, N. S., & Sudarshan, P. (1982). A method of classifying regions from multivariate data. *Econ. Pol. Weekly* 57: 2048–2052.

Joy, J., Kanga, S., & Singh, S. K. (2019). Kerala flood 2018: Flood mapping by participatory GIS approach, Meloor Panchayat. *Int. J. Emerging Techn.* 10(1): 197–205.

INDIA, P. (2011). Census of India 2011 provisional population totals. New Delhi: Office of the Registrar General and Census Commissioner.

Kanga, S., Meraj, G., Farooq, M., Nathawat, M. S. and Singh, S. K. (2021). Analyzing the risk to COVID-19 infection using remote sensing and GIS. *Risk Anal.* 41(5): 801–813.

Kanga, S., Sudhanshu, Meraj, G., Farooq, M., Nathawat, M. S., & Singh, S. K. (2020a). Reporting the management of COVID-19 threat in India using remote sensing and GIS based approach. *Geocarto Int.* 2020: 1–8.

Kumar, S., Kanga, S., Singh, S. K. and Mahendra, R. S. (2018). Delineation of Shoreline Change along Chilika Lagoon (Odisha), East Coast of India using Geospatial technique. *International J. Afr. Asian Stud.* 41: 9–22.

Kumar, S., Raizada, A., & Biswas, H. (2014). Prioritising development planning in the Indian semi-arid Deccan using sustainable livelihood security index approach. *Int. J. Sustain. Dev. World Ecol.* 21: 332–345.

Li, W. B., Du, Z. Q., Ling, F., Zhou, D. B., Wang, H. L., Gui, Y. M., Sun, B. Y., Zhang, X. M. (2013). A comparison of land surface water mapping using the normalized difference water index from TM, ETM plus and ALI. *Remote Sens.* 5: 5530–5549.

Lindenberg, M. (2002). Measuring household livelihood security at the family and community level in the developing world. *World Dev.* 30: 301–318.

Littig, B, & Grießler, E. (2005). Social sustainability: A catchword between political pragmatism and social theory. *Int. J. Sustain. Dev.* 8: 65–79.

Maikhuri, R. K., Nautiyal, S., Rao, K. S., & Saxena, K. G. (2001). Conservation policy – People conflicts: A case study from Nanda Devi Biosphere Reserve (a World Heritage Site), India. *For. Pol. Econ.* 2: 355–365.

McCaston, K. (2005). *Moving CARE's Programming Forward: Unifying Framework for Poverty Eradication & Social Justice and Underlying Causes of Poverty.* CARE International, Geneva.

McCracken, J. A., & Pretty, J. N. (1988). *Glossary of Selected Terms in Sustainable Agriculture.* International Institute of Environment and Development, London.

Meraj, G., et al. (2015). Assessing the influence of watershed characteristics on the flood vulnerability of Jhelum basin in Kashmir Himalaya. *Nat. Hazards* 77(1), 153–175.

Meraj, G., Farooq, M., Singh, S. K., Romshoo, S. A., Nathawat, M. S., & Kanga, S. (2020). Coronavirus pandemic versus temperature in the context of Indian subcontinent: A preliminary statistical analysis. *Environ. Dev. Sustain,* 23(4): 6524–6534.

Meraj, G., Romshoo, S. A., & Altaf, S. (2016). Inferring land surface processes from watershed characterization. In N. Janardhana Raju (ed.), *Geostatistical and Geospatial Approaches for the Characterization of Natural Resources in the Environment.* Springer, Cham, 741–744.

Meraj, G., Singh, S. K., Kanga, S., & Islam, M. N. (2021). Modeling on comparison of ecosystem services concepts, tools, methods and their ecological-economic implications: A review. *Modeling Earth Systems and Environment,* 1–20. https://doi.org/10.1007/s40808-021-01131-6

Meraj, G., Singh, S. K., Kanga, S. & Islam, M. N. (2021a). Modeling on comparison of ecosystem services concepts, tools, methods and their ecological-economic implications: a review. *Model. Earth Syst. Environ.* 1–20. https://doi.org/10.1007/s40808-021-01131-6

Meraj, G., Kanga, S., Kranjčić, N., Đurin, B., & Singh, S. K. (2021b). Role of Natural Capital economics for sustainable management of earth resources. *Earth,* 2(3): 622–634.

Moser, A. (1996). Ecotechnology in industrial practice: implementation using sustainability indices and case studies. *Ecol. Eng.* 7: 117–138.

Nathawat, M. S., et al. (2010). Monitoring & analysis of wastelands and its dynamics using multiresolution and temporal satellite data in part of Indian state of Bihar. *Int. J. Geomat. Geosci.* 1(3): 297–307.

Ouyang, W., Song, K., Wang, X., & Hao, F. (2014). Non-point source pollution dynamics under long-term agricultural development and relationship with landscape dynamics. *Ecol. Indic.* 45: 579–589.

Pekel, J. F., Cottam, A., Gorelick, N., & Belward, A. S. (2016). High-resolution mapping of global surface water and its long-term changes. *Nature* 540: 418–422.

Qu, F., Kuyvenhoven, A., Shi, X., & Heerink, N. (2011). Sustainable natural resource use in rural China: Recent trends and policies. *China Econ. Rev.* 22: 444–460.

Ranga, V., Pani, P., Kanga, S., Meraj, G., Farooq, M., Nathawat, M. S., & Singh, S. K. (2020b). National Health-GIS Portal – A conceptual framework for effective epidemic management and control in India. Preprints 2020, 2020060325. doi: 10.20944/preprints202006.0325.v1

Sajjad, H., Nasreen, I., & Ansari, S. A. (2014). Assessing spatiotemporal variation in agricultural sustainability using sustainable livelihood security index: Empirical illustration from Vaishali District of Bihar, India. *Agroecol. Sustain. Food Syst.* 38: 46–68.

Saleth, R. M., & Swaminathan, M. S. (1993). Sustainable livelihood security index; Towards a welfare concept and robust indicator for sustainability. In: Moser, F. (Ed.), *Proceedings of the International Workshop on Evaluation Criteria for a Sustainable Economy*, April 6–7: 42–58, Graz .

Scoones, I. (1998). *Sustainable Rural Livelihoods: A Framework For Analysis*. IDS Working Paper 72: 67–71. IDS, Brighton.

Scoones, I. (2009). Livelihoods perspectives and rural development. *J. Peasant Stud.* 36: 171–196.

Shaw, A., & Kristjanson, P. (2014). A catalyst toward sustainability? Exploring social learning and social differentiation approaches with the agricultural poor. *Sustainability* 6: 2685–2717.

Shaw, R., & Das, A. (2017). Identifying peri-urban growth in small and medium towns using GIS and remote sensing technique: A case study of English Bazar urban agglomeration, West Bengal, India. *Egypt. J. Remote Sens. Space Sci.* 20: 125–145.

Singh, P. K., & Hiremath, B. N. (2010). Sustainable livelihood security index in a developing country: A tool for development planning. *Ecol. Indic.* 10: 442–451.

Swaminathan, M. S. (1991a). Greening of the mind. *Indian J. Soc. Work* 52: 401–407.

Swaminathan, M. S. (1991b). From Stockholm to Rio de Janeiro: The road to sustainable agriculture. In: Monograph No. 4. MS Swaminathan Research Foundation, Madras.

Tao, S. L., Fang, J. Y., Zhao, X., Zhao, S. Q., Shen, H. H., Hu, H. F., Tang, Z. Y., Wang, Z. H., & Guo, Q. H. (2015). Rapid loss of lakes on the Mongolian Plateau. *Proc. Natl. Acad. Sci. U. S. A.* 112: 2281–2286.

Tomar, J. S., Kranjčić, N., Đurin, B., Kanga, S., & Singh, S. K. (2021). Forest fire hazards vulnerability and risk assessment in Sirmaur district forest of Himachal Pradesh (India): A geospatial approach. *ISPRS Int. J. Geo-Inf.* 10(7): 447.

Tulbure, M. G., & Broich, M. (2013). Spatiotemporal dynamic of surface water bodies using Landsat time-series data from 1999 to 2011. *ISPRS. J. Photogramm. Remote Sens.* 79: 44–52.

Tulbure, M. G., Kininmonth, S., & Broich, M. (2014). Spatiotemporal dynamics of surface water networks across a global biodiversity hotspot-implications for conservation. *Environ. Res. Lett.* 9: 114012.

United Nations Development Programme (UNDP) (2006). *Human Development Report 2006*. Oxford University Press, New York.

Wu, B. (2004). *Sustainable Development in Rural China: Farmer Innovation and Self-organisation in Marginal Areas*. Routledge, London.

You, H, & Zhang, X. (2017). Sustainable livelihoods and rural sustainability in China: Ecologically secure, economically efficient or socially equitable? *Resour. Conserv. Recycl.* 120: 1–13.

Zou, Z., Dong, J., Michael, A., Xiangming, M., Yuanwei, X., Russell, Q., Doughty, B., Katherine, V. H., & Hambright, D. K. (2017) Continued decrease of open surface water body area in Oklahoma during 1984–2015. *Sci. Total Environ.* 595: 451–460.

18 Carrying Capacity of Water Supply in Shimla City

A Study of Sustainability and Policy Framework

Rajesh Kumar
Governmen College Haripur Guler

M. L. Mankotia
Government Degree College Theog

CONTENTS

18.1 Introduction 348
18.2 Geographical Personality of the Study Area...349
18.3 Materials and Methods 349
18.4 Results and Discussion 350
    18.4.1 Water Source...350
        18.4.1.1 Dhalli Catchment...351
        18.4.1.2 Cherot and Jagroti...351
        18.4.1.3 Chair Nallah...352
        18.4.1.4 Source at Gumma at Nauti Khad.............................352
        18.4.1.5 Ashwani Khad 353
        18.4.1.6 Giri Khad...353
        18.4.1.7 Existing Distribution System of Water Supply in
                  Shimla City 353
    18.4.2 Distribution of Water Connection in Shimla City 354
    18.4.3 Water Storage Reservoirs in Shimla City 354
    18.4.4 Carrying Capacity, Demand and Deficits of Water Supply..........356
    18.4.5 Population Projection, Water Demand and Deficits 359
    18.4.6 Sustainable Water Supply 361
18.5 Conclusion 361
References..362

DOI: 10.1201/9781003147107-23

18.1 INTRODUCTION

Water is one of the basic necessities of life. It is very important as it is the major constituent of both plants and animals (Chakhaiyar, 2010). The demand for portable water supply and distribution increases with urbanization which is rapid in most of the developing countries. In recent times, the rate and dimension of urbanization has increased, which resulted in having more than 50% of the world population living in urban areas within which 64% of them are in developing countries (UNPD, 2012). The increasing rate of population and water demand has compounded the issue of water source depletion in many parts of the world (Farooq and Muslim, 2014). Towns and cities in developing countries are currently facing serious challenges of efficiently managing the scarce water resources, urbanization, and infrastructural decay, as well as the issue of sustainability of conventional water management (Zeraebruk, 2014; Gujree et al., 2020; Hassanin et al., 2020; Tomar et al., 2021; Meraj et al., 2021; Chandel et al., 2021; Bera et al., 2021).

To signify the importance of water, the year 2005 marked the beginning of the "International Decade for Action: Water for Life" and renewed the effort to achieve the Millennium Development Goal to reduce by half the proportion of the world's population without sustainable access to safe drinking water and sanitation by 2015 by the World Health Organization (NIHR, 2013).

Of the total water available on the Earth's surface, 97.3% is found in the oceans, 2.1% makes up the glaciers and polar ice, and 0.6% accounts for groundwater. The rivers, lakes, and streams contain only 0.02% of the total, which is immediately available for use (Snoeyink and Jenkins, 1980; Singh et al., 2017; Kanga et al. 2020a, b).

The present scale of water to rural or urban population is generally inadequate all over the country, and hence MC, Shimla is no exception. Increasing demand of rising population, nonavailability of potential source, inadequacy of infrastructure, etc. all lead to scarcity of drinking water in the study area. Beginning as a feeder of insignificant population of 16,000 souls in year 1875 AD, Shimla city water supply grew to larger proportion augmented and improved over the years. Despite the improvements carried out so far as aforesaid, scarcity of water prevails in many pockets of the city particularly in special areas amalgamated within the municipal boundary and the areas that are likely to be amalgamated. The depletion of yield of sources during lean period causes further increase in demand and supply gap, resulting in much hardships and miseries to the inhabitants of the city. City Development Plan proposed under JNNURM identifies water supply and sanitation as a major priority area to be tackled in order to lessen the growing gap between demand and supply to provide adequate quantum of portable water to the citizens, as a government norms 135 lpcd water supply to the city. The plan has estimated total quantity of water required till the end of planning year of 2039 as 71 MLD. Further, the plan reiterates the availability of 30 MLD of water, which falls too short of the present requirement of the city causing acute shortage.

The sustained strategy shall take into account the identification of source of supply, collection, transmission, distribution and other related aspects. The sustainability should be so framed that it would not only integrate the existing system of water supply but would also envisage a strategic plan consistent with future overall development of the city area. The American Water Works Association (AWWA) defines sustainability as "providing an adequate and reliable water supply of desired quality – now and for future generations – in a manner that integrates economic growth, environmental protection, and social development" (AWWA, 2009).

18.2 GEOGRAPHICAL PERSONALITY OF THE STUDY AREA

Shimla is situated in the Central Himalayas at 31°4′ to 31°10′ north latitude and 77°5′ to 77°15′ east longitude, and its mean elevation is 2,130 m above mean sea level. The topography of Shimla is characterized by rugged mountains, steep slopes and deep valleys. It experiences cold winters during December–February, with temperatures ranging from 0°C to 13°C. Shimla receives snowfall around Christmas or the last week of December. The summers (May–June) are mild with temperatures varying from 20°C to 30°C. The monsoon period extends from June to September and records moderate rainfall. The average rainfall recorded for the last 25 years (1980–2005) in Shimla is 1,437 mm. The Shimla region consists of a thin soil layer (0.15 m on ridges to 7 m deep in valleys), an intervening layer of detritus (mix of soil and fragments of weathered bedrock) and hard bedrock (Tandon, 2008; Altaf et al., 2013, 2014; Meraj et al., 2015, 2016). Shimla city was discovered in 1819 by the British and has evolved from a small hill settlement to one of the popular tourist destinations in India. Part of Himachal Pradesh was carved out of erstwhile the Punjab state in 1966, and Shimla became the capital of newly formed state of Himachal Pradesh in 1971. As per Census (2011), Shimla is the only Class I City in the State of Himachal Pradesh with a total population of 169,758 persons. Consequently, the population of the city has grown by 12 times (from 14,000 in 1901 to 169,758 in 2011) during the century. The total area under the jurisdiction of MC Shimla also has increased after merger of New Shimla, Totu (including some parts of Jutog) and Dhalli areas to 3,500 sq. km. According to 2011, the Municipal Corporation of Shimla (Shimla city) is divided into 25 wards, covering urban core and urban fringes (Figure 18.1).

The aim of this research is to assess the carrying capacity of water supply in Shimla city. This aim can be achieved through the following objectives;

- To analyze the water source and existing water distribution network in the Shimla City.
- To trace the ward wise distribution of water connection in study area.
- To determine the carrying capacity, demand and deficits of water supply in Shimla city.
- To study optimal level of distribution to achieve sustainability.

18.3 MATERIALS AND METHODS

The research design comprises four sections, namely, data collection, data input, data analysis and the results and findings.

Data collected for this study include water source, existing water distribution network, water connection and storage reservoirs obtained from SMC Shimla, and

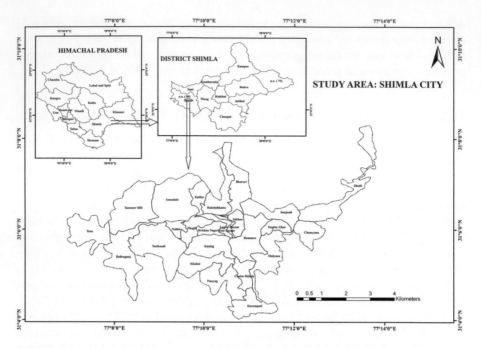

FIGURE 18.1   Location of the study area with respect to Himachal Pradesh

data relating to population and household have been collected from the census of India publications, viz Primary Census Abstracts of 2011 census years for Himachal Pradesh.

At the data input stage, the data collected were incorporated into the Arc GIS 10.3 software for processing. The existing distribution network, water source and storage reservoirs were digitized on the city map to delineate areas covered by the existing network. The distribution pattern was also established based on the water demand, as determined by population.

Having the data integrated and processed, the data was analyzed. A suitability model was developed with multiple inputs such as reclassified DEM and reclassified open spaces map. The model was run in order to determine suitable locations for reservoirs. Water demand for service areas was also calculated based on the population and number of reservoirs required.

In the results and findings section, the results and findings of this research are outlined. The findings include suitable location for reservoirs and suitable service (catchment) areas for water distribution based on the daily water demand and number of reservoirs.

18.4   RESULTS AND DISCUSSION

18.4.1   WATER SOURCE

Water supply facilities were introduced in Shimla more than 130 years ago by pumping water from nearby spring sources. The city grew over the years increasing its boundary as well as population with subsequent augmentation of water supply to match the needs of the growing population. With the commissioning of the last augmentation scheme of water supply in the year 1992, the total installed capacity of the system rose to 56.62 MLD.

The details and distribution of water source for Shimla city are provided in (Table 18.1 and Figure 18.2). Water supply system for the city developed over a period of time and the history of development are provided below in brief.

18.4.1.1   Dhalli Catchment

The first water supply scheme 4.54 MLD was implemented to utilize the water from the storage reservoir of 10.92 million liters (located at 12.85 km from Shimla), which stores water from spring sources from Dhalli Catchment Area, during 1875 to support a population of 16,000. An average of 0.45 MLD water is also received from this source under gravity condition at Dhalli filter. This is the source based on spring sources of the dense protected forest known as Dhalli catchment.

18.4.1.2   Cherot and Jagroti

Subsequently, to fulfill the growing need of the city and the tourists, the first augmentation of Shimla Water Supply Scheme was by provision of pump sets near Cherot Nallah (year 1889) and Jagroti Nallah (year 1914) to tap 4.80 MLD of water at the source. Part of this water is distributed in the adjoining area of Dhalli Zone, and balance is received at Sanjauli reservoir through gravity main from Dhalli filter.

TABLE 18.1
Source of Water for Shimla City

					Supply to SMC	
			Quantities Drawn in	Installed	None	
		Transmission	Year of	MLD	Capacity	Lean	Lean
S. No	Name of Source	Type	Start	(Average)	in (MLD)	Period	Period
1	Dhalli Catchment	Gravity	1875	1.80	4.54	0.23	0.20
2	Churat Nallah	Pumping	1914	3.86	4.80	3.86	2.65
3	Chair Nallah	Pumping	1914	3.00	2.50	2.50	1.42
4	Nauti Khad	Pumping	1924 and 1982	19.75	24.06	24.06	16.80
5	Ashwani Khad	Pumping	1992	10.80	10.06	10.80	6.30
6	Giri Khad	Pumping	2008	20	20	15	15
	Total			53.55	56.62	52.36	39.52
	Tube Wells – 10 nos						2.63
	Total						42.15

Source: MC Shimla.

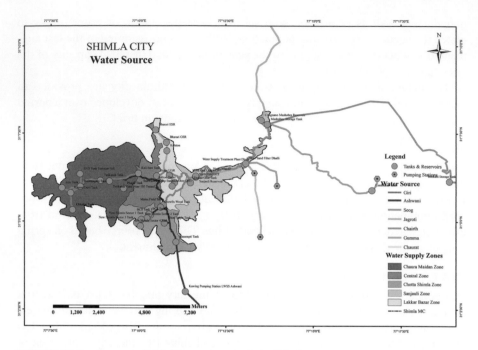

FIGURE 18.2   Sources of drinking water supply in Shimla

18.4.1.3   Chair Nallah

Shimla Water Supply Scheme (year 1914) was implemented by installation of two pump sets at Chair Nallah to tap 2.50 MLD of water at the source. From the source to storage tank at Lambidhar from where water is distributed in Charabra area, president house located at Charabra, the balance water gravitates to the Dhalli filter.

18.4.1.4   Source at Gumma at Nauti Khad

The third augmentation of Shimla Water Supply Scheme was commissioned during the year 1924 to tap 7.72 MLD of water from Nauti Khad with further upgradation of pumps at various stages.

The fourth augmentation of Shimla Water Supply Scheme was installation of pump sets at Gumma and Darabla to tap additional 16.34 MLD of water at the source. Today, the system is designed to lift 24.06 MLD of water at the source.

This is the main source of water supply, which provides approximately 16.75 MLD of water. This quantity of water is pumped from Gumma and received at Carignano reservoir from where it is gravitated to the Sanjauli reservoir. From Sanjauli, this water is further gravitated to the Ridge reservoir and also to Mans field reservoir. In addition to this, some of the sectoral tanks are also fed through feeder line from Sanjauli reservoir (Figure 18.2). These sectoral tanks are Engine Ghar, North Oak, Bharari, Advance Study, Sandal point, Chakkar and Totu.

18.4.1.5   Ashwani Khad

The fifth augmentation of Shimla Water Supply Scheme was commissioned in April 1992 designed to pump 10.80 MLD (Table 18.1) of water at two stages lifting at Ashwani Khad and Kawalag.

From this source, water is pumped through two stages and received at Kasumpti reservoir from where part of the water is distributed in the adjoining area and sectoral tanks located at sectors 1, 3 and 4 New Shimla are filled up. The balance water is again pumped from this reservoir to Mansfield tank. From Mansfield tank, the water is distributed in the adjoining area, and the sectoral tank at Knolls wood is also connected with this water tank (Figure 18.2). In addition to this, the sectoral tank at Tuti Kandi and sump well of Kamna Devi Pump house is also fed.

18.4.1.6   Giri Khad

Sixth Augmentation of Shimla Water Supply Scheme from Giri Khad commissioned in 2008 with the installed carrying capacity 20 MLD and also drawn about 20 MLD. The water supply from the Giri Khad source approximates 30 MLD to Shimla city.

18.4.1.7   Existing Distribution System of Water Supply in Shimla City

The city has five major water supply zone and 15 delineated water zones based on topography and location of feeder reservoirs as shown in (Figure 18.3). The supply of drinking water is not fairly in Shimla city. The water is drawn from five different gravity and pumping stations. The study reveals that water is not evenly distributed with the same quantity to all the wards in the city.

Figure 18.3 shows that Chaura Maidan Zone covered the western part i.e. ward nos. 3, 4, 5, 6, 7, 8, 9 and 10 of city and gets supply of drinking water once in three days. Under this zone, most feeder reservoirs are A G Office, Vice Regal Lodge, University, Totu and Chakkar, whereas in Central zone, around five wards namely, ward nos. 2, 11, 12, 13 and 25 get water supply every third day. Ridge and High Court are the two most feeder reservoirs found under this zone. Thirdly, the Lakkar Bazar Zone covered around three wards namely Bharari, Banmore and Jakhoo in northern and central part of the city. In southern part of Shimla city, Chotta Shimla Zone covered the four wards i.e. ward nos. 21, 22, 23 and 24; under the feeder reservoir Mansfield, Kasumpti and BCS also get water supply every third day. The fifth, water distributional Sanjauli Zone, mostly covered the eastern part of the city. Around five wards of the city have come under this zone namely ward nos. 16, 17, 18, 19 and 20. Sanjauli and Dhalli are the main feeder reservoirs of this zone. In this zone, four wards 16, 17, 18 and 19 get the water supply every third days, and ward no. 20 gets it weakly in peak season.

Table 18.2 also shows that there are about 12 reservoirs under the five water distributional zones which covered different areas of the Shimla city.

The study reveals that the water supply situation needs to be improved considerably. Due to inadequate and unplanned distribution network, the water distribution is not uniform with some areas getting excess water and many areas receiving very less quantity of water. At many places, the feeder mains have been tapped directly and used as distribution mains.

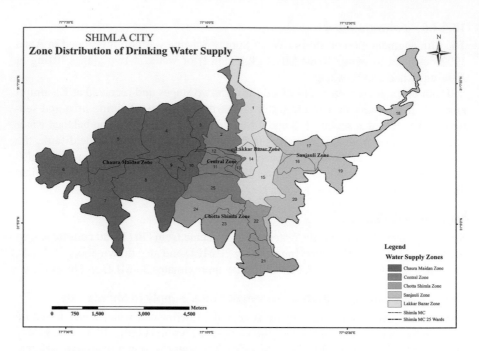

FIGURE 18.3   Zone wise drinking water distribution map of Shimla

18.4.2   DISTRIBUTION OF WATER CONNECTION IN SHIMLA CITY

At present, 30,995 water connections have been provided by the department
(Table 18.3). About 63% connections are private, while the rest account for public connections (Table 18.4). The total number of domestic connections is 24,209
followed by commercial 4,403, construction 1,395, Govt. institution 384, religious
places 29 and hotel 575 in 2018. A perusal at Table 18.5 leads one to see that the
growth rate variation 11.95% points from the year 2013 (27,765) to 2017 (30,995).
Variations in the distribution of water connection were still more pronounced at ward
level (Table 18.6). From these connections, water is supplied for around 60–90 minutes every day in nonlean period and for around 45 minutes on alternate days during
lean period.

18.4.3   WATER STORAGE RESERVOIRS IN SHIMLA CITY

Storage reservoirs are also called conservation reservoirs because they are used to
conserve water. The SMC is being served water from 11 major reservoirs having a
total capacity of 36.95 ML. In addition to the above 11 major reservoirs, 28 minor
reservoirs having a total capacity of 5.8 ML are also available for distribution of
water. The details of the storage reservoirs and their distribution network are shown
in Table 18.7 and Figure 18.4.

Water from the sources are treated and pumped barring Dhalli catchment source,
which operates under gravity conditions. Water thus pumped or gravitated is stored
in 10 number of service reservoirs located at suitable sites covering the MC, Shimla.

TABLE 18.2
Zone Distribution of Drinking Water Supply in Shimla City

S. No	Name of Zone	Reservoirs	Area Covered
1	Sanjauli Zone	Sanjauli	Sanjauli Bazaar, Engine Ghar, Nav Bahar, Snowdown, Jakhoo, Pumping Station, Grand Hotel, Shankli, Scandal, Sangti
		Dhalli	Dhalli Bazaar, HRTC workshop, Dhingu Devi, Pumping Station for higher belt of Sanjauli area
2	Lakkar Bazar Zone	Bharari	Bharari, Harvington, Kuftu, Anu, Bermu, etc.
3	Central Zone	Ridge	Telegraph office, Krishna Nagar, Sabzi Mandi, Ripon, Lalpani, Western Command, Ram Bazaar, Middle bazaar
		High Court	Lower High Court area, Paradas Garden, Kanlog, Talland
4	Chaura Maidan Zone	AG office	Kaithu, Annandale, Kavi, AG Office, Ram Nagar Vidhan Sabha, Chaura Maidan, Tuti Kandi, Kumar House, Raj Bhavan, Ava Lodge, Labour Bureau, Kenndy House, Win Gate
		Vice Regal lodge	Institute of Advanced Studies, Tilak Nagar, Ghora Chowk, Hanuman Temple
		University	University Complex, Summer Hill, Govt. Quarters, Shiv Mandir
		Tutu	Tutu Bazaar, Jutogh, Dhamida, Fatenchi
		Chakkar	Sandal Hill, Tara Devi, Shoghi
		Kamna Devi	Hill Spur of Kamna Devi, Forest Colony
5	Chotta Shimla Zone	Mansfield	Mansfield to Marina, Secretariat, Chotta Shimla bazaar, Brock Hurst upto Govt. School
		Kasumpti	Kasumpti Colony, Lower Brock Hurst, Patti Rehana, Patina Mehli, Pantha Ghati, Patelog
		BCS	BCS, Khalini, Forest Colony

Source: MC Shimla.

TABLE 18.3
Distribution of Total Water Connections by MC Shimla

S. No	Head	No. of Connections
1	Domestic	24,209
2	Commercial	4,403
3	Construction	1,395
4	Govt. Institution	384
5	Religious places	29
6	Hotel	575
	Total running connection	30,995

Source: MC Shimla.

TABLE 18.4

Distribution of Water Connections by MC Shimla

S. No	Head	No. of Connections
1	Private	23,000
2	Public	200
3	Street hydrants	525
4	Tank and others	100
	Total	23,825

Source: MC Shimla.

TABLE 18.5

Distribution of Total Water Connection by MC Shimla in 2013 and 2018

S. No	Head	No. of Connections (2013)	No. of Connections (2018)	Growth Rate (%)
1	Domestic	21,276	24,209	13.78
2	Commercial/hotels	5,161	4,978	−3.54
3	Street hydrants/ construction connection	900	1,395	55
4	Govt. institution/religious places	428	413	−3.50
Total		27,765	30,995	11.63

Source: MC Shimla.

The total capacity of existing reservoirs as shown above is 42.75 MLD. The reservoir located at Sanjauli is the largest one having a capacity to store 8.78 MLD, and the smallest one is situated at Kasumpti having a capacity of 0.223 MLD. Newly developed areas of BCS, Chakkar, Tohu, etc. do not have separate service reservoirs; instead, they are fed from existing ones causing considerable loss in pressure at the tail end of the network.

18.4.4 Carrying Capacity, Demand and Deficits of Water Supply

As may be noted from the table above, I&PH and MC Shimla departments are responsible for water supply in Shimla. There is metric distribution system of water in place. There are six water stations that provide supply to entire Shimla: Sanjauli, Chhota Shimla, New Shimla, Central Zone, Lakkar Bazar and Chaura Maidan. There are two big water storage tanks at Ridge and Sanjauli which are used for bulk water storage and supply.

TABLE 18.6
Distribution of Total Water Connection Ward Wise in Shimla City

S. No	Wards	Total Water Connection	Percentage to Total Water Connection
1	Bharari	1,165	3.9
2	Ruladu Bhatta	331	2.8
3	Kaithu	1,068	3.5
4	Anndale	957	3.2
5	Summerhill	613	2
6	Totu	429	1.4
7	Boileuganj	1,500	5
8	Tutti Kandi	793	2.6
9	Nabha	947	3.1
10	Phagli	693	2.3
11	Krishna Nagar	628	2.1
12	Ram Bazaar	1,628	5.4
13	Mall Road	1,041	3.5
14	Jakhoo	864	2.9
15	Banmori	735	2.4
16	Engine Ghar	823	2.7
17	Sanjauli	1,541	5.1
18	Dhali	1,138	3.8
19	Chamyana	992	3.3
20	Malyana	802	2.7
21	Kasumpti	3,604	11.9
22	Chhota Shimla	9,079	6.5
23	Pateyog	3,237	10.7
24	Khalini	993	3.3
25	Kanlog	1,170	3.9

Source: MC Shimla.

The water supply system in Shimla, which pumps water (around 1,470 m head) from the nearby streams, is about 135 years old. At present, the city relies on six key water sources for its daily water needs. The combined designed capacity of the sources is 61 MLD, but the present yield is limited to 37–38 MLD only. During lean periods, i.e., May–June, the quantum of water available at source reduces to 30–31 MLD. During normal periods, the water supplied from the reservoirs is approximately 38 MLD and post losses, 28–30 MLD is available to the resident population for daily use. During lean periods, only 23–25 MLD is available for usage after transmission losses. This translates into average per capita water availability in the range of 100–115 LPCD during normal and lean periods (after transmission losses). The demand for water during normal days in Shimla is around 38 MLD and during tourist season reaches as high as 50 MLD. The city receives water supply daily in

TABLE 18.7

Water Storage Reservoirs of Shimla City

S. No	Reservoirs	Capacity in MLD
1	Carignano	3.00
2	Sanjauli	8.78
3	Ridge	4.63
4	Mansfield	3.63
5	Kasumpti	2.00
6	Kasumpti	0.22
7	Viceregal lodge	0.23
8	Jakhu	0.32
9	Boileaugunj	0.24
10	Seog	10.90
11	Masobara	3.00
12	Others	5.8
Total		42.75

Source: Water Supply and Sewerage Division, New Shimla.

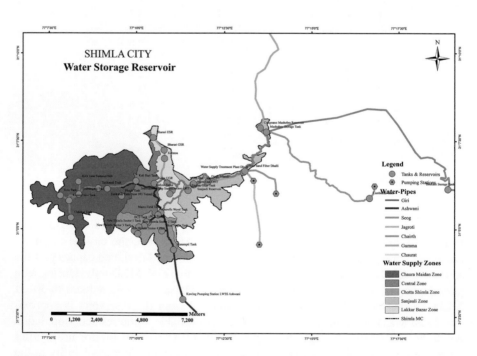

FIGURE 18.4 Drinking water storage distribution map of Shimla

the morning for around 1–1.5 hours. The total demand of water for Shimla city with a population of 169,758 and a floating population of 75,000 comes to about 35 MLD, considering as a government norms 135 LPCD of water supply. During lean periods, water is supplied every alternate day. In some of the areas, the supply is only every 3–4 days.

18.4.5   POPULATION PROJECTION, WATER DEMAND AND DEFICITS

Graph 18.2 provides details of the population of city and the per capita water availability. At present, the per capita surface water availability is about 32 MLD. Rise in population may require the per capita water availability to 72 MLD by 2071. As the availability of water has wide spatial and temporal variations (including inter-annual variations), the general availability situation is more alarming than that depicted by the averages.

The population projection (Graph 18.1) shows the permanent and floating population of Shimla city to be at the level of 493,632 by 2071. It is also assumed that the population of Shimla city is continuously growing from 1971. In Shimla city, Graph 18.2 shows that population growth and water demand totally going to beyond carrying capacity continuously. At present, total demand of water for total population and floating population is 35.24 MLD, considering a government norms 135 LPCD of water supply. The carrying capacity of water supply in Shimla is limited, which can sustain only about 2.15 million populations which will be insufficient for future. The study analyzed that the deficits between water supply and demand continuously increase. If the population of the city continuously increases at the present growth rate, the future requirement of water will be 72 MLD in 2071 (Table 18.8). The assessment of the intermediate demand of water supply will be 51 MLD for the population 369,195 in 2041 (Table 9.9).

The total ultimate water demand is 72 MLD by the year 2071, and the intermediate demand by the year 2041 is 51.11 MLD. The deficits between water supply and demand continuously increase. If the population of the city continuously increases at the present growth rate, the future requirement of water will be 72 MLD in 2071.

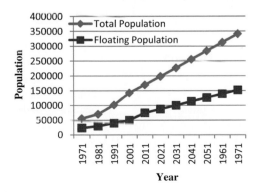

GRAPH 18.1   Population projection.

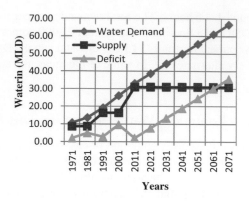

GRAPH 18.2   Water crisis.

TABLE 18.8
Appraisal of Future Water Demand

S. No	Description	Total Population	Rate of Supply (L/day)	Water Requirement (L/day)
1	Permanent population for the year 2071	341,316	135	46,077,660
2	Floating population	**152,316**	135	20,562,660
3	Add 15% on account of losses i.e. unaccounted			4,442,688
			Grand total	71,083,008
				(72 MLD)
	Average water available from existing sources			30.00 MLD
	Balance water requirement for the year 2071			**43.00 MLD**

TABLE 18.9
Intermediate Water Demand

S. No	Description	Total Population	Rate of Supply (L/day)	Water Requirement (L/day)
1	Permanent population for the year 2041	255,537	135	34,497,495
2	Floating population	113,658	135	15,343,830
3	Add 15% unaccounted			1,278,459
Total				51,119,784
				51.11 MLD

18.4.6 SUSTAINABLE WATER SUPPLY

An important aspect of sustainable water supply systems is thus the concept of providing water of a specific quality needed for a specific use, rather than simply providing adequate quantities of water for all uses. The AWWA defines sustainability as "providing an adequate and reliable water supply of desired quality – now and for future generations – in a manner that integrates economic growth, environmental protection, and social development" (AWWA, 2009).

The sustainable water requirement by Shimla city will be met from newly proposed water supply scheme from Pabbar River, the DPR for which already stands submitted for external funding to the Government of India. In Shimla city, more newly proposed water supply schemes should be established by the government. The government should be so framed that sustainability would not only integrate the existing system of water supply but would also envisage a strategic plan consistent with future overall development of the city area. The strategies of sustainable development shall take into account the identification of source of supply, collection, transmission, distribution and other related aspects. It should also be streamlined to fit into the regional development plans, long-term sector plan, land use plan and other open space planning. This scenario may also include the supporting activities like health, education, staff training and infrastructural improvements, etc.

18.5 CONCLUSION

The study explores that with inadequate and unplanned distribution network, the water distribution is not uniform with some areas getting excess water and many areas receiving very less quantity of water. At many places, the feeder mains have been tapped directly and used as distribution mains. At present, the Shimla water system is a nonstorage type and has been designed only on the basis of lean period stream flow. Inadequacy of sources is responsible for water scarcity, hampering developmental and tourists' activities. The study observed that the main reasons for the suboptimal functioning of the existing water supply system in Shimla city are inappropriate augmentation of source, inadequate storage facility, aged and leaking pipeline network, illegal tapping of transmission water supply pipelines, unauthorized house connections, faulty metering, lack of operation and maintenance of system components, adoption of inappropriate design methodology, etc. Certain parts of the city are experiencing severe water supply crisis, and the level of services offered by service providers has reduced drastically. Study highlights the population growth and water demand totally going to beyond carrying capacity continuously. At present, the total demand of water for total population and floating population is 35.24 MLD, considering a government norms 135 LPCD of water supply. The deficits between water supply and demand continuously increase. If the population of the city continuously increases at the present growth rate, the future requirement of water will be 72 MLD in 2071. Sustainable water requirement by Shimla city will be met-out from newly proposed water supply scheme from Pabbar River, the DPR for which already stands submitted for external funding to the Government of India.

REFERENCES

Altaf, F., Meraj, G., & Romshoo, S. A. (2013). Morphometric analysis to infer hydrological behaviour of Lidder watershed, Western Himalaya, India. *Geography Journal* 2013. https://doi.org/10.1155/2013/178021

Altaf, F., Meraj, G., & Romshoo, S. A. (2014). Morphometry and land cover based multi-criteria analysis for assessing the soil erosion susceptibility of the western Himalayan watershed. *Environmental Monitoring and Assessment* 186(12), 8391–8412.

AWWA (2009). Sustainability and the American Water Works Association: Recommended activities and implementation strategy. Denver. http://www.awwa.org/files/Resources/SustainableWater/AdHocCommittee SustainabilityReport2009.pdf

Bera, A., Taloor, A. K., Meraj, G., Kanga, S., Singh, S. K., Durin, B., & Anand, S. (2021). Climate vulnerability and economic determinants: Linkages and risk reduction in Sagar Island, India; A geospatial approach. *Quaternary Science Advances* 100038. https://doi.org/10.1016/j.qsa.2021.100038

Chakhaiyar, H. (2010). *Periwinkle Environmental Education Part IX*. Jeevandeep Prakashan Pvt Ltd, Mumbai.

Chandel, R. S., Kanga, S., & Singh, S. K. (2021). Impact of COVID-19 on tourism sector: A case study of Rajasthan, India. *AIMS Geosciences* 7(2), 224–242.

Farooq, M., & Muslim, M. (2014). Dynamics and forecasting of population growth and urban expansion in Srinagar city – A geospatial approach. *The International Archives of Photogrammetry, Remote Sensing and Spatial Information Sciences* 40(8), 709.

Gujree, I., Arshad, A., Reshi, A., & Bangroo, Z. (2020). Comprehensive spatial planning of Sonamarg resort in Kashmir Valley, India for sustainable tourism. *Pacific International Journal* 3(2), 71–99.

Hassanin, M., Kanga, S., Farooq, M., & Singh, S. K. (2020). Mapping of Trees outside Forest (ToF) from Sentinel-2 MSI satellite data using object-based image analysis. *Gujarat Agricultural Universities Research Journal* 207–213.

Kanga, S., et al. (2020a). Modeling the spatial pattern of sediment flow in Lower Hugli Estuary, West Bengal, India by quantifying suspended sediment concentration (SSC) and depth conditions using geoinformatics. *Applied Computing and Geosciences* 8, 100043.

Kanga, S., Sheikh, A. J., & Godara, U. (2020b). GIscience for groundwater prospect zonation. *Journal of Critical Reviews* 7(18), 697–709.

Meraj, G., et al. (2015). Assessing the influence of watershed characteristics on the flood vulnerability of Jhelum basin in Kashmir Himalaya. *Natural Hazards* 77(1), 153–175.

Meraj, G., Romshoo, S. A., & Altaf, S. (2016). Inferring land surface processes from watershed characterization. In Janardhana Raju, N. (ed.), *Geostatistical and Geospatial Approaches for the Characterization of Natural Resources in the Environment*. Springer, Cham, 741–744.

Meraj, G., Singh, S. K., Kanga, S., & Islam, M. N. (2021). Modeling on comparison of ecosystem services concepts, tools, methods and their ecological-economic implications: A review. *Modeling Earth Systems and Environment*, 1–20. https://doi.org/10.1007/s40808-021-01131-6

NIHR (2013). *Impact of Sewage Effluent on Drinking Water Sources of Shimla City and Suggesting Ameliorative Measures (PDS Under Hydrology Project-II)*. National Institute of Hydrology Jal Vigyan Bhawan, Roorkee.

Singh, S. K., Mishra, S. K., & Kanga. S. (2017). Delineation of groundwater potential zone using geospatial techniques for Shimla city, Himachal Pradesh (India). *International Journal for Scientific Research and Development* 5(4), 225–234.

Snoeyink, V. L., & Jenkins, D. (1980). *Water Chemistry*. John Wiley & Sons, New York.

Tandon, M. (2008). Catchment sensitive planning for sustainable cities. *ITPI Journal* 5, 21–26.

Tomar, J. S., Kranjčić, N., Đurin, B., Kanga, S., & Singh, S. K. (2021). Forest fire hazards vulnerability and risk assessment in Sirmaur district forest of Himachal Pradesh (India): A geospatial approach. *ISPRS International Journal of Geo-Information* 10(7), 447.

UNPD (2012). *World Urbanization Prospects, the 2011 Revision.* Department of Economic and Social Affairs Population Division United Nations, New York.

Zeraebruk, K. N. (2014). Assessment of level and quality of water supply service delivery for development of decision support tools: Case study Asmara water supply. *International Journal of Sciences: Basic and Applied Research (IJSBAR)* 14(1), 93–107.

Index

Analytical Hierarchy Process (AHP) 5
ArcGIS spatial analyst tool 115
Artificial Neural Network (ANN) 97
ASTER DEM 115, 120, 132

calibration 33, 40, 41, 100, 177, 184
carrying capacity 334, 353
climate variability 224
crop simulation model (CSM) 182
CROPWAT model 182

debris flow 147
Decision Support System for Agro-Technology
        Transfer (DSSAT) model 182
deficit 359
demand 348
digital numbers (DN) 103
Dynamic Vegetation Models (DVMs) 237

Ecological Security Indicators 332
Environmental Policy Integrated Climate (EPIC)
        model 182

flood disaster management 4
Focused Group Discussions (FGDs) 292

geographical information system (GIS) 310
global climate model (GCM) 183
global positioning system (GPS) 91, 292, 295
Google Earth Pro 115, 120

habitat mapping 316
HadCM3 46
hazard 5, 114
hazard identification 10
HEC-RAS 10
Hindu Kush-Karakoram-Himalayan
        (HKH) 196
Home-Based Industries (HBI) 333
hydrological cycle 204, 225
hydro-morphological 19

Kashmir Himalayas 202

land surface temperature (LST) 241
Landslide Hazard Zonation (LHZ) 133
Leaf Area Index (LAI) 185
LiDAR 91

lithology 133, 160
logistic regression 161
long-lived greenhouse gases (LLGHGs) 224
LTEM (Long Term Ecological monitoring) 225

map correlation method 20
MIKE FLOOD 5
morphometric analysis 60

Nash-Sutcliffe efficiency 23
National Bureau of Soil Survey & Land Use
        Planning (NBSS & LUP) 132
NDWI 242

overlay analysis 314

Pearson Type III distribution 8
percent bias (PBIAS) 40
Permanent Observational Plots (POPs) 226
Permanent Preservation Plots (PPPs) 225
Problem, Plan, Data, Analysis, and Conclusion
        (PPDAC) model 94

rainfall erosivity index 64
regional climate model (RCM) 183
remote sensing (RS) 3
Representative Concentration Pathways
        (RCPs) 176

satellite-derived bathymetry (SDB) 88
sedimentation 85
Seed Cell Area Index (SCAI) 163
Shyama Prasad Mukherjee Rurban Mission
        (SPMRM) 326
skewness 9
snow and glacier melt 196
snow water equivalent (SWE) 44
soil erodibility factor 61, 63
spatial probability mapping 161
spatial risk analysis 6
Special Report on Emission (SRES) scenarios
        179, 180
Species Distribution Models (SDMs) 237
SRTM DEM 5, 16
Standard Deviation Ratio 40
streamflow 41
SWAT 41
SWAT-CUP 50

temporal probability 163

ungauged catchment 20
UNISDR 127
United States Geological Survey (USGS) 266, 288
urban heat island (UHI) 246
USLE 45

validation 45, 100, 104, 147, 162
variogram 22
vegetation line 216
vulnerability analysis 10

WASMOD 42
water yield 47